Biochemical Techniques: Theory and Practice

Biochemical Techniques: Theory and Practice

Kylie Joseph

MURPHY & MOORE
www.murphy-moorepublishing.com

Biochemical Techniques: Theory and Practice
Kylie Joseph
ISBN: 978-1-63987-074-5 (Hardback)

MURPHY & MOORE

Published by Murphy & Moore Publishing,
1 Rockefeller Plaza,
New York City, NY 10020, USA

Cataloging-in-Publication Data

Biochemical techniques : theory and practice / Kylie Joseph.
        p. cm.
Includes bibliographical references and index.
ISBN 978-1-63987-074-5
1. Biochemistry--Technique. 2. Biochemistry. I. Joseph, Kylie.
QP519.7 .B56 2022
572.36--dc23

For more information regarding Murphy & Moore Publishing and its products, please visit the publisher's website www.murphy-moorepublishing.com

Table of Contents

Preface   **VII**

Chapter 1 **An Introduction to Biochemical Techniques**   **1**
    a. Biochemical Parameters   3
    b. Analytical and Bioanalytical Techniques   4

Chapter 2 **Polymerase Chain Reaction**   **16**
    a. Optimization and Troubleshooting in Polymerase Chain Reaction   27
    b. PCR Product Detection Methods   36

Chapter 3 **Chromatographic Techniques**   **48**
    a. High-performance Liquid Chromatography   56
    b. Paper Chromatography   62
    c. Ion-exchange Chromatography   66
    d. Gel Filtration Chromatography   72
    e. Affinity Chromatography   79
    f. Gas Chromatography   105

Chapter 4 **Spectroscopic Techniques**   **111**
    a. Ultraviolet and Visible Light Spectroscopy   118
    b. Fluorescence Spectroscopy   126
    c. Circular Dichroism Spectroscopy   132
    d. Atomic Spectroscopy   144
    e. Fourier Transform Infrared Spectroscopy   151
    f. Raman Spectroscopy   160
    g. Mass Spectrometry   194

Chapter 5 **Electrophoretic Techniques**   **200**
    a. Polyacrylamide Gel Electrophoresis   205
    b. Agarose Gel Electrophoresis   213
    c. Capillary Electrophoresis   218
    d. Microchip Electrophoresis   224

Permissions

Index

Preface

Techniques and methods that are used to analyze substances which govern the chemical reactions underlying various life processes are studied under biochemical techniques. It includes investigative procedures such as spectroscopy and gel staining which help in determining the concentration and purity of various proteins and nucleic acids. Most biomolecules occur in very minute quantities inside living cells. Their analysis requires their purification and freeing them from contamination. This is achieved using various techniques including centrifugation, gel electrophoresis, precipitation and chromatography. These are interrelated methods which are based on different physical and chemical properties of biomolecules like shape, size, net charge, etc. Chromatography is the most widely used biochemical technique which can be further classified into gel filtration chromatography, affinity chromatography, gas chromatography and paper chromatography. This book is compiled in such a manner, that it will provide in-depth knowledge about the theory and practice of different biochemical techniques. It is an upcoming field of science that has undergone rapid development over the past few decades. This textbook is appropriate for those seeking detailed information in this area.

Given below is the chapter wise description of the book:

Chapter 1- The set of methods, procedures and assays used to analyze the biomolecules and understand the chemical reactions underlying life processes is known as biochemical analysis techniques. This is an introductory chapter which will introduce briefly all the significant aspects of biochemical techniques such as biochemical parameters, and analytical and bioanalytical techniques.

Chapter 2- Polymerase chain reaction is a process in which a small sample of DNA is amplified to a large amount. By using polymerase chain reaction, billions of copies from a single molecule of DNA can be made. The topics elaborated in this chapter will help in gaining a better perspective about polymerase chain reaction as well as PCR product detection methods.

Chapter 3- Chromatography is a technique used for the separation of components or solutes by using a moving solvent on filter paper. There are various types of chromatography techniques such as high-performance liquid chromatography, paper chromatography, ion-exchange chromatography, gel filtration chromatography, affinity chromatography and gas chromatography. All these types of chromatography are discussed in detail in this chapter.

Chapter 4- Spectroscopic techniques make use of light and its interaction with the matter to evaluate the consistency or structure of the sample. The different types of spectroscopic techniques include ultraviolet and visible light spectroscopy, fluorescence spectroscopy, and circular dichroism spectroscopy. All these types of spectroscopic techniques have been carefully analyzed in this chapter.

Chapter 5- Electrophoretic techniques are based on the motion of charged particles under the influence of electric field. There are four types of electrophoretic techniques: polyacrylamide gel electrophoresis, agarose gel electrophoresis, capillary electrophoresis and microchip electrophoresis. The chapter closely examines the key concepts of these electrophoretic techniques to provide an extensive understanding of the subject.

At the end, I would like to thank all those who dedicated their time and efforts for the successful completion of this book. I also wish to convey my gratitude towards my friends and family who supported me at every step.

Kylie Joseph

An Introduction to Biochemical Techniques

The set of methods, procedures and assays used to analyze the biomolecules and understand the chemical reactions underlying life processes is known as biochemical analysis techniques. This is an introductory chapter which will introduce briefly all the significant aspects of biochemical techniques such as biochemical parameters, and analytical and bioanalytical techniques.

Biochemical analysis techniques refer to a set of methods, assays, and procedures that enable scientists to analyze the substances found in living organisms and the chemical reactions underlying life processes. The most sophisticated of these techniques are reserved for specialty research and diagnostic laboratories, although simplified sets of these techniques are used in such common events as testing for illegal drug abuse in competitive athletic events and monitoring of blood sugar by diabetic patients.

To perform a comprehensive biochemical analysis of a biomolecule in a biological process or system, the biochemist typically needs to design a strategy to detect that biomolecule, isolate it in pure form from among thousands of molecules that can be found in an extracts from a biological sample, characterize it, and analyze its function. An assay, the biochemical test that characterizes a molecule, whether quantitative or semi-quantitative, is important to determine the presence and quantity of a biomolecule at each step of the study. Detection assays may range from the simple type of assays provided by spectrophotometric measurements and gel staining to determine the concentration and purity of proteins and nucleic acids, to long and tedious bioassays that may take days to perform.

The description and characterization of the molecular components of the cell succeeded in successive stages, each one related to the introduction of new technical tools adapted to the particular properties of the studied molecules. The first studied biomolecules were the small building blocks of larger and more complex macromolecules, the amino acids of proteins, the bases of nucleic acids and sugar monomers of complex carbohydrates. The molecular characterization of these elementary components was carried out thanks to techniques used in organic chemistry and developed as early as the nineteenth century. Analysis and characterization of complex macromolecules proved more difficult, and the fundamental techniques in protein and nucleic acid and protein purification and sequencing were only established in the last four decades.

Most biomolecules occur in minute amounts in the cell, and their detection and analysis require the biochemist to first assume the major task of purifying them from any contamination. Purification procedures published in the specialist literature are almost as diverse as the diversity of biomolecules and are usually written in sufficient details that they can be reproduced in different laboratory with similar results. These procedures and protocols, which are reminiscent of recipes in cookbooks have had major influence on the progress of biomedical sciences and were very highly rated in scientific literature.

The methods available for purification of biomolecules range from simple precipitation, centrifugation, and gel electrophoresis to sophisticated chromatographic and affinity techniques that are constantly undergoing development and improvement. These diverse but interrelated methods are based on such properties as size and shape, net charge and bioproperties of the biomolecules studied.

Centrifugation procedures impose, through rapid spinning, high centrifugal forces on biomolecules in solution, and cause their separations based on differences in weight. Electrophoresis techniques take advantage of both the size and charge of biomolecules and refer to the process where biomolecules are separated because they adopt different rates of migration toward positively (anode) or negatively (cathode) charged poles of an electric field. Gel electrophoresis methods are important steps in many separation and analysis techniques in the studies of DNA, proteins and lipids. Both western blotting techniques for the assay of proteins and southern and northern analysis of DNA rely on gel electrophoresis. The completion of DNA sequencing at the different human genome centers is also dependent on gel electrophoresis. A powerful modification of gel electrophoresis called twodimensional gel electrophoresis is predicted to play a very important role in the accomplishment of the proteome projects that have started in many laboratories.

Chromatography techniques are sensitive and effective in separating and concentrating minute components of a mixture and are widely used for quantitative and qualitative analysis in medicine, industrial processes, and other fields. The method consists of allowing a liquid or gaseous solution of the test mixture to flow through a tube or column packed with a finely divided solid material that may be coated with an active chemical group or an adsorbent liquid. The different components of the mixture separate because they travel through the tube at different rates, depending on the interactions with the porous stationary material. Various chromatographic separation strategies could be designed by modifying the chemical components and shape of the solid adsorbent material. Some chromatographic columns used in gel chromatography are packed with porous stationary material, such that the small molecules flowing through the column diffuse into the matrix and will be delayed, whereas larger molecules flow through the column more quickly. Along with ultracentrifugation and gel electrophoresis, this is one of the methods used to determine the molecular weight of biomolecules. If the stationary material is charged, the chromatography column will allow separation of biomolecules according to their charge, a process known as ion exchange chromatography. This process provides the highest resolution in the purification of native biomolecules and is valuable when both the purity and the activity of a molecule are of importance, as is the case in the preparation of all enzymes used in molecular biology. The biological activity of biomolecules has itself been exploited to design a powerful separation method known as affinity chromatography. Most biomolecules of interest bind specifically and tightly to natural biological partners called ligands: enzymes bind substrates and cofactors, hormones bind receptors, and specific immunoglobulins called antibodies can be made by the immune system that would in principle interact with any possible chemical component large enough to have a specific conformation. The solid material in an affinity chromatography column is coated with the ligand and only the biomolecule that specifically interact with this ligand will be retained while the rest of a mixture is washed away by excess solvent running through the column.

Once a pure biomolecule is obtained, it may be employed for a specific purpose such as an enzymatic reaction, used as a therapeutic agent, or in an industrial process. However, it is normal in a research laboratory that the biomolecule isolated is novel, isolated for the first time and, therefore, warrants full characterization in terms of structure and function. This is the most difficult part in a biochemical analysis of a novel biomolecule or a biochemical process, usually takes years to accomplish, and involves the collaboration of many research laboratories from different parts of the world.

Recent progress in biochemical analysis techniques has been dependent upon contributions from both chemistry and biology, especially molecular genetics and molecular biology, as well as engineering and information technology. Tagging of proteins and nucleic acids with chemicals, especially fluorescent dyes, has been crucial in helping to accomplish the sequencing of the human genome and other organisms, as well as the analysis of proteins by chromatography and mass spectrometry. Biochemical research is undergoing a change in paradigm from analysis of the role of one or a few molecules at a time, to an approach aiming at the characterization and functional studies of many or even all biomolecules constituting a cell and eventually organs. One of the major challenges of the post-genome era is to assign functions to the entire gene products discovered through the genome and cDNA sequencing efforts. The need for functional analysis of proteins has become especially eminent, and this has led to the renovated interest and major technical improvements in some protein separation and analysis techniques. Two-dimensional gel electrophoresis, high performance liquid and capillary chromatography as well as mass spectrometry are proving very effective in separation and analysis of abundant change in highly expressed proteins. The newly developed hardware and software, and the use of automated systems that allow analysis of a huge number of samples simultaneously, is making it possible to analyze a large number of proteins in a shorter time and with higher accuracy. These approaches are making it possible to study global protein expression in cells and tissues, and will allow comparison of protein products from cells under varying conditions like differentiation and activation by various stimuli such as stress, hormones, or drugs. A more specific assay to analyze protein function in vivo is to use expression systems designed to detect protein-protein and DNA-protein interactions such as the yeast and bacterial hybrid systems. Ligand-receptor interactions are also being studied by novel techniques using biosensors that are much faster than the conventional immunochemical and colorimetric analyzes.

The combination of large scale and automated analysis techniques, bioinformatic tools, and the power of genetic manipulations will enable scientists to eventually analyze processes of cell function to all depths.

Biochemical Parameters

Albumin

As a measure in a nutritional assessment, albumin is useful because a fast diminishing albumin concentration is a sign for an inflammatory reaction. An increasing albumin level can be interpreted as an improvement; the patient becomes anabolic. The albumin level only increases when the inflammation decreases. Nutrition has no influence on that.

Creatinine

In certain cases creatinine can be used to get an impression of the quantity of muscle mass. When the kidney works well, a decreased creatinine can indicate decreased muscle mass. Creatinine arises by the conversion of creatine to creatinine in the muscle mass.

CRP

An increased CRP is a result of inflammation. CRP can increase up to a thousand times as a reaction to inflammation, sepsis or infection. It can be used to monitor stressresponse during the acute fase. Hemoglobin parameter can be used to determine the response to illness. During illness Hb decreases very fast.

Lymphocytes

The total lymphocytes count is not a sensitive index for malnutrition, because it reacts very slowly to recovery from malnutrition. The total amount of lymphocytes can increase in case of inflammation, radiation therapy and chemotherapy. Screening on point deficiencies with the aid of medical and food questionnaires. Only in case of severe deficiencies this is shown in the blood.

Prealbumin

This is a sensitive indicator for protein-deficiency. Prealbumin increases with nutritional therapy, even when the disease condition is not getting better. It decreases fast in case of a low energy intake, even if protein intake is adequate. However it also decreases in case of inflammation and is dependent on the level of hydration. Careful interpretation of prealbumin values in the clinical setting is advised.

Transferrin

Transferrin is produced in the liver. It is a transport protein for iron and zinc, and can also be used as an indicator for the iron status in the body. In iron-deficiency, the serum transferrin increases. In illness, transferrin is low because the liver produces less transferrin. Anaemia and nephrosis also influence transferrin.

Urea

The liver produces urea if amino acids break down. The urea production is more after a protein rich meal and when endogenous catabolism is increased (in case of infections, internal bleedings, intoxication, fever, and after tissue damage).

Analytical and Bioanalytical Techniques

Analytical chemistry is concerned with the chemical characterization of matter and refining the qualitative and quantitative problem about that matter. It plays a vital role in almost all the aspects of scientific research and development, for example, clinical, forensic, environmental, and pharmaceutical sciences. In medicine, analytical chemistry is the key for clinical laboratory tests which imparts basis of disease diagnosis and chart progress for recovery to the physicians. Figure describes the scheme through which physicians ruled out or analyze the disease prognosis and therapeutic drug monitoring in patients. In accord with this, an analytical chemist also explores the idea of developing advanced technique for betterment of human healthcare and in sorting out the problems related to the disease diagnosis. Implementation of an analytical technique mainly depends on the varying degree of selectivity, sensitivity, accuracy, precision, cost, and rapidity of that particular technique. The techniques employed may be based either on physical property or chemical property of an analyte. An analyte is defined as a constituent which has to be determined in a given sample type. The classical analytical techniques include gravimetric, volumetric, and titrimetric methods; on the other hand, instrumental techniques involve ultraviolet-visible (UV-Vis), infrared (IR), and near-infrared (NIR) spectrophotometry fluorimetry, atomic spectroscopy (absorption/emission), electroanalytical chromatography, and radioimmunoassay. Instrumental techniques are usually more sensitive and selective than classical techniques but are less precise. Precision of techniques means the repeatability of a result and is expressed as standard deviation. Selectivity of an analytical method defines the measurement of a particular analyte from sample solution to a certain degree, in the presence of other analytes, without any interference. However, sensitivity of a method describes the ability to recognize two different concentrations.

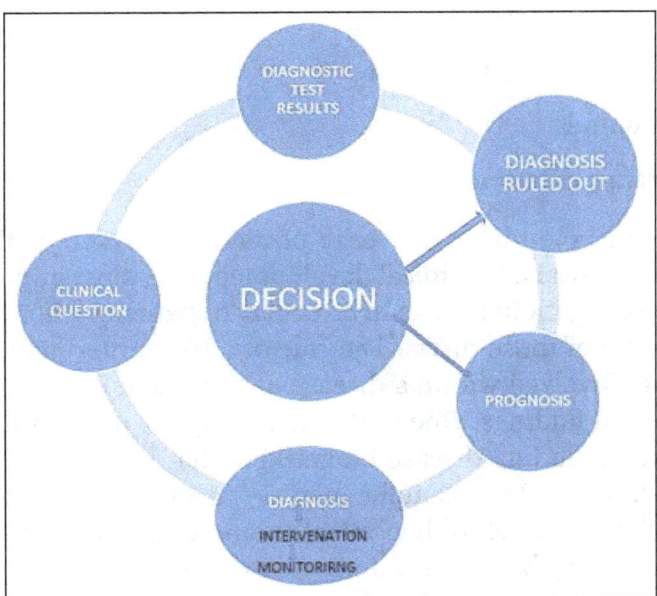

Figure: Schematic representation of four common decision making steps in which the result of an investigation is involved.

However, medical and clinical analyses are undergoing the greatest extension of instrumental methods. Interest in identifying biologically active compounds is growing rapidly and providing new challenges for the analytical chemists. These challenges have been resolved by the introduction of bioanalytical technique as a modern approach to disease diagnosis and therapy. A bioanalytical method is a combination of different procedures which are (i) collection, (ii) processing, (iii) storage, and (iv) analysis of a biological sample (blood-cerebrospinal fluid (CSF), serum, plasma, or urine, tissue, and skin). This method is also useful for quantitative determination of drugs and metabolites in biological samples. For that reason, technologies used to perform bioanalytical methods vary according to the analyte's nature. Hence, to find out the appropriate technologies involved in a bioanalytical method for the purpose of quantification of an analyte, the method validation is important. This procedure is termed as bioanalytical method validation (BMV). Few techniques commonly applied in bioanalytical studies include hyphenated (combination of two techniques) techniques like liquid chromatography (LC), gas chromatography (GC), capillary electrophoresis (CE) coupled with mass spectrometry (MS), and advanced automated chromatographic techniques, for example, high-performance liquid chromatography (HPLC).

During the past decades, the analytes that have been targeted in bioanalytical studies include amino acid, peptides, proteins, serum enzyme, tumor and cancer genes, carbohydrates, vitamins, catecholamines, cardiac risk factors, etc. With the recurrent analysis of biomolecules, numerous analytical techniques and instrumentation have been evolved and applied in the field of medical sciences which are as follows:

- Sensors
- Electrophoresis
- Chromatography
- Mass spectrometry
- Optical techniques (microscopy)
- Radio and immunochemical techniques
- Hyphenated techniques
- Point-of-care instrumentation

Biological samples have the potential to deliver important biomarkers in the clinic due to accessibility of these biological materials. In clinical development, the most important benefit offered by biomarkers is to limit investigational drugs to critical care patients who would gain the therapies to observe the effectiveness of those drugs. The role of a biomarker is to give information about the biological mechanism involved within a disease or treatment of disease having the capability to correlate with the clinical findings. One of the most tangible problems that research scientists are facing in recent years is finding disease biomarkers that are translated well from animal or computer simulation to humans. For example, increase in enzyme activity in computer or animal model may have a significant impact in theoretical computer or animal model, whereas same enzyme activity enhancement may have a very limited or no clinical impact.

There is no denying that "analytical and bioanalytical technique" is a broad topic, incorporating technologies from classical chromatography to point-of-care instrumentation. But unifying and doing those as quickly, accurately, and inexpensively as possible are drives to make chemical or biochemical measurements.

Researchers are interested in mapping the neural connectivity of the brain through scanning electron microscopy (SEM). This could be now employed with more powerful microscopes, such as focused ion beam and multi-beam SEM, to collect serial images of ultrathin brain slices. They can now build surface plasmon resonance substrates out of silver rather than the more typical gold and an SPR microscope to image and quantify 1296 binding events in parallel. Those scientists who are interested in surface properties can now scan those surfaces faster than ever, thanks to high-speed atomic force microscopy.

Sample Collection and Storage

We have to keep in mind that biological samples, collected from the patients, must be transported to the initial assessment center as soon as possible. The type of preservatives should be known to protect the samples from degradation prior to cryopreservation at a reasonable cost. Cryopreservation is a process to store biological samples at very low temperature for prevention of damage. The purpose is to find readily accessible and data-rich biological samples. The stability of a wide range of bioanalytes and cells as a component of whole blood should be estimated, taking into account different anticoagulant (inhibition of coagulation of blood) media, at different temperatures and under varying transport conditions. Bioanalytes can be known biochemicals, such as DNA, defined proteins, and specific metabolites, or unknown analytes, such as the constituent plasma/serum proteome and metabonome.

Design and testing of the sample handling protocol considered as key factors that affect the stability of biological samples, including anticoagulants, stabilizing agents, and temperature, elapsed time from collection to initial processing and endogenous degrading properties (enzymes, cell death). We also aim for cost-efficiency by avoiding collecting multiple sources of material for the same analyte. The samples undergo minimal processing locally in the assessment centers before being shipped to the main laboratory for processing with the aim of cryopreservation within 24 h of collection. Samples are protected against degradation during shipping by being chilled at 4 °C (only peripheral blood lymphocytes, at 18 °C). Once the samples get processed in the laboratory, they are placed in cabinets maintained at −80 °C for the working archive or in nitrogen vapor at −180 °C or below for the backup archive.

Sample Preparation

Biological samples involve plasma, serum, CSF, bile, urine, tissue homogenates, saliva, seminal fluid, and frequently whole blood. Quantitative analysis of drugs and metabolites containing huge amounts of proteins and large numbers of endogenous compounds within these samples is very complicated. Direct injection of drug containing biological sample into a chromatographic column results in the precipitation or absorption of proteins on the column packing material, resulting in an immediate loss of column performance. A number of advances have made to convert sample preparation techniques, used for the cleanup of drugs in biological samples into formats that are acceptable for high-volume processing with or without automation. The most widely used cleanup methods for separation of biomolecules from biological samples are summarized in table.

Table: Sample preparation techniques used for biological samples.

Sample preparation techniques	Advantages
Liquid phase extraction (LLE)	LLE is one of the first methods used for extraction. It depends on the partitioning of analytes between two immiscible liquids. The resulting extract may be directly analyzed or further purified and concentrated by subsequent LLE and evaporations.
Solid phase extraction (SPE)	SPE is a method for the isolation and concentration of selected analytes from a fluid sample by their transfer on a solid phase. The analytes are recovered by elution or thermal desorption. This method has high recovery, uses less organic solvent than LLE.

Affinity separation: molecularly imprinted polymers (MIPs)/ antibodies	The affinity sorbent may consist of an immobilized antibody or a molecularly imprinted polymer. This technique is highly specific and very sensitive, but the sorbent is difficult to prepare; it suffers from cross-reactivity and leaking of template.
Solid phase micro extraction (SPME)	SPME, a solvent free extraction method, consists of a single extraction step, but the experimental variables must be well controlled. It reduces solvent and sample volume needs and sample preparation time. Improves detection limits i.e., parts per trillion level detection.
Ultrafiltration and microdialysis (MD)	Ultrafiltration consists of filtering the sample through a special size-excluding filter, either by applying pressure (10–100 psi) or by centrifugation. The method is widely used, simple, efficient, but suffers from ligand binding to the filter and shift of equilibrium. Dialysis and MD can be used to separate an analyte by diffusion through a semi-permeable membrane.

Because of this, sample preparation became a prominent step in the analysis of biological samples. In recent years, the necessity of new developed method is largely required. Frequently, it was earlier considered as a separate procedure prior to the analysis, while it nowadays has become a more or less integrated part of the analytical procedure. It is necessary to lay the foundation of their development on a systematic and scientific approach. Thus, fundamental understanding of the different processes involved in a sample preparation method is served as a basis for its optimization. We should select the appropriate sample preparation method on the basis of requirements of the assay and time allowed to run sample preparation method.

Spectrophotometry

The spectrophotometric technique is used to study interactions between electromagnetic radiations and analyte. The concentration of an analyte is determined by using a graph which is called standard analytical curve. An example is determination of iron in blood serum. The iron content of blood serum is determined after deprotonation (by precipitating protein) with trichloroacetic acid and reduction with hydroxyl ammonium sulfate. Iron (II) ions are reacted in the medium buffered with ammonium acetate and with diphenyl-1,10-phenanthroline-disulfonic acid disodium salt (bathophenanthroline disulfonate–Na), and the absorbance of the complex formed is measured at 535 nanometer. The concentration belonging to the absorbance data of the test solution is read from the standard analytical curve and multiplied by three for threefold dilution.

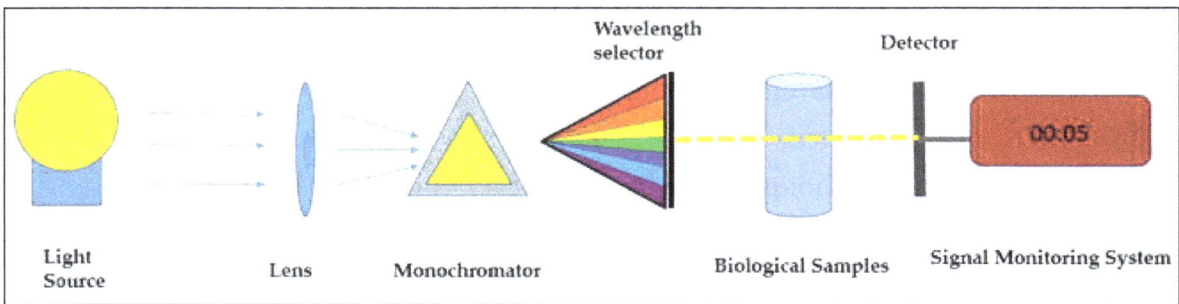

Figure: Schematic representation of biological sample determination using spectrophotometric technique.

Optical Spectroscopic Techniques of Human Models

Optical spectroscopy for biomedical applications covers up the plethora of medical technological and fundamental research areas. This includes screening and early detection of diseases which remain clinically silent over long periods. It is a noninvasive, fast spectroscopic technique. The technique is also capable of observations from femtosecond time scale at nanometer spatial resolution, so it can be applied in all areas of life sciences. This technique is to make an early, noninvasive, and patient-specific diagnosis near the source of the disease and then to treat the disease at primary stage, for example, Alzheimer's disease and coronary disease. Thus, optics offers a wide variety of diagnostic methods and products of biomedical spectroscopy.

Absorptions and Reflectance Spectroscopy

This method involves investigations of brain dysfunction and mental health problems like depression, epilepsy, and Alzheimer's disease. The direct absorption like near-infrared (NIR) techniques and instrumentation are particularly suitable in routine neonatal care applicatons. However, diffused reflectance spectroscopy, in the UV-Vis-NIR region, can be used for biomedical applications, like studies on skin condition (vitiligo, psoriasis, skin cancer) and glucose concentration measurement.

Photoacoustic Spectroscopy (PAS)

This includes measurement of concentration of biomolecular species. Examples are glucose determination, characterization of tissue status (biopsy tissue), and imaging application. PAS can provide information on three-dimensional distribution of specific molecular species in a specimen, by appropriate choice of excitation wavelength.

Raman and Infrared Spectroscopy

The "fingerprint" molecular specific technique will be of great advantage in understanding the biochemical interactions involved in induction, progression, therapeutic invention, and regression. This technique is suitable for biomedical applications such as breath analysis, drug-cell interaction, microscopy, and imaging of biopsy sample (tissue, fine needle aspiration).

Fluorescence Spectroscopy

Depending on excitation wavelength, the fluorescence peak has been observed in blood and urine spectra. A few examples are spectra observed from epithelial tissues, proteins, NAH, FAD, and hemoglobin.

Mass Spectrometry

Mass spectroscopy (MS) measures masses within the sample. In mass spectroscopy, chemical species get ionized and ions get sorted on the basis of their mass-to-charge ratio. Major application of MS includes confirmation of immunoassay-positive drug screens, identification of inborn errors of metabolism, and analysis of steroid hormones. Conclusive identification of molecules that range in size from tens of daltons (small molecules) to hundreds of thousands of daltons (biomolecules) is based on different principles.

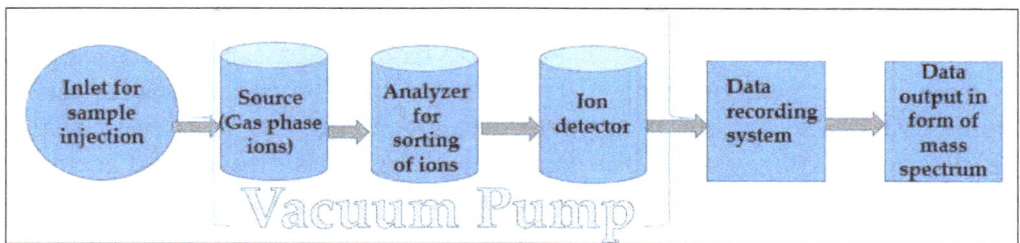

Figure: Schematic representation of mass spectrometry detection of sample.

Discovery of highly sensitive polymerase chain reaction (PCR) was a major step forward in the biomedical research and diagnostics. This technique is used for analysis of small quantities of short sequences of DNA and RNA without cloning. PCR can detect the presence of pathogens earlier than the culture tests. The miniaturization of MS systems allows a transportable device that minimizes the need of highly skilled operators and allows for rapid and accurate MS analysis in a point-of-care format (near the physician's clinic).

Imaging Techniques

Advances in medical imaging present a great opportunity in drug development. A number of different imaging technologies are available. These include computed tomography (CT), magnetic resonance imaging (MRI), magnetic resonance spectroscopy (MRS), positron-emission tomography (PET), and single photon emission computed tomography (SPECT). If adequately qualified, imaging biomarkers can be very helpful in the early stages of clinical development.

Sensors

An electrochemical sensor consists of a diffusion barrier, a sensing electrode (working electrode, measuring electrode, or anode), a counter electrode (cathode), and an electrolyte. Their fabrication includes various types of systems such as conductometry, voltammetry, potentiometry, and capacitance, and it is an important tool to detect various analytes in environmental, clinical, and biological fields due to their high sensitivity, cheapness, and miniaturization. These sensors have the potential to achieve sensitive, specific, and low-cost detection of biomolecules which is relevant to the diagnosis and monitored treatment of disease. A few are listed below:

Potentiometric Sensors

Techniques based on measurement of potential sensor are termed as potentiometry, for example, determination of potassium in blood serum by direct potentiometry with an ion-selective electrode. The determination of urea is a frequent task of clinical laboratories. The basis of enzyme electrode function is the selective recognition of urea by urease enzyme. The following reaction takes place:

$$\text{Urea} + 3\text{H}_2\text{O} \xrightarrow{\text{Urease}} 2\text{NH}_4^+ + \text{HCO}_3^- + \text{OH}^-$$

This reaction can be followed using different potentiometric electrodes.

Molecularly Imprinted Polymer (MIP) Sensors

MIP could be one of the important tailor-made systems for targeted analyte recognition exclusively even their presence in complex real biological samples in parts per million to parts per billion levels. The general idea is to create the cavity in the presence of the guest. The guest organizes and promotes energy-minimized interactions with polymer forming around it. Thus, after washing the guest out, the polymers retain a template cavity of the guest's size and shape which subsequently display binding selectivity toward the guest just like induced-fit model of enzyme. The MIP-modified sensors can be used for biological and pharmaceutical analyte determination from biological samples. An example is the development of a polyscopoletin-based MIP nanofilm for the electrochemical determination of elevated human serum albumin (HSA) in urine. The results suggest that MIP-based sensors may be applicable for quantifying high-abundance proteins in a clinical setting.

Biosensors

The need for rapid, simple handheld testing devices in medicine paves the way for introduction of biosensor. Biological sensors are optical, electrical, and piezoelectrical devices that have the ability to detect biological compounds, such as nucleic acids and proteins. Early diagnosis of inherited disease is important for effective treatment and is sometimes lifesaving. Methods, like enzyme-linked immunosorbent assay and PCR, can require highly skilled professionals and expensive chemicals and can be time-consuming. In this area, many biosensor schemes had developed as an alternative to classical methods.

The latest advancements in nanotechnology result in its application for cancer biomarker recognition. Several other biosensors include nanomaterial-based biosensors, peptide nucleic acid-based biosensors, biosensors for medical mycology, optical DNA biosensors, and last but not least, the biosensors for the diagnosis of heart disease.

Chromatographic Separation Techniques

In chromatographic separation technique, various constituents of the mixture in the given sample travel at different speeds, causing them to separate. This technique involves two phases: a mobile phase and a stationary phase. The separation mainly depends on the differential partitioning between these two phases. A bonded phase is a stationary phase that is covalently bonded to the support particles or to the inside wall of the column tubing. The stationary phase is the substance fixed in place for the chromatography procedure, and the mobile phase is moving in a definite direction. Examples include the silica layer in thin layer chromatography (TLC). Archer John Porter Martin and Richard Laurence Millington Synge won a Nobel Prize in chemistry for chromatography invention. Their work encouraged the rapid development of several advanced chromatographic methods such as paper chromatography, gas chromatography, and HPLC. The differences in a compound's partition coefficient bring about differential retention on the stationary phase and thus affect the separation process. Chromatography may be preparative or analytical. The preparative chromatography separates the components of a mixture for later use and is thus a form of purification. Analytical chromatography is done normally with smaller amounts of material and is for establishing the presence or measuring the relative proportions of analytes in a mixture. Figure describes a chromatogram for a biological system where the signal is proportional to the concentration of the specific analyte separated.

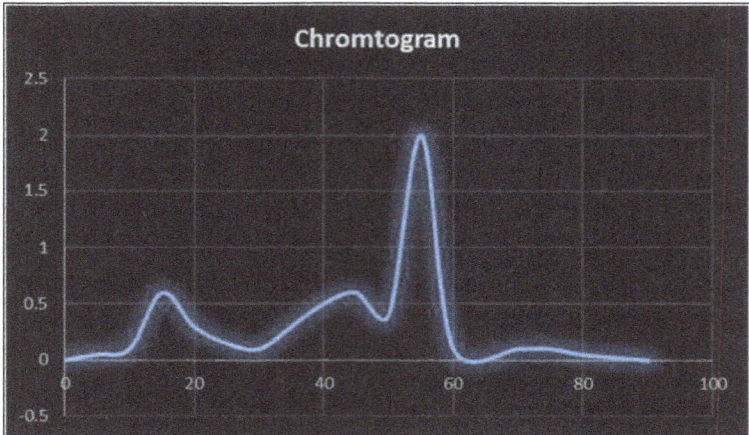

Figure: Chromatogram response of a biological sample. The retention time is plotted on X-axis and signal on Y-axis obtained from detector corresponding to the response created by the analytes exiting the system.

Depending upon the shape of stationary phase, chromatography may be (i) planar chromatography, having one-dimensional bed support such as paper or TLC, or (ii) column chromatography with three-dimensional bed support. TLC is useful for separating mixtures of organic compounds and is often used to monitor the progress of organic reactions and to check the purity of products.

On the basis of physical state of mobile phase, chromatographic technique may be GC or LC. GC can be used to separate mixtures of volatile organic compounds. A GC consists of a flowing mobile phase, an injection port, a separation column containing the stationary phase, a detector, and a data recording system. LC is useful for separating mixtures of ions or molecules that are dissolved in a solvent. If the matrix support, or stationary phase, is polar (e.g., paper, silica, etc.), it is normal-phase chromatography; and if it is nonpolar (C-18), it is reversed-phase chromatography.

Chromatography is a method of separating the constituents of a solution, based on one or more of its chemical or physical properties. This could be charge, polarity, or a combination of these traits and pH balance. The solution is passed through a medium which will hinder the movement of some particles more than others. These principles are used to isolate and analyze enzymes, pigments, amino acids, constituents of DNA, and almost any other molecule you can imagine. A wide variety of chromatography techniques had developed to allow mixed substances to be separated.

Capillary Electrophoresis (CE)

Jorgenson and Lukacs in 1981 invent capillary electrophoresis (CE) most often termed as capillary zone electrophoresis (CZE). It is a type of electrophoresis in which analytes are separated by applying an electric field across buffer solution-filled capillary tubes. The proposed instrumental technique was later implemented in a number of applications such as bioanalytical and forensic drug analysis. This technique is an alternative to the gel electrophoresis or LC.

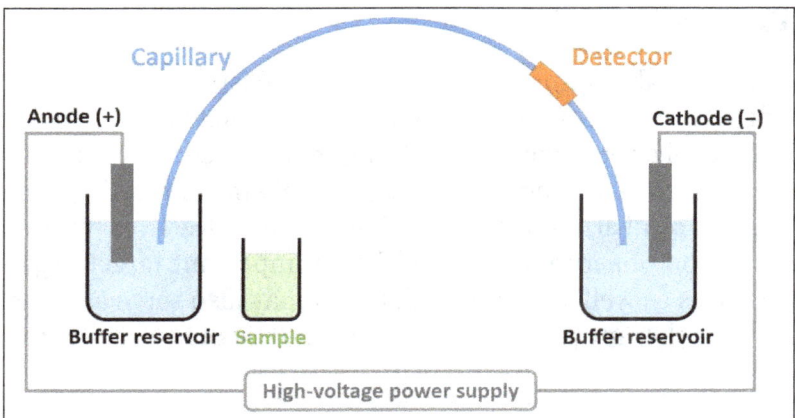

Figure: Capillary electrophoresis separation method.

CZE method correlated well with an automated kinetic fluorescent assay. An example is analysis of NAG by CE after incubation of urine samples using synthetic substrate, methylumbelliferyl-β-d-glucosaminide.

Microscopy

In the simplest microscopic methods, a specimen is illuminated by visible light and observed either against a bright background (bright-field microscopy) or a dark background (dark-field microscopy). The presence of cells that are not expected in the healthy person may be an indicator of disease. For example, a simple microscopic analysis of blood sample identifies the sickle cell anemia, and analysis of urine quantifies the presence of pus cells, which is an indicator of infection.

Light microscopy uses light as the illumination radiation. This is used to identify the microorganisms based on their morphology. An application of microscopy is to count the number of different cells per unit volume of blood or any other sample using a hemocytometer. Fluorescence microscopy has emerged as a very powerful tool for studying molecular processes owing largely to the advancement in optics and discovery of the green fluorescent protein and development of its analogs with different spectral properties. Several advancements in the field of fluorescence microscopy have been achieved that includes the following techniques:

- Confocal laser scanning microscopy (CLSM): CLSM is a type of fluorescence microscopy that allows imaging of the samples at different focal planes that light emitting from below or above the desired focal plane is eliminated. This results in very high lateral resolution and allows determining the spatial localization of the molecules.

- Total internal reflection fluorescence (TIRF) microscopy: TIFR is another type of fluorescence microscopy wherein the optics allows imaging of the molecules that are almost like microscopic slide. The resolution of light microscopes depends on the wavelength of the light used. The smaller the wavelength of the light used, the better the resolution obtained. Wavelength of the visible light imposes a resolution limit of ~0.2 μm on the light microscopes. Hemocytometer (Neubauer chamber) is a glass slide which has a counting chamber at the center. A glass cover is placed on the hemocytometer, and the sample is gently introduced into the chamber. The sample chamber has a grid which allows counting of cells in a defined region using a microscope.

Gene Therapy Protocol

Gene therapies are considerable improvements over the existing therapy because of the advantage in dosing schedule, patient compliance, toxicity, immunogenicity, and cost. Owing to this, gene therapy provides novel approaches for the treatment of inherited and acquired diseases. The development of a nonviral gene delivery vehicle capable of efficient, cell-specific delivery will be a valuable addition to the clinical armamentarium. For example, the liver places a central role in the metabolism and production of serum proteins; it is an important target organ for gene therapy. Hepatic metabolic diseases as well as acquired diseases may also serve as targets for hepatic gene therapy. Most recent gene therapy protocols describe delivery of foreign genes by means of injecting lentiviral particles.

Immunological and Radioisotope Techniques

Immunoassays are the quantitation of bioanalyte that depends on the reaction of an antigen (analyte) and an antibody. These methods are based on a competitive binding reaction between a fixed amount of labeled form of an analyte and a variable amount of unlabeled sample analyte for a limited amount of binding sites on a highly specific anti-analyte antibody. When immunoanalytical reagents (analyte or antibody) are mixed and incubated, the analyte is bound to the antibody forming an immune complex. This complex is separated from the unbound reagent fraction by physical or chemical separation technique. Analysis is achieved by measuring the label activity (e.g., radiation, fluorescence, or enzyme) in either of the bound or free fraction.

Immunoassay methods have been widely used in many important areas such as diagnosis of diseases, therapeutic drug monitoring, clinical pharmacokinetic and bioequivalence studies in drug discovery, and pharmaceutical industries. A few immunoassays based on different labels are as follows:

- Radioimmunoassay (RIA) methods have been used successfully for the determination of a limitless number of pharmaceutically important compounds in biological fluids. RIA is used to analyze thyroid hormone testing in patients after iodine-131 therapy.

- Enzyme immunoassay (EIA) is analogous to RIA except that the label is an enzyme rather than a radioisotope.

- Fluoroimmunoassay (FIA) is analogous to RIA except that the label is a fluorophore rather than a radioisotope.

- Chemiluminescence immunoassay (CLIA) involves a chemiluminescent substance as a label.

- Liposome immunoassay (LIA) is the assay involving a liposome-encapsulating marker.

- Cloned enzyme donor immunoassay (CEDIA) methodology is a novel approach which uses the DNA technology to produce homogenous enzyme immunoassays for drugs.

- Flow injection immunoassay (FIIA) methods were recently introduced to enhance the efficiency of immunochemical reaction, as well as to increase the performance of the analysis.

- Capillary electrophoresis immunoassay (CEIA) has been recently introduced as a sensitive analytical technique, particularly when combined with a sensitive detection method.

Hyphenated Separation Techniques and its Application in Clinical Chemistry

The hyphenated techniques improved detection limits, sample identification capability, and miniaturization potential; hence, about 60% of the application of electrochemical detection (ED) has been found in the field of bioanalysis. The principle of ED used in biomedical analysis is a transfer of charge between substances in a column effluent and a working electrode; mainly, two types of ED either coulometric detection or amperometric detection are frequently used. The main advantages of using ED are the selectivity and sensitivity over UV-Vis detection. In HPLC, most of the application has been carried out by using UV-Vis detector, and ED is only used in small portion. The development of HPLC with ED facilitated highly sensitive and selective determination of homovanillic acid (HVA) and vanillylmandelic acid (VMA) in urine for the differential diagnosis of neuroblastoma pheochromocytoma and related tumors. HPLC, coupled with UV-Vis using photodiode array as a detector, is widely used for determination of different drugs in serum. Other applications include determination of vitamins, antioxidants, and other components in biological samples.

CE coupled with MS provides an advantage of the sensitivity (parts per million ranges) and selectivity of these detection systems. A detector that is becoming more frequently attached to CE is inductively coupled plasma mass spectrometry (ICP-MS). To date, CE-ICP-MS has been performed using a quadrupole detector within the MS allowing a small number of elements to be analyzed at any one time.

Lab on a Chip (LoC)

Lab on a chip is defined as a microform of analytical devices that assimilate numerous laboratory operations such as PCR and DNA sequencing into a single chip on a very small scale. Miniaturized version of LoC provides cost-efficiency, use of low-volume reagents, high parallelization, high diagnostic speed, high sensitivity, and high expandability.

On the other hand, chronic disease (CD) healthcare is experiencing few limitations owing to lengthy and costly diagnosis procedures. Rapid, reliable, and low-cost diagnostic tools at point-of-care (PoC) instrumentation are therefore on high demand. LoC technology has a high potential to enable improved biomedical applications. In this regard, research toward developing new LoC-based PoC systems for CD diagnosis is fast growing into a nascent area such as chronic respiratory diseases (CRD), diabetes, and chronic kidney diseases (CKD).

Polymerase Chain Reaction

Polymerase chain reaction is a process in which a small sample of DNA is amplified to a large amount. By using polymerase chain reaction, billions of copies from a single molecule of DNA can be made. The topics elaborated in this chapter will help in gaining a better perspective about polymerase chain reaction as well as PCR product detection methods.

Polymerase chain reaction (PCR) can be recognized as a revolutionary technology for which the inventor Kary B. Mullis was awarded with the prestigious Nobel Prize in 1993 for chemistry along with Michael Smith. The secret story of any living organism which gives it unique characteristics is coded within the large and complexed DNA of that individual. To understand the story, it was needed to bring the DNA outside the living cell into the laboratory; this gave birth to the concept of PCR. Kary Mullis conceptualized of a machine through which from a single molecule of DNA we can get multiples of its copy after the end, like a Xerox machine does. With this in 1983 Kary Mullis developed PCR which is now a most generalized technique used in medical or biological science research labs for a range of applications. Examples are its use in DNA cloning, phylogeny, functional analysis of genes, diagnosis of diseases, genetic fingerprints for forensic sciences as paternity test where the aim is to amplify a single or a few copies of a specific DNA segment in generating thousands to millions of copies of that specific DNA sequence. The PCR techniques currently available can amplify DNA from a wide range of samples as small as hair, blood spot, and ancient biological samples.

Principle of Polymerase chain Reaction

The principle behind PCR is based on using the ability of DNA to replicate into a new strand of DNA complementary to the previous template strand. PCR differs from the replication in the sense that here only a fragment of DNA is multiplied several time to get a multiple copy of the fragment.

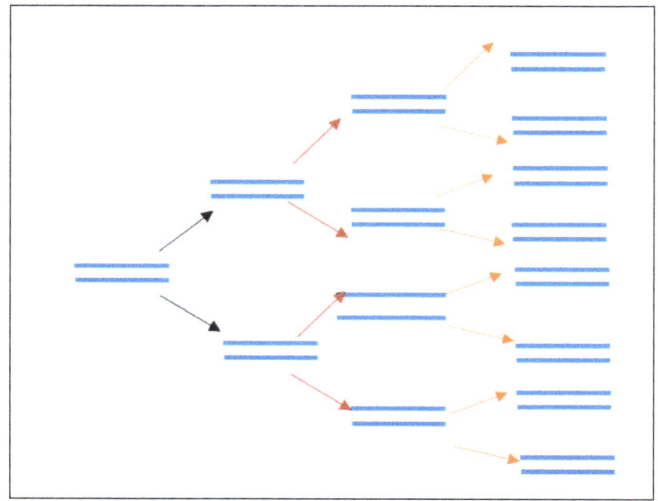

The PCR is a chain reaction of thermal cycling, which consists of repeated heating and cooling steps of reaction involving 3 basic steps:

- Denaturation

- Annealing

- Elongation

Denaturation

Denaturation is a structural change in a protein that results in the loss (usually permanent) of its biological properties. Because the way a protein folds determine its function, any change or abrogation of the tertiary structure will alter its activity. Denaturation of proteins can usually be caused by two key conditions – temperature and pH.

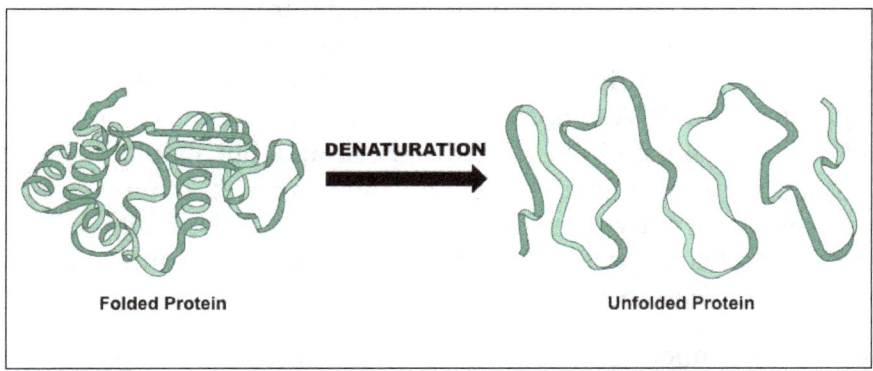

Denaturation of a Protein.

Temperature

- High levels of thermal energy may disrupt the hydrogen bonds that hold the protein together.

- As these bonds are broken, the protein will begin to unfold and lose its capacity to function as intended.

- Temperatures at which proteins denature may vary, but most human proteins function optimally at body temperature (~37 °C).

pH

- Amino acids are zwitterions, neutral molecules possessing both negatively (COO^-) and positively (NH_3^+) charged regions.

- Changing the pH will alter the charge of the protein, which in turn will alter protein solubility and overall shape.

- All proteins have an optimal pH which is dependent on the environment in which it functions (e.g. stomach proteins require an acidic environment to operate, whereas blood proteins function best at a neutral pH).

Effect of pH on Protein Structure.

Annealing

In the next step gradually the temperature of reaction mixture is lowered down to around 50–65 °C for 15–60 seconds which allows the DNA molecules to reanneal but during this step the short stretches of ss oligonucleotides (primers) added in the reaction mixture incorporates and get annealed to the single-stranded DNA template. The temperature selected for this step must be low enough to allow for hybridization of the primer to the strand, but should be higher enough to let the binding be sequence specific between template and primer. In other words, primer can only bind to a perfectly complementary sequence of the template. If the temperature is selected very low, the primer would bind non-specifically. When annealing temperature is kept too high, primer will fail to anneal as H-bonds cannot form.

In general, the annealing temperature is adjusted about 3–5 °C below the Tm of the primers. Tm is the temperature at which 50% of the oligonucleotide and its perfect complement are in duplex. It is dependent on the length of the DNA molecule and its specific nucleotide sequence. Theoretically, Tm = {4(G+C) + 2(A+T)}. Now the polymerase can bind to the primer-template hybrid and further starts formation of new strand complimentary to the template.

Elongation

This is the step where the polymerase attached to primer-template hybrid starts extending the new strand of DNA by adding specific nucleotides dNTPs which are complementary to the template strand into the developing strand. The temperature is different for different DNA polymerase. Most commonly used polymerase is Taq polymerase which is derived from a thermostable bacterium (Thermus aquaticus) and has its optimum activity at temperature 75–80 °C. However, chain elongation is carried out at 72 °C. The time required for this step depends both on the length of the target DNA to be amplified and on the efficiency of DNA polymerase used and. In general, DNA polymerase can add a thousand bases per minute. At each extension step, the amount of DNA target is doubled.

These three steps are repeated in a cyclic process which gives this technique the name Polymerase Chain Reaction which means a chain of reaction involving polymerase. PCR are carried out for 25 to 30 cycles in an automated thermal cycler, which can heat and cool the reaction mixture speedily. It is theoretically calculated that a single molecule of DNA can give 1 billion copies of the template after 30 cycles of PCR.

Requirements for PCR

Any PCR reaction requires the following components for reaction:

DNA Template

DNA template is the sample DNA that contains the target sequence to be amplified by PCR. At beginning of reaction a high temperature is applied to this template DNA to separate the strand, called as initial denaturation. For PCR a very small quantity of DNA is required as more amount of DNA template will increase the amount of contaminants and reduce the efficiency of PCR.

DNA Polymerase

DNA polymerase is the enzyme required for amplification of template DNA. These enzymes should have: 1) Capability to generate new strands of DNA by using the DNA template and primers. 2) The enzyme should be resistant to high temperature. In these parameters the most common enzyme used for PCR is Taq DNA polymerase (isolated from Thermus aquaticus) as it is better than other polymerase giving good results whereas it is having no 3' to 5' proof reading activity and results to low replication fidelity showing an error in 9,000 nucleotides. Other polymerase Pfu DNA polymerase (obtained from Pyrococcus furiosus) is widely used due to its higher fidelity when copying DNA. In addition, Taq also adds an extra A at the 3' end (it is found useful for TA cloning).

Reaction Conditions

The volumes of reaction medium vary between 10 and 100 µl. There are a multitude of reaction medium formulas. However, it is possible to define a standard formula that is suitable for most polymerization reactions. This formula has been chosen by most manufacturers and suppliers, who, moreover, deliver a ready-to-use buffer solution with Taq polymerase. Concentrated 10 times, its formula is approximately the following: 100 mM Tris-HCl, pH 9.0; 15 mM $MgCl_2$, 500 mM KCl.

It is possible to add detergents (Tween 20, Triton X-100) or glycerol in order to increase the conditions of stringency that make it harder and therefore more selective hybridization of the primers. This approach is generally used to reduce the level of nonspecific amplifications due to the hybridization of the primers on sequences without relationship with the sequence of interest. We can also reduce the concentration of KCl until eliminated or increase the concentration of $MgCl_2$. Indeed, some pairs of primers work better with solutions enriched with magnesium. On the other hand, with high concentrations of dNTP, the concentration of magnesium should be increased because of stoichiometric interactions between magnesium and dNTPs that reduce the amount of free magnesium in the reaction medium. dNTPs (deoxyribonucleoside triphosphates) provide both the energy and the nucleotides needed for DNA synthesis during the chain polymerization. They are incorporated in the reaction medium in excess, that is, about 200 µM final. Depending on the reaction volume chosen, the primer concentration may vary between 10 and 50 pmol per sample. Matrix DNA can come from any organism and even complex biological materials that include DNAs from different organisms. But to ensure the success of a PCR, it is still necessary that the DNA matrix is not too degraded. This criterion is obviously all the more crucial as the size of the sequence of interest is large. It is also important that the DNA extract is not contaminated with inhibitors of the polymerase chain reaction (detergents, EDTA, phenol, proteins, etc.). The amount of template DNA in the reaction medium initiate that the amplification reaction can be reduced to a single copy. The maximum quantity may in no case exceed 2 μg. In general, the amounts used are in the range of 10–500 ng of template DNA. The amount of Taq polymerase per sample is generally between 1 and 3 units. The choice of the duration of the temperature cycles and the number of cycles depends on the size of the sequence of interest as well as the size and the complementarity of the primers. The durations should be reduced to a minimum not only to save time but also to prevent risk of nonspecific amplification. For denaturation and hybridization of primers, 30 seconds are usually sufficient. For elongation, it takes 1 minute per kilobase of DNA of interest and 2 minutes per kilobase for the final cycle of elongation. The number of cycles, generally between 20 and 40, is inversely proportional to the abundance of DNA matrix.

DNA Polymerases Requires a Template and a Primer

DNA polymerases catalyze the formation of polynucleotide chains through the addition of successive nucleotides derived from deoxynucleoside triphosphates. The polymerase reaction takes place only in the presence of an appropriate DNA template. Each incoming nucleoside triphosphate first forms an appropriate base pair with a base in this template. Only then does the DNA polymerase link the incoming base with the predecessor in the chain. Thus, DNA polymerases are template-directed enzymes.

DNA polymerases add nucleotides to the 3′ end of a polynucleotide chain. The polymerase catalyzes the nucleophilic attack of the 3′-hydroxyl group terminus of the polynucleotide chain on the α-phosphate group of the nucleoside triphosphate to be added. To initiate this reaction, DNA polymerases require a primer with a free 3′-hydroxyl group already base-paired to the template. They cannot start from scratch by adding nucleotides to a free single-stranded DNA template. RNA polymerase, in contrast, can initiate RNA synthesis without a primer.

All DNA Polymerases have Structural Features in Common

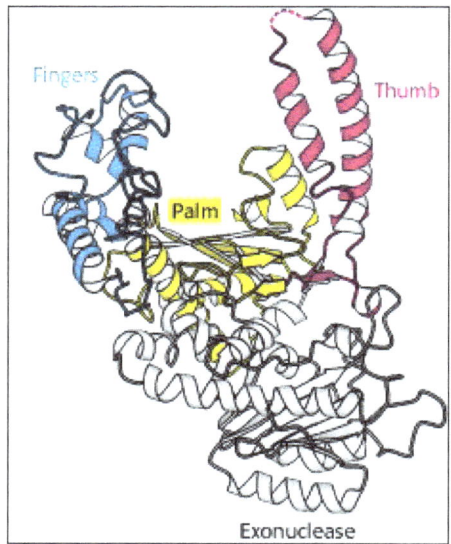

Figure shows DNA polymerase structure. The first DNA polymerase structure determined was that of a fragment of E. coli DNA polymerase I called the Klenow fragment. Like other DNA polymerases, the polymerase unit resembles a right hand with fingers (blue), palm (yellow), and thumb (red). The Klenow fragment also includes an exonuclease domain.

The three-dimensional structures of a number of DNA polymerase enzymes are known. The first such structure to be determined was that of the so-called Klenow fragment of DNA polymerase I from E. coli. This fragment comprises two main parts of the full enzyme, including the polymerase unit. This unit approximates the shape of a right hand with domains that are referred to as the fingers, the thumb, and the palm. In addition to the polymerase, the Klenow fragment includes a domain with $3' \rightarrow 5'$ exonuclease activity that participates in proofreading and correcting the polynucleotide product.

DNA polymerases are remarkably similar in overall shape, although they differ substantially in detail. At least five structural classes have been identified; some of them are clearly homologous, whereas others are probably the products of convergent evolution. In all cases, the finger and thumb domains wrap around DNA and hold it across the enzyme's active site, which comprises residues primarily from the palm domain. Furthermore, all the polymerases catalyze the same polymerase reaction, which is dependent on two metal ions.

Two Bound Metal Ions Participate in the Polymerase Reaction

Like all enzymes with nucleoside triphosphate substrates, DNA polymerases require metal ions for activity. Examination of the structures of DNA polymerases with bound substrates and substrate analogs reveals the presence of two metal ions in the active site. One metal ion binds both the deoxynucleoside triphosphate (dNTP) and the 3'-hydroxyl group of the primer, whereas the other interacts only with the 3'-hydroxyl group. The two metal ions are bridged by the carboxylate groups of two aspartate residues in the palm domain of the polymerase. These side chains hold the metal ions in the proper position and orientation. The metal ion bound to the primer activates the 3'-hydroxyl group of the primer, facilitating its attack on the α-phosphate group of the dNTP substrate in the active site. The two metal ions together help stabilize the negative charge that accumulates on the pentacoordinate transition state. The metal ion initially bound to dNTP stabilizes the negative charge on the pyrophosphate product.

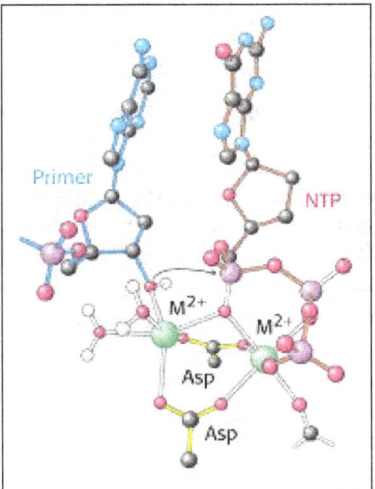

Figure shows DNA polymerase mechanism. Two metal ions (typically, Mg^{2+}) participate in the DNA polymerase reaction. One metal ion coordinates the 3'-hydroxyl group of the primer, whereas the phosphate group of the nucleoside triphosphate bridges between the two metal ions. The hydroxyl group of the primer attacks the phosphate group to form a new O-P bond.

Specificity of Replication Dictated by Hydrogen Bonding and the Complementarity of Shape between Bases

DNA must be replicated with high fidelity. Each base added to the growing chain should with high probability be the Watson-Crick complement of the base in the corresponding position in the template strand. The binding of the NTP containing the proper base is favored by the formation of a base pair, which is stabilized by specific hydrogen bonds. The binding of a non-complementary base is unlikely, because the interactions are unfavorable. The hydrogen bonds linking two complementary bases make a significant contribution to the fidelity of DNA replication. However, DNA polymerases replicate DNA more faithfully than these interactions alone can account for.

The examination of the crystal structures of various DNA polymerases indicated several additional mechanisms by which replication fidelity is improved. First, residues of the enzyme form hydrogen bonds with the minor-groove side of the base pair in the active site. In the minor groove, hydrogen-bond acceptors are present in the same positions for all Watson-Crick base pairs. These interactions act as a "ruler" that measures whether a properly spaced base pair has formed in the active site. Second, DNA polymerases close down around the incoming NTP. The binding of a nucleoside triphosphate into the active site of a DNA polymerase triggers a conformational change: the finger domain rotates to form a tight pocket into which only a properly shaped base pair will readily fit. The mutation of a conserved tyrosine residue at the top of the pocket results in a polymerase that is approximately 40 times as error prone as the parent polymerase.

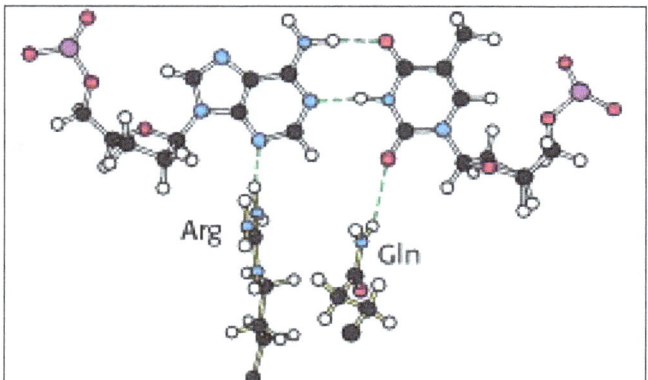

Figure: Minor-Groove Interactions. DNA polymerases donate two hydrogen bonds to base pairs in the minor groove. Hydrogen-bond acceptors are present in these two positions for all Watson-Crick base pairs including the A-T base pair shown.

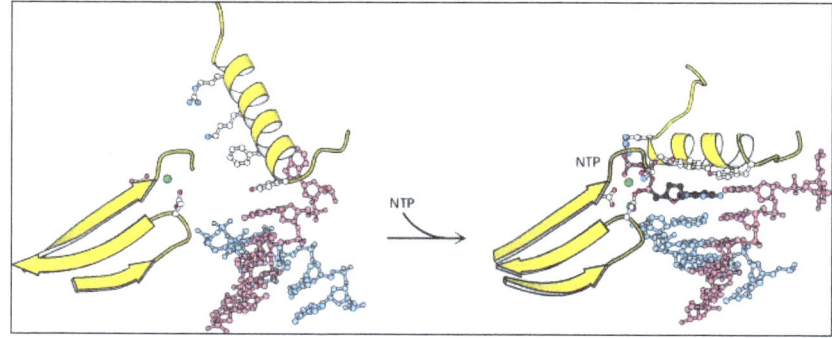

Figure shows shape selectivity. The binding of a nucleoside triphosphate (NTP) to DNA polymerase induces a conformational change, generating a tight pocket for the base pair consisting of the NTP and its partner on the template strand. Such a conformational change is possible only when the NTP corresponds to the Watson-Crick partner of the template base.

Many Polymerases Proofread the Newly Added Bases and Excise Errors

Many polymerases further enhance the fidelity of replication by the use of proofreading mechanisms. As already noted, the Klenow fragment of E. coli DNA polymerase I include an exonuclease domain that does not participate in the polymerization reaction itself. Instead, this domain removes mismatched nucleotides from the 3′ end of DNA by hydrolysis. The exonuclease active site is 35 Å from the polymerase active site, yet it can be reached by the newly synthesized polynucleotide chain under appropriate conditions. The proofreading mechanism relies on the increased probability that the end of a growing strand with an incorrectly incorporated nucleotide will leave the polymerase site and transiently move to the exonuclease site.

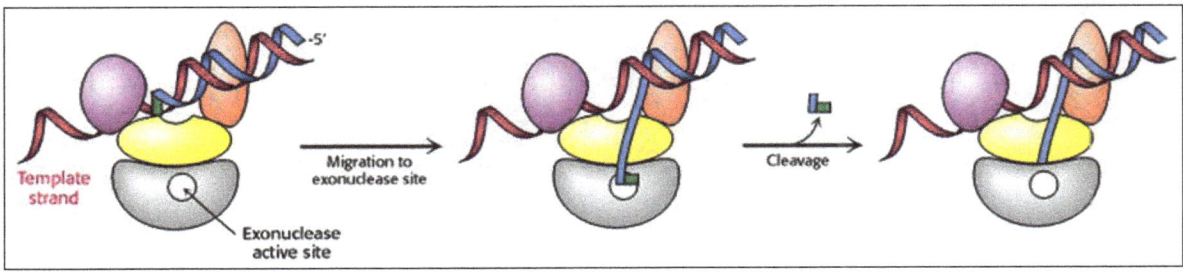

Figure shows proofreading. The growing polynucleotide chain occasionally leaves the polymerase site of DNA polymerase I and migrates to the exonuclease site. There, the last nucleotide added is removed by hydrolysis. Because mismatched bases are more likely to leave the polymerase site, this process serves to proofread the sequence of the DNA being synthesized.

How does the enzyme sense whether a newly added base is correct? First, an incorrect base will not pair correctly with the template strand. Its greater structural fluctuation, permitted by the weaker hydrogen bonding, will frequently bring the newly synthesized strand to the exonuclease site. Second, after the addition of a new nucleotide, the DNA translocates by one base pair into the enzyme. The newly formed base pair must be of the proper dimensions to fit into a tight binding site and participate in hydrogen-bonding interactions in the minor groove similar to those in the polymerization site itself. Indeed, the duplex DNA within the enzyme adopts an A-form structure, allowing clear access to the minor groove. If an incorrect base is incorporated, the enzyme stalls, and the pause provides additional time for the strand to migrate to the exonuclease site. There is a cost to this editing function, however: DNA polymerase I removes approximately 1 correct nucleotide in 20 by hydrolysis. Although the removal of correct nucleotides is slightly wasteful energetically, proofreading increases the accuracy of replication by a factor of approximately 1000.

The Separation of DNA Strands Requires Specific Helicases and ATP Hydrolysis

For a double-stranded DNA molecule to replicate, the two strands of the double helix must be separated from each other, at least locally. This separation allows each strand to act as a template on which a new polynucleotide chain can be assembled. For long double-stranded DNA molecules, the rate of spontaneous strand separation is negligibly low under physiological conditions. Specific enzymes, termed helicases, utilize the energy of ATP hydrolysis to power strand separation.

The detailed mechanisms of helicases are still under active investigation. However, the determination of the three-dimensional structures of several helicases has been a source of insight. For example, a bacterial helicase called PcrA comprises four domains, hereafter referred to as domains A1, A2, B1, and B2. Domain A1 contains a P-loop NTPase fold, as was expected from amino acid sequence analysis. This domain participates in ATP binding and hydrolysis. Domain B1 is homologous to domain A1 but lacks a P-loop. Domains A2 and B2 have unique structures.

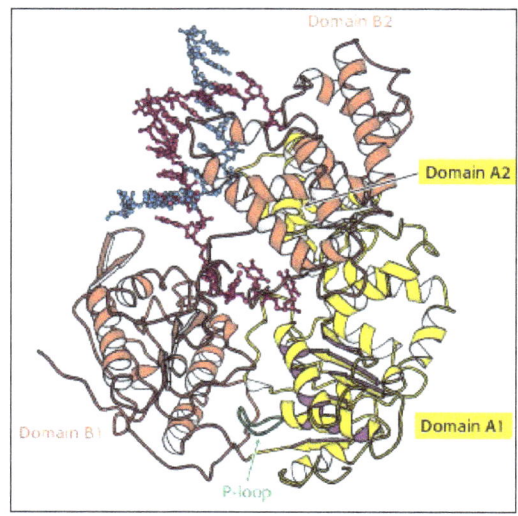

Figure shows Helicase structure. The bacterial helicase PcrA comprises four domains: A1, A2, B1, and B2. The A1 domain includes a P-loop NTPase fold, whereas the B1 domain has a similar overall structure but lacks a P-loop and does not bind nucleotides. Single-stranded DNA binds to the A1 and B1 domains near the interfaces with domains A2 and B2.

From an analysis of a set of helicase crystal structures bound to nucleotide analogs and appropriate double- and single-stranded DNA molecules, a mechanism for the action of these enzymes was proposed. Domains A1 and B1 are capable of binding single-stranded DNA. In the absence of bound ATP, both domains are bound to DNA. The binding of ATP triggers conformational changes in the P-loop and adjacent regions that lead to the closure of the cleft between these two domains. To achieve this movement, domain A1 releases the DNA and slides along the DNA strand, moving closer to domain B1. The enzyme then catalyzes the hydrolysis of ATP to form ADP and orthophosphate. On product release, the cleft between domains A and B springs open. In this state, however, domain A1 has a tighter grip on the DNA than does domain B1, so the DNA is pulled across domain B1 toward domain A1. The result is the translocation of the enzyme along the DNA strand in a manner similar to the way in which an inchworm moves. In regard to PcrA, the enzyme translocates in the $3' \rightarrow 5'$ direction. When the helicase encounters a region of double-stranded DNA, it continues to move along one strand and displaces the opposite DNA strand as it progresses. Interactions with specific pockets on the helicase help destabilize the DNA duplex, aided by ATP-induced conformational changes.

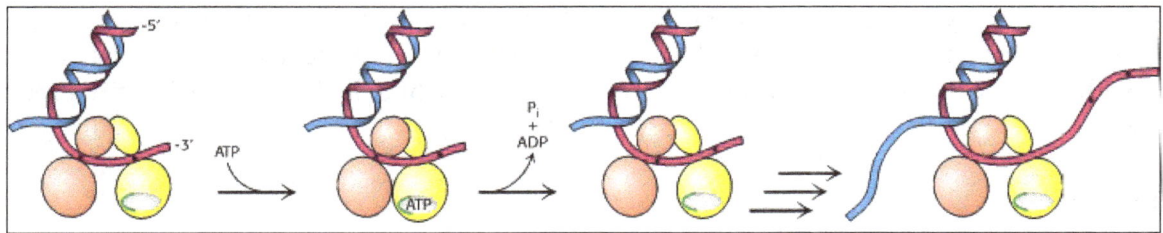

Figure shows Helicase mechanism. Initially, both domains A1 and B1 of PcrA bind single-stranded DNA. On binding of ATP, the cleft between these domains closes and domain A1 slides along the DNA. On ATP hydrolysis, the cleft opens up, pulling the DNA from domain B1 toward domain A1. As this process is repeated, double-stranded DNA is unwound.

Helicases constitute a large and diverse class of enzymes. Some of these enzymes move in a $5' \rightarrow 3'$ direction, whereas others unwind RNA rather than DNA and participate in processes such as RNA splicing and the initiation of mRNA translation. A comparison of the amino acid sequences of hundreds of these enzymes reveals seven regions of striking conservation. Mapping these regions onto the PcrA structure shows that they line the ATP-binding site and the cleft between the two domains, consistent with the notion that other helicases undergo conformational changes analogous to those found in PcrA. However, whereas PcrA appears to function as a monomer, other members of the helicase class function as oligomers. The hexameric structures of one important group are similar to that of the F1 component of ATP synthase, suggesting potential mechanistic similarities.

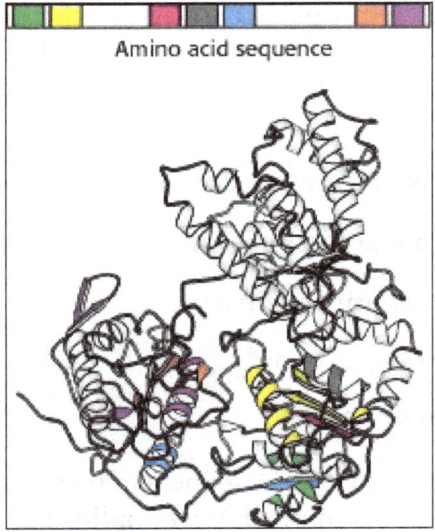

Figure shows conserved residues among Helicases. A comparison of the amino acid sequences of hundreds of helicases revealed seven regions of strong sequence conservation (shown in color). When mapped onto the structure of PcrA, these conserved regions lie along the interface between the A1 and B1 domains and along the ATP binding surface.

Buffer

For any enzymatic reaction an optimal reaction condition is needed which is provided by the Buffer solution. For example, buffer used for PCR reaction provides a suitable chemical environment for optimum activity and stability of the DNA polymerase. PCR buffer contains tris-HCl, KCl or MgCl$_2$ and glycerol or Triton X-100 and water. In most cases the PCR reaction buffer is provided along with the DNA polymerase and in 10X concentration, which is diluted to 1X in final reaction volume.

dNTPs

dNTP (Deoxynucleoside triphosphates) is the basic substrate for PCR reaction which is a mixture of dATP, dTTP, dGTP and dCTP. It acts as the building block for polymerization of new DNA strand. Excess of dNTP can increase rate of error and can also inhibit the Taq. Lower amount of dNTP may reduce rate of reaction. Thus an optimum concentration of dNTP is required to be taken for PCR. Usually dNTP are used at 10-50 µM concentration. If we need to amplify large size of DNA, higher concentration of dNTP is required.

Primers

Primers are the last but the most important in PCR. They are short pieces of single-stranded DNA about 10 base pair that are complementary to the target sequence. DNA polymerase starts synthesizing new DNA from the end of the primer. Up to 3µM of primers is sufficient. High template to primer ratio may result to non-specific amplification and formation of dimers of primer. In most of PCR a set of 2 primers (forward and reverse) is used for amplifying a specific region of gene. But a number of PCR have been designed which can run with a single primer as RAPD (Rapid Amplification of Polymorphic DNA) where the primer binds to several polymorphic sites in genome and amplifies all of these regions resulting into PCR product of variable sizes. RAPD is successfully used for differentiating plant varieties, animal breeds, and microbial strains.

While designing any primer few points are to be considered:

- It should be of a size around 20-30 nucleotide.

- It should avoid formation of primer-dimer.

- GC content should be between 40-60%.

- Tm should be between 55-65 °C.

Increasing number of information about DNA sequences over various public databases has made it easy to design specific primer sets. A number of free online bioinformatics tools can be used for designing primers and PCR conditions as Primer3, Primer-BLAST etc.

Instrument

Any PCR reaction is most commonly carried out in a reaction tube (volume 0.2–0.5 ml) covering reaction volume of 10-200 µl in a thermal cycler. The thermal cycler is the machine which heats and cools the reaction tubes to provide the temperatures required for each PCR step of the reaction (denaturation, annealing, and extension). At the end of PCR the reaction is given a hold at 4 °C to stop the reaction and save the PCR product from degradation.

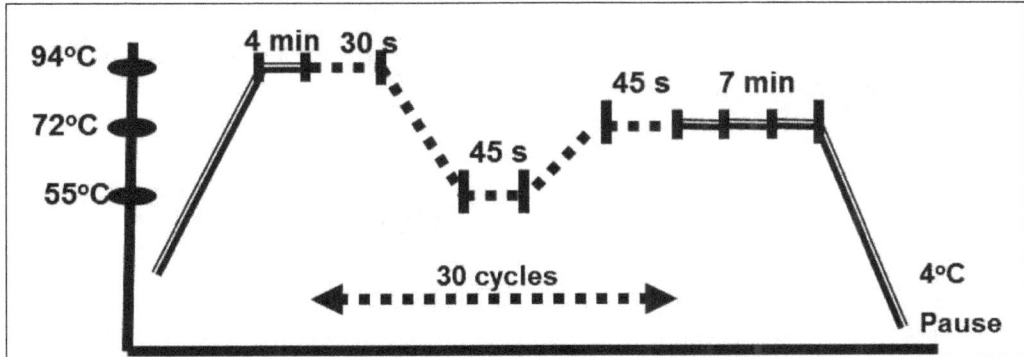

After each PCR reaction the result is visualized by agarose gel electrophoresis of the PCR product which should show a band of specific size for positive result. Absence of band or band of different size shows nonspecific result.

Thermal cyclers now use Peltier chips for both heating and cooling of the block. PCR reaction tubes are made thin to allow rapid thermal equilibration. Thermal cyclers also heated lids to preventing condensation at top of PCR tubes.

Optimization and Troubleshooting in Polymerase Chain Reaction

Optimization of PCR is essential to achieve complete reaction in each step of each cycle. If elongation step results in complete and un-complete extension of primers, it will result in synthesis of undesired products of variable sizes. The yield of desired product which is derived from completion of strand in extension step will be low and at times yield could be too low and will require performing PCR again. Optimization is done with respect to concentration of template, concentration of primers, concentration of Mg2+ and annealing temperature. Low annealing temperature results in mispairing while higher temperature may lead to poor binding of primer to template. When nonspecific products persist in spite of all attempted variations in reaction conditions, nested PCR can be attempted. In this approach, PCR is carried out in two rounds. In first round of PCR, primers are designed to give product which must have target DNA fragment. In second round, specific primers are used to amplify regions of target DNA.

The use of PCR to generate large amounts of a desired product can be a double-edged sword. Failure to amplify a sample under optimum conditions can lead to the generation of multiple undefined and unwanted products - even to the exclusion of the desired product. At the other extreme, no product may be amplified. A typical response at this point is to vary one or more of the many parameters that are known to contribute to primer-template fidelity and primer extension. High on the list of optimization variables are Mg^{2+} concentrations, buffer pH, and cycling conditions. With regard to the last, the annealing temperature is most important. The situation is complicated further by the fact that some of the variables are quite interdependent. For example, because dNTPs directly chelate a proportional number of Mg^{2+} ions, an increase in the concentration of dNTPs decreases the concentration of free Mg^{2+} available to influence polymerase function.

Enhancing Agents

Various additives such as DMSO (2%-5%), PEG 6000 (5%, 15%), glycerol (500-20%), nonionic detergents, and formamide (5%) can also be incorporated into the reaction to increase specificity. Some reactions may amplify only in the presence of such additives. Several optimization kits incorporating these and other enhancing agents, and a variety of buffers, are currently being marketed.

Matrix Analyses

The basic challenge is to devise an optimization strategy that is efficient in both time and cost. A full matrix analysis in which several values for each of the variables are tested in combination with each of the other variables can quickly become overwhelmingly cumbersome and costly. The size of the matrix can be pared down significantly by application of the Taguchi method, in which several key variables are altered simultaneously. A more typical strategy is to run a simple matrix analysis focused on those parameters most likely to have the greatest impact on hybridization fidelity (i.e., Mg^{2+} concentration and annealing temperature).

Mg^{2+} Concentration

Mg^{2+} concentration is the easiest to manipulate because all concentration variations can be run in separate tubes simultaneously. Suppliers of Taq polymerase now provide $MgCl_2$ solution separate from the rest of the standard reaction buffer to simplify its adjustment. A typical two-step optimization series might first include Mg^{2+} at 0.5 mM increments from 0.5 to 5.0 mM and, after the range is narrowed, at several 0.2 or 0.3 mM increments.

Annealing Temperature

Optimization of annealing temperature begins with calculation of the melting temperatures (T_m) of the primer-template pairs by one of several methods, the simplest being $T_m = (G + C) 4 + (A + T)2$. A single base mismatch lowers the T_m by ~5 °C. More complex formulas can be used also, but in practice, because the T_m is affected variously by the individual buffer components and even the primer and template concentrations, any calculated T_m value should be regarded as an approximation. Several reactions run at temperature increments (2-5° straddling a point 5 °C below the calculated T_m will give a first approximation of the optimum annealing temperature for a given set of reaction conditions. It should be noted that some primers, for reasons that are not entirely apparent, are refractory to optimization. One possible explanation may be that unique characteristics of the target amplicon give a T_m above the temperature of the denaturation cycle segment. In such situations, it may be more time and cost efficient simply to design a second set of primers that hybridize to neighboring DNA.

Primer Design

Ideally, PCR primers should have a 40-60% GC content and a 3'-terminal "G/C clamp" (at least one or two 3' G's and/or C's); be similar in size (18-25 bases), T_m values, and nucleotide ratios; and be free of repetitive motifs, palindromes, excessive degeneracy, and long stretches of polypurines or polypyrimidines. Fortunately, deviation from any or all of these guidelines is tolerable but may necessitate greater attention to other optimization parameters. Several useful computer programs have been developed to aid in efficient primer design.

Cycle Number, Reamplification and Product Smearing

Increasing the number of cycles may enhance an anemic reaction, but this modification can also lead to the generation of spurious bands and to smears composed of high molecular weight products rich in single-stranded DNA. Similar smearing can occur under normal conditions if the level of starting template is too high, as often occurs in attempts to reamplify from a previous PCR. A general rule of thumb is to use 1 μl of a $1:10^4$-10^5 dilution of a PCR if a gel band is detectable.

Pipetting and Thermal Cycler Variables

To minimize pipetting variables, a master mix containing all of the reactants except the Taq polymerase can be made in advance and stored at 4°C for the several days (or even longer) that it might take to run the reactions at each of the selected annealing temperatures. If multiple thermal cyclers are available, it may be tempting to run the reactions concurrently, but small differences in calibration and performance among machines could render the results useless, particularly for marginal reactions. Also, when running multiple sequential reactions, it is prudent to use the same wells on the sample block to avoid possible well-to-well variations. At least one manufacturer of thermal cyclers (Stratagene) offers a machine that can run eight different temperature profiles simultaneously in the same block to aid in annealing temperature optimization.

Touchdown PCR

Touchdown (TD) PCR represents a fundamentally different approach to PCR optimization. Rather than multiple reaction tubes, each with different reagent concentration and/or cycling parameters, a single tube or a small set of tubes is run under cycling conditions that inherently favor amplification of the desired amplicon, often to the exclusion of artifactual amplicons and primer-dimers. Multiple cycles are programmed so that the annealing segments in sequential cycles are run at incrementally lower temperatures. As cycling progresses, the annealing segment temperature, which was selected to be initially above the suspected T_m, gradually declines to, and falls below, this level. This strategy helps ensure that the first primer-template hybridization events involve only those reactants with the greatest complementarity, (i.e., those yielding the target amplicon). Even though the nealing temperature may eventually drop down to the T_m of nonspecific hybridizations, the target amplicon already will have begun its geometric amplification and is thus in a position to outcompete any lagging (nonspecific) PCR products during the remaining cycles. Because the aim is to avoid low Tm priming during the earlier cycles, it is imperative that the hot start modification be utilized with TD PCR. TD PCR should be viewed not as a method of determining the optimum cycling conditions for a specific PCR but as a potential one-step method for approaching optimal amplification. We have found that a variety of otherwise satisfactory single amplicon-yielding reactions are rendered more robust (i.e., yield more product) when subjected to TD PCR.

TD PCR is of particular value when the degree of identity between the primer and template is unknown. This situation often arises as primers are designed on the basis of amino acid sequences, members of a multigene family are amplified, or evolutionary PCR is attempted (i.e., amplification of DNA from one species using primers with identity to a homologous segment of another species). In such cases, the mismatches among the primers and template may have lowered the T_m s of the target amplicons enough to approach those of the spurious priming sites. Degenerate primers with multiple base variation or inosine residues are often used in such situations, but the greater variety of sequences in the former case and the relaxed stringency in the latter case might tend to increase the chances of nonspecific priming. Moreover, in some cases the locations of potential base mismatches will be unknown. Although TD PCR can be used with degenerate primers, we have shown that non-degenerate primers displaying a significant degree of template - sequence mismatches can yield single-target amplicons of single copy genes from genomic DNA under standard buffer conditions. Even mismatches clustered near the 3' end of the primer are tolerated.

Nested PCR

Nested and seminested PCRs are often quite successful in reducing or eliminating unwanted products while at the same time dramatically increasing sensitivity. An initial set of primers straddling the DNA segment of interest is first amplified under standard conditions. Spurious products are frequently primed with one or both primers and contain irrelevant sequences internally. An aliquot of the reaction product mixture is then subjected to an additional round of amplification using primers complementary to the sequences internal to the first set of primers. Only the legitimate product should be amplified in this second round. This approach is often successful even if the desired product is initially below the level of detection by ethidium bromide staining and in the presence of visible spurious bands. Seminested PCR, in which a second primer is internal to only one end of the target segment, can be equally effective. This variation is often required for gene walking or attempts at 5' or 3' RACE (rapid amplification of cDNA ends) in which the template DNA sequence internal to only one of the primers is known.

A second form of artifact, known as jumping PCR, may not be eliminated by nested PCR. Incompletely extended products can occasionally rehybridize to an adjacent segment of DNA, perhaps to a similar gene element, to prime an unintended product. In such instances, sequences internal to one or both primers will still be present, but the amplicon size will differ.

If it is anticipated that nested PCR methods will be employed, better results may be obtained if the first and second rounds of amplification are terminated after 20 or so cycles rather than the usual 30-35. This modification will minimize the chances of generating unwanted high molecular weight bands and smears. Such artifacts often contain considerable single-stranded DNA and appear to be the result of mispriming by DNA products amplified in earlier cycles. Nested PCR is extremely sensitive, and as little as a single copy of a viral gene has been detected in a background of 10^6 genomes.

Hot Start PCR

Even brief incubations of a PCR reaction mix at temperatures significantly below the T_m can result in primer-dimer and nonspecific priming. Hot start PCR methods can dramatically reduce these problems. The aim is to withhold at least one of the critical components from participating in the reaction until the temperature in the first cycle rises above the T_m of the reactants. For example, in smaller assays incorporating an oil overlay, one of the components common to all tubes (e.g., Taq polymerase) can be initially withheld and added only after the temperature rises above 80~ during the first denaturing stage. Alternatively, a wax bead can be melted over the bulk of the reaction mix in each tube and allowed to solidify and the withheld component pipetted on top of the wax cap. These beads can be made in the laboratory or purchased (Ampliwax PCR Gems, Perkin-Elmer Cetus). During the temperature ramp into the first denaturation segment, the wax will melt and the final component will become incorporated and mixed by convection in each tube, a great convenience in dealing with large numbers of tubes. A recent hot start variation involves adding specific anti-Taq polymerase antibody (TaqStart Antibodies, Clontech) to the PCR reaction tubes prior to the addition of Taq polymerase. The antibody prevents polymerase activity from beginning until the rising temperature dissociates and denatures the blocking antibody. This modification is compatible with newer thermal cyclers and techniques that seek to avoid the extra handling and purification steps accompanying oil and wax overlays.

Controls

Negative controls are mandatory in each PCR run to rule out false-positives caused by contamination. All primer sets and each new aliquot of buffer component must be tested. Fortunately, often this can be done in one or two tubes by incorporating all of the reagents except the DNA template. Even if multiple primer sets are being used, all can be run in the same tube. It is often desirable to run positive controls as well to ensure that master mixes have been formulated properly and to aid in optimization. When amplification from genomic DNA or some other low copy number source is attempted, it may be advisable first to optimize with higher-copy dilutions ($10^3 - 10^5$ copies) of the target DNA segment from plasmid or phage prep if available. Of course, great care must be exercised not to create contamination problems in the process. Another useful control is to purify irrelevant DNA in parallel with the template-containing DNA. PCR of this template-free DNA will serve as a process control to monitor for contamination picked up during the purification procedure.

Contamination

The best way to deal with the ever-present threat of contamination is to take every precaution to prevent it from happening in the first place. If the intended target template is from a human source or a common laboratory vector, the chance of contamination is considerably increased. All reagents including primers, templates, and H_2O should be stored in multiple small (ideally, single-use) aliquots and freely discarded if suspected of contamination. The use of positive displacement micropipettes and tips or standard micropipettes with filter-plugged tips may be advantageous. Pipetting should be slow and deliberate to minimize the generation of aerosols. Hands should be gloved at all times, and gloves should be changed frequently. A likely source of much contamination occurs when reaction tubes are opened and closed and can take the form of aerosol generation or inadvertent contact with the inner lip of the cap. Loose-fitting gloves tend to aggravate the problem. Tube-opening devices may be useful but can themselves be sources of contamination. Contamination also can be introduced during the purification of the template through equipment and reagents (e.g., homogenizers, reused tube, electrophoresis apparatus, centrifuges, common laboratory buffers, and phenol).

The primary source of contamination in many laboratories is carryover from previous PCRs. It is prudent to keep the workstations, including the pipettes, racks, and reagents used for setting up

PCR physically separate from those used for analyzing the reactions (in separate rooms if possible). A particular dilemma arises when secondary amplification is required (i.e., additional cycles are needed, a reamplification of a previous reaction is desired, or two-step protocols such as nested PCR are conducted). In such cases, sterile, uncapped tubes containing all of the reactants except the diluted template and Taq polymerase (withheld for hot start) are placed on a freshly prepared "sterile" field (i.e., a fresh sheet of plastic wrap or foil) to receive an aliquot of the first amplification product. Care is taken not to touch the tubes or rack with anything (including gloved hands) but the pipette tip during the transfer process. With a fresh pair of gloves, the tubes and rack can be transferred to the thermal cycler for further processing.

No matter how much care is exercised, contamination may occur on occasion. Several methods for dealing with this eventuality have been described. Environmental surfaces, including pipettes, can be decontaminated by wetting with a 0.07 M sodium hypochlorite (i.e., 10% Clorox bleach) solution. This method has been found to be more efficient and less corrosive than decontamination with HC1. UV irradiation of the work station is also beneficial, though dried DNA is much less susceptible to UV damage than hydrated forms. Containers, racks, and micropipettors, if permissible, should be autoclaved. Also, reagents can be decontaminated with UV light. The various components of a reaction mix display dramatically different sensitivities to UV-induced irreversible denaturation. Because dNTPs in the concentrations used in PCR are strong absorbers of UV irradiation and can effectively block decontamination efforts of final reaction mixes, it is advisable to decontaminate individual stock reagents or at least to withhold the dNTPs (perhaps in conjunction with hot start) until after reaction mix decontamination. Primers vary widely in their susceptibility to UV-induced denaturation depending, in part, on the likelihood of thymine dimer formation. $MgCl_2$, stock $10 \times$ buffers, and dNTPs are UV insensitive, whereas Taq polymerase is UV sensitive. The nature of the contaminant also influences the effectiveness of decontamination efforts. DNA of < 250 bases is neutralized less effectively. To decontaminate, place reaction components, less Taq polymerase and template, in clear reaction tubes on a Fotodyne 1000 transilluminator containing 254- and 300-nm UV bulbs or Photodyne Foto/Prep I (without transparent cover) with 300-nm bulbs or similar devices for 10-20 min. The 300-nm irradiation alone can be used but is somewhat less effective. A 10-min treatment in a Stratalinker 1800 (Stratagene) set to deliver 100 mJ per min at 254 nm can also be used.

Other approaches to decontamination include γ irradiation, digestion with restriction endonucleases, DNaseI, exonuclease III, isopsoralen treatment, incorporation of dU into PCR products followed by uracil-N-glycosylase (UNG) digestion, and centrifugal ultrafiltration. None of these additional approaches is ideal. An appropriate γ-ray source is frequently not available. Enzymatic digestions usually entail additional manipulations, which increase the risk of contamination.

The last two approaches are primarily directed at treating PCR products so as to render them incapable of serving as templates (i.e., preventing amplicons from serving as sources of carryover contamination for future PCRs). Isopsoralen derivatives are thermostable molecules capable of forming polymerase-inhibiting monoadducts upon photoactivation in the 320-nm to 400-nm light range. Isopsoralens can be conveniently added during PCR setup and subsequently activated by UV exposure to sterilize PCR products after amplification but before tube opening. The amount of isopsoralen added per tube depends on experimental conditions.

Taq polymerase readily incorporates uracil in place of thymine into PCR amplicons. As a consequence, UNG can be used to digest previously amplified uracil-containing amplicons that might otherwise be contaminating a fresh PCR reaction mix. UNG depyrimidinates uracil residues to block polymer extension and increase the rate of strand hydrolysis. UNG is not reactive with free dUTP and is thermolabile. Consequently, 5 ng of UNG can be added to a PCR mix and incubated for 15 min to digest uracil-containing DNA. The UNG is denatured during a 5-min incubation at 94°C preceding a normal PCR cycling program.

No Product

We have adjusted the Mg^{2+} concentration, buffer pH, and cycling parameters; added more cycles; tried lower annealing temperatures and TD PCR; and still no product is seen on ethidium bromide-stained gels (acrylamide gels are considerably more sensitive than agarose gels). What is the next step? Lengthening the initial denaturation step and/or increasing temperature will increase the likelihood that the template DNA is fully denatured to provide the maximal number of priming sites. Standard conditions for this optional step are 5 min at 95 °C. An in-tube thermocouple can be used to predetermine that the indicated temperature will correspond to the actual sample temperature.

Amplification may have occurred but may have been inefficient. If so, the amplicons can be revealed by a probe of the dried gel or a blot. A secondary amplification using the same primers or, preferably, nested primers may be all that is needed to generate a specific product. Serial 10-fold dilutions ranging from 1:100 to 1:10,000 should be used.

Inhibitors

Poor or nonexistent amplification may indicate the presence of inhibitors in the DNA sample. Numerous inhibitors of PCR have been described. These include ionic detergents (e.g., SDS and Sarkosyl), phenol, heparin, xylene cyanol, and bromphenol blue. Test for inhibitor in the template preparation by spiking original PCR mix with dilutions of known positive (demonstrably amplifiable) template. Re-extraction, ethanol precipitation, and/or centrifugal ultrafiltration may resolve the problem. Proteinase K carryover can serve to digest the Taq polymerase but is readily denatured by a 5-min incubation at 95 °C.

Suggested Optimization Strategy

The example given is TD PCR, but the same principles apply to conventional PCR.

- Design an optimal primer pair on the basis of considerations.

- Calculate or estimate approximate Tin. Program the thermal cycler for TD PCR.

- Set up several standard hot starts PCR mixes incorporating a range of Mg^{2+} concentrations and including appropriate positive and negative controls. Use $10^4 - 10^5$ copies of the template.

- Amplify and analyze products.

  - If weak or no product detected:

    - Subject reaction tubes to 10 additional cycles at constant annealing temperature (i.e., 55 °C and recheck.

- Reamplify 1 μl of 10-fold dilutions (1:10--1:1,000) of TD PCR at fixed annealing temperature.

- Use more template and check for inhibitor in template preparation by spiking original PCR mix with dilutions of known positive (demonstrably amplifiable) template.

- Increase initial template denaturation time and/or temperature.

- Vary concentrations of other buffer components (pH, Taq polymerase, dNTPs, and primers).

- Add enhancers to PCR mix.

- Reamplify dilutions (1:10-1:1,000) of first reaction using nested primers.

- Abandon this primer set, design new primers, and begin again. Depending on the researcher's degree of impatience and tolerance for frustration, this step might supercede any of the above.

○ If multiple products or a high molecular weight smear is observed:

- Raise the maximum and minimum annealing temperatures (i.e., move the range upward) in the TD PCR program.

- Remove some cycles from the bottom of the range and/or from the terminal constant temperature cycles.

- Increase the number of cycles per degree annealing temperature by one cycle (i.e., to three cycles/degree). Doing so may necessitate removing terminal cycles to prevent smearing.

- Vary concentrations of other buffer components (pH, Taq polymerase, dNTPs, and primers).

- Attempt band purification followed by reamplification. Target bands can be cut from gels and allowed to diffuse out or be liberated by freeze/thaw cycles or gel digestion. Alternatively, a small plug of gel can be removed with a micropipette tip, or most simply, by stabbing the band directly in the gel with an autoclaved toothpick (plastic may be preferable) and inoculating a fresh reaction tube.

- Reamplify $1:10^4$ and 1:10 s dilutions of first reaction using nested primers.

- Abandon primer set, design new primers, and begin again.

Touchdown PCR Programming

The goal in programming for TD PCR is to produce a series of cycles with progressively lower annealing temperatures. The annealing temperature range should span ~15 °C and extend from at least a few degrees above the estimated T_m to ~10 °C below. For example, for a calculated primer template T_m of 62 °C with no degeneracy, program the thermal cycler to decrease the annealing temperature 1~ every second cycle (i.e., run two cycles per degree) from 65 °C to 50 °C followed by

15 additional cycles at 50 °C Some thermal cyclers (Perkin-Elmer model 9600 and MJ Research model PTC-100) readily accommodate TD PCR and are easily programmed to decrease the temperature of a segment automatically by a fixed amount per cycle (e.g., 0.5° cycle). For others, a long series of files must be linked or extensive strings of commands entered. In these latter cases, it may be more convenient to create a generic TD PCR program covering a broader temperature range (~20 °C than to reprogram every time the range needs to be modified by a few degrees. Another alternative to programming restrictions and inconvenience is to use fewer but more abrupt steps (e.g., seven 2 °C steps, or five 3 °C steps); doing so may decrease the chances for discriminating among products with two closely spaced T_m S.

The continued presence of spurious bands following TD PCR indicates that the initial annealing temperature was too low, that there is a relatively small gap between the T_m S of the target and unwanted amplicons and/or that the unwanted amplicon is being more efficiently amplified. Raising the number of cycles per 1 °C descending step to three or four will give the target amplicon added competitive advantage before the initiation of the spurious amplification. A proportional number of cycles should be removed from the end of the program to prevent excess cycling and the concomitant degradation of the amplicon and generation of high molecular weight smears.

Modifications of TD PCR for use with degenerate and mismatched primers include lowering the annealing temperature range (e.g., 50 °C declining to 35 °C while keeping the last 15 cycles at > 50 °C (once priming has begun, the primers are fully complementary to the newly formed amplicons, have a much higher T_m, and do not benefit from excessively low annealing temperatures).

Conditions Favoring Enhanced Specificity

Adjusting conditions in the direction opposite to that listed below usually favors increased sensitivity (i.e., more products) and the concomitant risk of nonspecific amplification. The aim is to strike a balance between these two opposing tendencies. (↑ and ↓ signify increase and decrease, respectively).

- Use hot start

- Use TD PCR (favors enhancer specificity and sensitivity)

- Optimize primer design

  ↓ Mg^{2+}

  ↓ dNTP (also favors higher fidelity)

- Optimize pH

  ↓ Taq polymerase

  ↓ Cycle segment lengths

  ↓ Number of cycles

  ↑ Annealing temperature

↓ Inhibitors

↑ Ramp speed

↑ Chance that target temperature is achieved in each tube

Add and optimize enhancer(s)

↓ Primer concentration

↓ Primer degeneracy

↑ Template denaturation efficiency

PCR Product Detection Methods

To utilize PCR to its fullest potential, the identification and measurement of PCR products, or amplicons, must be improved compared with traditional methods. Amplicons must be identified by their specific nucleotide sequence to prevent false or ambiguous results caused by primer-dimer formation, nonspecific amplification, or target sequence variation. Quantitation of amplicon levels is an important first step for the estimation of the relative number of initial target molecules.

Until several years ago, PCR was usually performed on clean model systems or on specimens that had been highly purified. Gel-based detection methods worked well for these applications when sample throughput, time, and persample costs were not important issues. Over the past several years, PCR has become a mainstream technique, moving from the realm of the molecular biology laboratory to a wide variety of fields that are testing large numbers of samples and a wide variety of sample types. New detection methods are needed that meet the requirements of these users.

Traditional Detection Methods

Gel-based methods for the detection of PCR products have a number of drawbacks. Agarose gel electrophoresis with ethidium bromide staining is simple and inexpensive but suffers from a lack of sensitivity and specificity. Typically, at least 10 ng of full-length PCR product in a 5-μl volume must be loaded onto a gel so that a distinct band will be visible against a clean background. For a 200-bp PCR product, this amount translates to nearly 5×10^{10} full-length product molecules and would require an amplification of 5×10^{9} for detection of 10 initial target molecules in a 100-1 μl amplification reaction. Achieving this level of amplification would require 32 amplification cycles of 100% efficiency, a level of amplification that may be nearly achievable in a clean model system but is rarely achievable in practice with experimental or clinical samples.

Specificity may also be a significant problem with ethidium bromide-based detection methods. Often, PCR products from clinical or biological samples appear on the gel as a number of bands or as a smear. In these cases, it is difficult to determine whether the correct size band has been generated. To overcome this problem, two alternative gel-based methods have been used.

In one method, PCR products in an agarose gel are transferred to a membrane by standard blotting techniques and are subsequently detected with a labeled probe of specific sequence. (1) With proper hybridization conditions, this technique is sequence specific and will not detect nonspecific amplification products. However, in some instances, a significant amount of the PCR product is not full length, and the results may be ambiguous because the detected products will still appear as a smear. The sensitivity of blotting methods may vary, depending on the specific label and the detection method used. Radioactive labels or enzymatic labels coupled with chemiluminescent detection usually give the most sensitive results, typically, one to three orders of magnitude better than ethidium bromide staining. Blotting methods are also laborious and time consuming, and radioactive labels pose significant disposal problems.

An alternative to blotting-based methods is nested-primer amplification, followed by agarose gel electrophoresis and ethidium bromide staining. With this technique, specificity and sensitivity are enhanced by the amplification conditions while the detection method itself remains insensitive and nonspecific. With the nested-primer technique, a first round of amplification is performed in a normal way. A portion of the first-round amplification reaction is then used as the sample in a second round of amplification using primers that are internal, or nested, to the first set. Although the sensitivity of agarose gel detection is low, the overall sensitivity is high, because the target has been subjected to two rounds of amplification. The specificity of amplification is also enhanced because the primers used in the second round of amplification will only amplify the first-round amplicons. The disadvantages of this method are the lengthy and tedious amplification procedures, the cost of amplification reagents, the higher probability of amplicon contamination during the seeding of the second round of amplification, and the inability to utilize the dUTPAJNG procedure to prevent carryover contamination.

Advanced Detection Methods

Advanced detection methods must provide significant advantages over traditional methods to justify the time, effort, and cost involved in converting to a new technology. Implementation of a new detection method into a laboratory will typically require significant investments in time to learn the new technology and to validate the performance of the new method against the currently used method. Moreover, advanced detection methods may require significant investment in capital equipment and the purchase of expensive consumables and reagents.

To overcome these drawbacks, advanced detection methods must exhibit superior sensitivity and specificity compared with traditional methods. Ten or fewer input target copies should be detectable after one round of amplification, and the detection should be sequence specific. Primer-dimers and other nonspecific amplification products should not be detected, and to ensure against carryover contamination, all detection methods should be capable of incorporating the dUTP/ UNG decontamination procedure.

Advanced detection methods should be easy to use, require little or no specialized training, and minimize tedious and laborious procedures. The detection method should require no purification of the PCR reaction products, require as little hands-on time as possible, and be amenable to automation with currently available equipment. Flexible throughput is desirable so that small or large batch sizes can be run without wasting reagents or consumables.

The total cost of an advanced detection method should be competitive with traditional detection methods, notwithstanding performance considerations. One-time costs are usually limited to capital equipment expenditures such as readers, shakers, and incubators--equipment that may already be available in many laboratories. Recurring costs include reagents, disposables, and labor. With advanced detection methods, however, cost savings may be realized from increased productivity, reduced hands-on time, higher sample throughput, less repeat amplification because of ambiguous results, and reduced cost of amplification reagents if converting from nested-primer methods.

Universality is an important property of a detection method for most laboratories. A universal detection system uses common equipment, reagents (except sequence-specific components such as primers and probes), and a common procedure for detecting PCR products. This feature minimizes the equipment and number of different reagents that laboratories must purchase and store. A universal detection procedure makes detection of new analytes quick and facile. Another important characteristic of a universal detection system is that multiple targets can be detected in the same assay, a feature that is especially important to small-volume users.

Advanced detection methods can be developed in-house from numerous reports found in the literature. Advantages of using in-house methods are lower reagent costs and a higher level of technical knowledge about the assay that comes from hands-on development. Conversely, the laboratory may be left with little or no technical expertise in the production, validation, or performance of the in-house assay if the individuals responsible for the development leave the laboratory or institution. Moreover, the reagents prepared for use in an in-house assay require additional in-house quality control and quality assurance programs on a routine and on-going basis.

Commercially available detection methods offer several distinct advantages over in-house assays. Reagents in commercial kits have already been validated for their intended use, the laboratory does not need to develop a reagent quality control or quality assurance program, and the supply of reagents is usually reliable and consistent. The manufacturer often supplies defined protocols, and lot-to-lot or kit-to-kit variability is controlled, giving enhanced reproducibility. An additional convincing argument for using a commercially available detection method is the technical support that is available with the purchase of a commercial advanced detection system. Support should include detailed protocols, hands-on training, support literature, applications notes, and troubleshooting.

ELISA-based detection methods offer the most promising alternative to gel-based detection methods because ELISA assays are relatively standardized, and ELISA techniques are familiar to most laboratory personnel throughout the world. In addition, a wide range of materials, equipment, and supplies are commercially available to support ELISA-based technology, and many laboratories are already thoroughly equipped to run ELISA-based assays. Other advanced detection methods based on alternative technologies, such as high performance liquid chromatography (HPLC) or capillary electrophoresis, are currently in limited use or under development. In this chapter, a number of in-house and commercially available advanced detection systems will be discussed with emphasis placed on ELISA methods.

Elisa Detection Methods

Oligonucleotide-based Detection Methods

Numerous new methods for detection of PCR products are found in the literature. Two common detection methods use specific oligonucleotide sequences to either capture or detect the PCR amplicons. In both formats, the PCR mixture contains one biotinylated primer and one unmodified primer. During amplification, the biotin-labeled primer is incorporated in the amplicon. In the oligonucleotide probe format the amplicon is denatured and hybridized to an enzyme-labeled oligonucleotide probe. The reaction mixture is simultaneously captured onto a streptavidin-coated plate, excess labeled oligonucleotide is washed away, and the enzyme label generates a signal with an appropriate substrate.

In the oligonucleotide capture format the amplicon is denatured and hybridized to a sequence-specific capture oligonucleotide that is prebound to a solid phase. Sequences that are not captured are washed away, and captured sequences are detected through the biotin label with enzyme-labeled streptavidin.

Both of these formats are straightforward and relatively simple to design and construct in either a blot or microplate format. Both formats are limited in sensitivity by the fact that only a few enzyme labels are bound per amplicon. Moreover, optimal hybridization conditions may vary significantly among different oligonucleotide probes, because of sequence variations within the hybridization region. As a consequence, genotypically similar strains of an organism may be captured or detected with dissimilar efficiency because of minor sequence variation, thus making relational quantitation difficult.

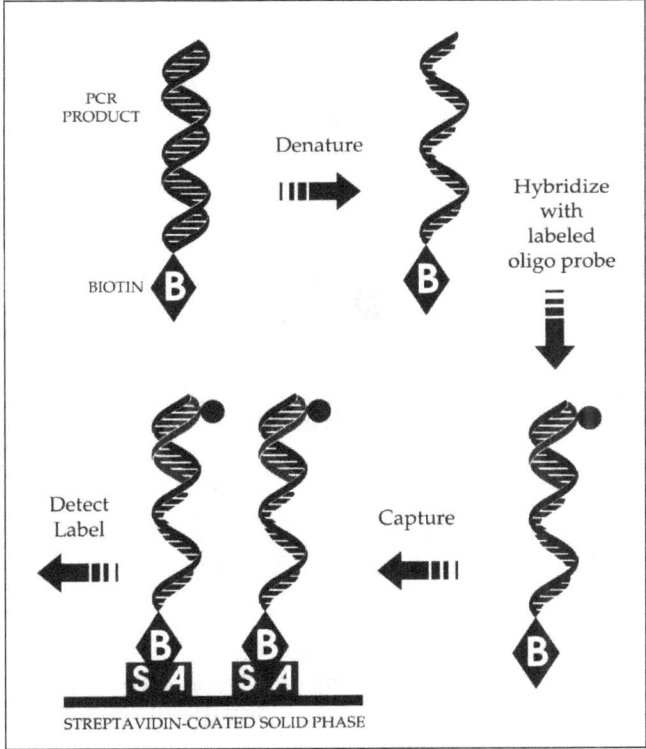

Figure: Oligonucleotide probe format. After amplification, the PCR product is denatured and hybridized to a labeled oligonucleotide. Hybrids are captured onto a streptavidin-coated plate, and the label generates a detectable signal. This format is utilized in many detection assays including the QPCR System 5000.

Immunological Detection Methods

An elegant ELISA method for the detection of PCR products has been described by Yolken that utilizes full-length RNA probes, solution hybridization, and a monoclonal antibody specific for RNA/DNA hybrids. Biotinylated RNA probes complementary to the sequence of interest are transcribed from template DNA and labeled by the addition of biotinylated nucleotide triphosphate to the transcription reaction. PCR is performed with unmodified primers, and a portion of the PCR reaction is hybridized in solution with the biotinylated RNA probe. After hybridization, RNA/DNA hybrids are captured onto a microplate coated with anti-biotin antibodies. The captured hybrids are subsequently detected with a β-galactosidase conjugate of a monoclonal antibody specific for RNA/DNA hybrids, and the β-galactosidase label is detected with a substrate that produces a fluorescent signal.

This detection system is sensitive because multiple antibodies, and hence multiple enzyme labels, can bind to each captured hybrid. Specificity is also enhanced because the formation of RNA/DNA hybrids is favored at the hybridization temperature of 75 °C Because the RNA probes are complementary to the whole length of amplicon, small sequence variations that occur between different strains of the same species will have negligible effect on hybridization or detection efficiency.

SHARP Signal System

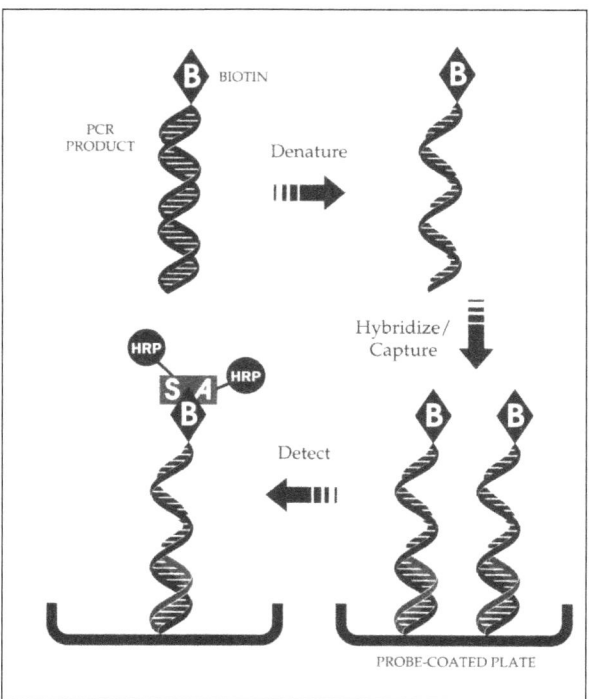

Figure: Oligonucleotide capture format. After amplification, PCR products are denatured and captured by an oligonucleotide probe bound to a plate (solid phase). Captured nucleic acids are detected by a label incorporated into the PCR product.

The SHARP Signal System is a capture ELISA assay that also utilizes an antibody to RNA/DNA hybrids. With the SHARP Signal System, PCR is performed using one biotinylated and one unmodified primer. After amplification, a portion of the reaction mixture is denatured and then hybridized in solution to a complementary unlabeled RNA probe. The RNA/DNA hybrids thus formed are captured onto a streptavidin-coated microplate and detected with an alkaline phosphatase-conjugated antibody specific for RNA/DNA hybrids. After washing, signal is generated with a colorimetric substrate and read on a conventional microplate reader at 405 nm.

The SHARP Signal assay procedure is summarized in Figure. In most applications, overall time to results is < 4 hr, with hands-on time of ~ 1 hr. In one laboratory, the SHARP signal assay has been automated with standard ELISA robotic equipment and is currently being used for rapid PCR detection of human papillomavirus (HPV) in cervical specimens with excellent results.

The SHARP Signal assay is quite sensitive because multiple antibodies, and hence multiple enzyme labels, can react with each captured hybrid. In a model system, the assay has been shown to be at least 100 times more sensitive than ethidium bromide staining of PCR products in agarose gels and can detect ~ 10 pg of biotinylated PCR product per well. The detection assay is extremely flexible and easy to adapt to any target sequence because the RNA probes are unlabeled and can be easily produced by transcription of plasmid DNA containing a T7, SP6, or T3 RNA polymerase promoter. Alternatively, a promoter can be incorporated into a PCR primer, and after amplification with target DNA, the resulting PCR product can be used as the transcription template.

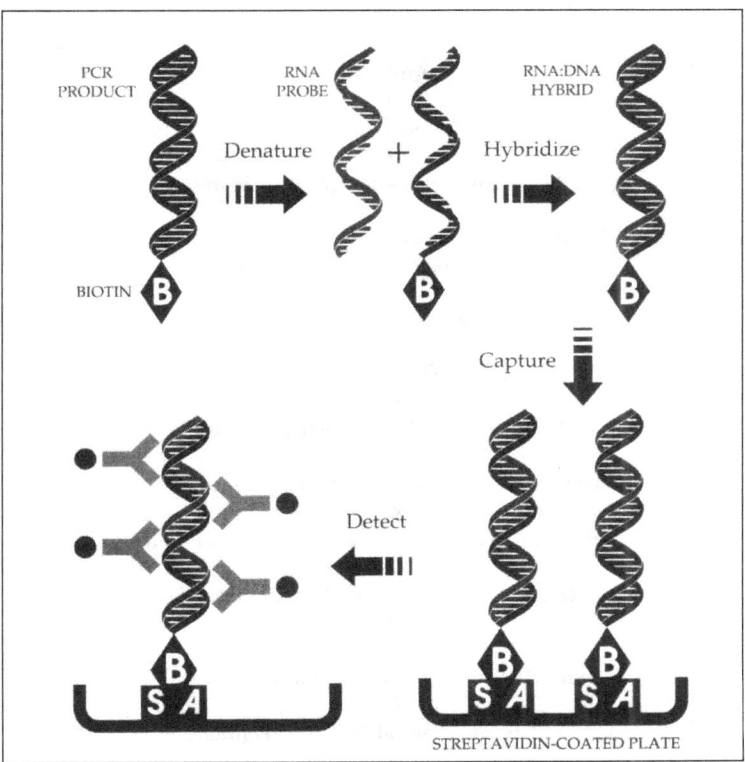

Figure: The SHARP Signal System. PCR products are denatured and hybridized to unlabelled RNA probes. RNA/DNA hybrids are captured onto a streptavidin-coated plate and detected with an enzyme-conjugated antibody specific for RNA/DNA hybrids.

The performance of the SHARP Signal System has been evaluated in a model system with several target analytes. In one study, plasmid DNA containing the 5'-noncoding region of hepatitis C virus (HCV) was serially diluted in buffer containing sheared herring sperm DNA. Aliquots containing 1 µg of sheared herring sperm DNA and 0, 10, 10^2, 10^3, 10^4, and 10^5 copies of the HCV sequence were PCR amplified for 35 cycles with primers directed against the HCV 5' noncoding region (the sense primer was biotinylated on the 5' end). A 5-µl aliquot of each amplification reaction was analyzed by ethidium bromide gel analysis and by the SHARP Signal System. The results are summarized in figure. The data show that 10 input copies are detectable and that the signal generated is proportional to the number of input copies.

In a recent study using clinical specimens, the SHARP Signal System was compared with a nested-primer method for the detection of HCV in serum. Specimens were processed by standard methods to preserve the RNA and to remove PCR inhibitors, cDNA copies of the RNA were prepared by a reverse transcriptase reaction using an antisense primer and the GeneAmp RNA PCR kit (Perkin-Elmer). One portion of the samples was then amplified with a single primer set containing one primer labeled with a 5' biotin, and another portion of the samples was amplified using the nested primer method. The samples amplified with the single primer set were analyzed by the SHARP Signal System after one round of amplification. The samples amplified with the nested-primer technique were analyzed by ethidium bromide gel after the second round of amplification. Of the 99 clinical specimens tested, 47 were positive and 50 were negative by both methods. The remaining two samples were nested-primer positive, SHARP Signal System negative.

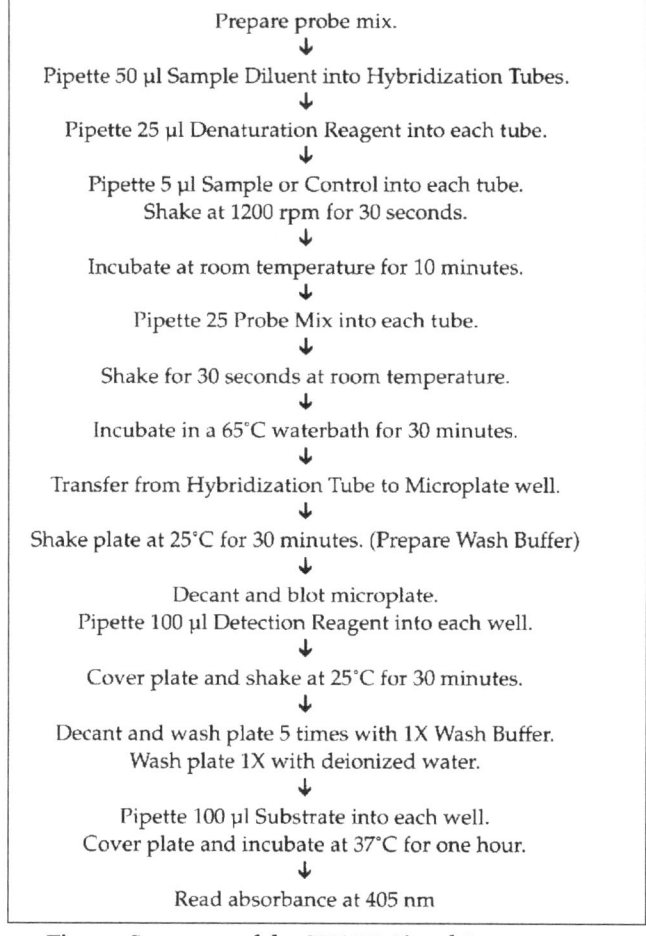

Figure: Summary of the SHARP Signal System assay.

Overall, the SHARP Signal System exhibited 98% agreement with the nestedprimer method. The SHARP Signal System assay for PCR products provides sufficient reagents to perform 192 tests in an 8-well strip, 96-well microplate format. Minimal reagent preparation is necessary. A control probe and positive and negative assay controls are provided so that the user can verify assay performance. RNA probes, primers, and detection controls for the amplification and detection of specific infectious disease targets such as human immunodeficiency virus (HIV), HCV, hepatitis B virus (HBV), HPV, Mycobacterium tuberculosis (Mtb), and cytomegalovirus (CMV) are currently available, and additional targets are under development. The methodology of the SHARP Signal System allows for maximum flexibility because the same reagents, except for the RNA probe, are used for all assays and the detection procedure is universal for all targets. Thus, in a single assay on a single capture plate, a user can test for several analytes simultaneously.

Use of unlabeled RNA probes may be superior to conventional DNA probes for several reasons. (1) RNA/DNA hybrids are more stable than DNA/DNA hybrid structures, thus allowing the use of higher temperatures to increase the stringency of probe/target association and to favor probe/target hybridization over reannealing of the target strands. (2) A posthybridization RNase digestion increases the specificity of detection and reduces background by removing free or nonspecifically bound RNA probe. (3) Strand-specific single-stranded probes are more sensitive and efficient in detecting target sequences because of a lack of competition from probe reassociation.

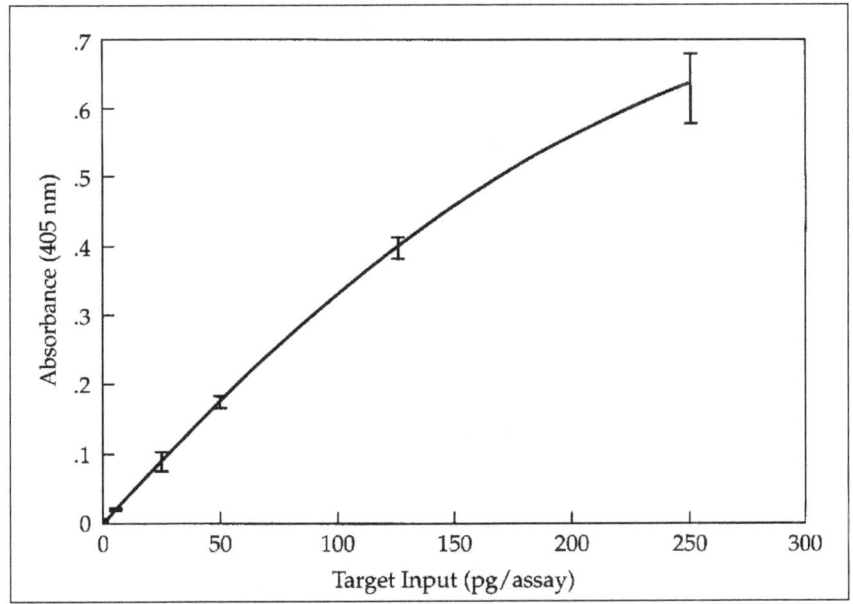

Figure: Sensitivity of the SHARP Signal System with 95% confidence intervals. PCR product from the L1 region of HPV-16 was quantitated, diluted, and tested in the SHARP Signal System.

The flexibility of the system is further increased because the assay uses easily produced unlabeled RNA probes allowing users to develop their own probe/primer set as outlined by the protocol shown in figure. A PCR product containing the target sequence of interest can be directly cloned into a plasmid containing a T7 RNA polymerase promoter. After plasmid DNA has been purified, RNA is easily (and inexpensively) transcribed from the plasmid DNA producing a high yield. The RNA is purified by a simple lithium chloride precipitation and is then ready to use. Biotinylated and unmodified primers may be ordered from commercial suppliers; however, the strand of amplification product containing the biotinylated primer must be the opposite sense of the RNA probe or detection will not occur. Alternatively, a custom RNA probe and primer set for specific targets of interest can be ordered commercially.

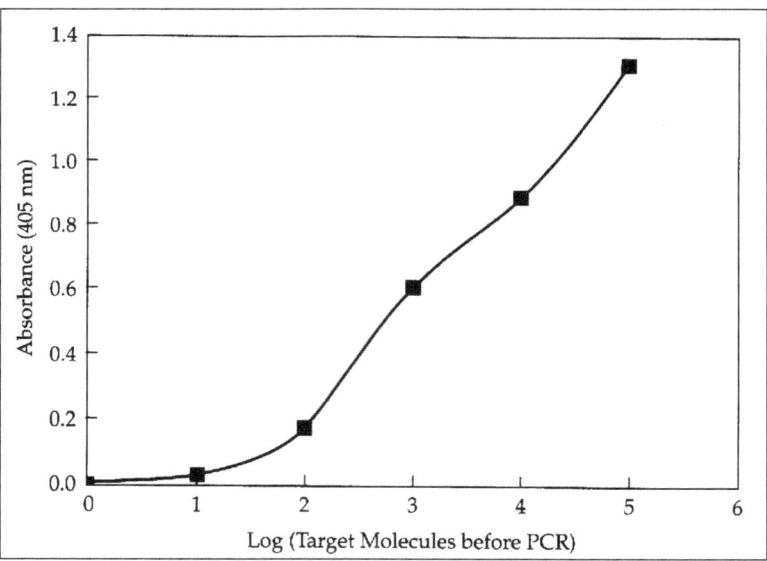

Figure: SHARP Signal System. Detection of amplified plasmid DNA containing the HCV 5'- noncoding region.

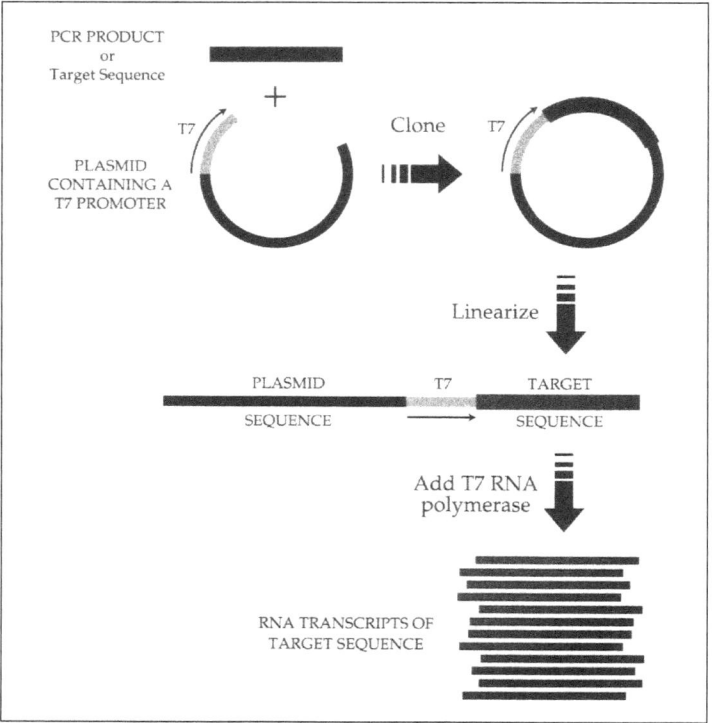

Figure: Preparation of RNA probes from PCR products or other DNA fragments. PCR products or other DNA fragments are cloned into a plasmid containing a T7 RNA polymerase promoter. After propagation and purification of the plasmid DNA, the DNA is linearized and RNA transcripts are produced.

For even greater flexibility, standard cloning techniques can be used to prepare much longer RNA probes. For instance, the HBV probe is a genomic, full-length RNA probe of ~ 3200 bases. By choosing the appropriate primer sets, many different genes can be amplified and then detected with the same probe in the same assay. The "extra" RNA that is not complementary to the much shorter PCR product does not interfere with the assay.

Other Elisa Detection Methods

AMPLICOR

The AMPLICOR PCR assay is a complete sample preparation, amplification, and detection system for specific analytes. PCR amplification is performed with one biotinylated and one unmodified primer. As in the oligonucleotide capture format, the PCR product is denatured and hybridized to a specific capture probe that has been prebound to the microplate. After capture, the biotin label on the capture DNA strand is detected with a streptavidin-horseradish peroxidase conjugate following a standard immunoassay procedure. Results are obtained in 4-5 hr. The AMPLICOR system contains all of the reagents necessary for sample collection, preparation, PCR amplification, and detection. The assay utilizes the PCR System 9600 thermal cycler and incorporates the dUTP/UNG procedure for control of carryover contamination. An FDA-approved version of the AMPLICOR PCR assay is currently available for detection of Chlamydia trachomatis. Because the AMPLICOR assay is a complete package designed to detect a single target, the assay is convenient but lacks the flexibility of other detection systems. Also, the cost of the AMPLICOR system may not be feasible for a research laboratory.

Gen-Eti-K Deia

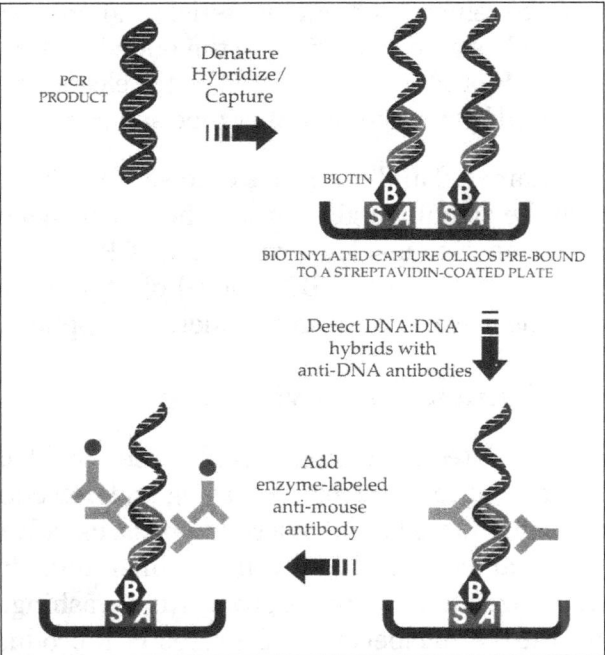

Figure: DEIA. PCR products are denatured and hybridized with a solid-phase capture probe. A mouse monoclonal antibody specific for DNA/DNA hybrids is added, followed by an enzyme labeled anti-mouse antibody.

DNA enzyme immunoassay (DEIA) is also an antibody-based detection system. Before detection, a biotin-labeled, target-specific capture probe is bound to a streptavidin-coated plate with an overnight incubation. Prepared capture plates may then be stored dry for 1 month. PCR amplification is performed with unmodified primers. After amplification, the PCR products are denatured by heating to 100 °C and a portion of the denatured PCR reaction is hybridized to the immobilized capture probe. Following hybridization and wash steps, a mouse monoclonal antibody specific for double-stranded DNA is added to the wells and reacts with the captured sequences that have formed double-stranded DNA complexes with the capture probe. After a second washing, an enzyme labeled anti-mouse antibody is added followed by a third washing and colorimetric detection. Drawbacks of this method include lack of flexibility because once the capture probe is bound to the plate; only a single analyte can be detected. In addition, multiple antibody additions and wash steps make the detection procedure more labor intensive than other methods.

Other Detection Methods

Electrochemiluminescence

The QPCR System 5000 applies a recently developed detection technology to PCR product detection. Electrochemiluminescence is the ability of a substance to emit light when stimulated by an electric field. With this detection method, PCR amplification is performed with one biotinylated and one unmodified primer. After amplification, the PCR products are hybridized to oligonucleotide probes containing the electrochemiluminescent label, Tris (2, 2'-bipyridine)-ruthenium(II) chelate (TBR). After hybridization, the reaction mixture is mixed with streptavidin-coated magnetic beads that bind the biotin-labeled hybrids. The magnetic beads are then loaded into the QPCR instrument, which contains an electrochemiluminescent detection chamber. The magnetic beads are bound magnetically to an electrode and are washed to remove unbound materials and unhybridized TBR-labeled probe. Subsequently, the electrode supplies an electric field to the bound material, and TBR-labeled probe that has hybridized to capture PCR product is stimulated by the electric field and produces light. The emitted light intensity is measured by a photomultiplier tube and converted to a signal output.

The QPCR system is highly automated and can process 50 samples in ~ 1 hr but requires 2.5 hr to process 96 samples. Because the signal is light output, the linear dynamic range is greater than most colorimetric-based detection systems. However, the QPCR system requires a significant investment in capital equipment, and although TBR-labeled phosphoramidite is available, TBR-labeled primers are not yet routinely available from commercial suppliers.

Reverse Dot Blots: HybriQuick and EnviroAmp

The HybriQuick and EnviroAmp detection systems employ the same basic reverse dot blot detection technology, but utilize different formats for processing and detection. In a reverse dot blot, a target-specific capture oligonucleotide is bound to a solid phase, usually a filter or membrane. PCR amplification is performed with at least one biotinylated primer, and after amplification, the PCR product is denatured and hybridized to the capture probe. After washing, enzyme-labeled avidin or streptavidin is used to detect the biotin label on the captured PCR product. A colorimetric reaction then produces a colored spot on the filter or membrane.

The EnviroAmp assay utilizes filter strips that are prebound with capture probe specific for Legionella. All processing, including hybridization, washing, and detection, is performed manually. The HybriQuick system utilizes a probe analysis card (PAC) that contains HIV-specific capture probes prebound to immobilize beads. After denaturation, the sample is simply spotted onto the appropriate site on the PAC. The MicroProbe Affirm processor then processes the cards through the remaining assay steps and color intensity is read by visual inspection. The Affirm processor can process six samples in 38 min.

Both of these methods allow only a limited sample throughput and provide only qualitative results. The manual methodology required by the EnviroAmp system is tedious and time consuming while the HybriQuick method requires a capital equipment investment in the Affirm processor and the use of expensive PACs. Neither method allows flexibility in assay design because there is no provision for allowing the user to bind alternate capture probes to the solid phase.

HPLC

HPLC can be used to separate DNA fragments by size-exclusion (SE) or ionexchange (IE) chromatography. IE chromatography is used more commonly to detect PCR products because of the high resolution obtained with this approach. HPLC detection of PCR products is a simple technique that requires few reagents or manipulations. After amplification, a portion of the reaction mix is loaded onto the column. With an anion-exchange column and gradient elution, DNA fragments are separated according to size and are eluted from smallest to largest. A specific PCR product is detected by the appearance of a peak corresponding to the expected product size. The technique is similar to gel electrophoresis, except that the resolution and sensitivity of the HPLC separation are much greater than gel methods. Quantitative measurement of the amounts of PCR product is also performed easily and accurately by the detector integrator; therefore, HPLC can be an especially useful technique for optimization of PCR conditions. An additional advantage of HPLC detection is that purified PCR products can be recovered for future manipulation such as cloning or sequencing. Because a typical HPLC separation takes from 10 to 60 min, sample throughput with HPLC analysis is low. However, for analysis of only a few samples per batch, HPLC might be an appropriate method. Because HPLC systems are quite expensive, HPLC detection is probably only appropriate for those laboratories that already have an HPLC system in place.

Chromatographic Techniques

Chromatography is a technique used for the separation of components or solutes by using a moving solvent on filter paper. There are various types of chromatography techniques such as high-performance liquid chromatography, paper chromatography, ion-exchange chromatography, gel filtration chromatography, affinity chromatography and gas chromatography. All these types of chromatography are discussed in detail in this chapter.

The chromatography can simply be defined as "a separation technique that uses the size, shape, chemical properties or charge of molecules in a sample to separate the sample into its constituent components". It is often used to detect one, or a number of components in a complex mixture. The visual picture obtained after the separation of molecules or constituents of a mixture is called chromotogram.

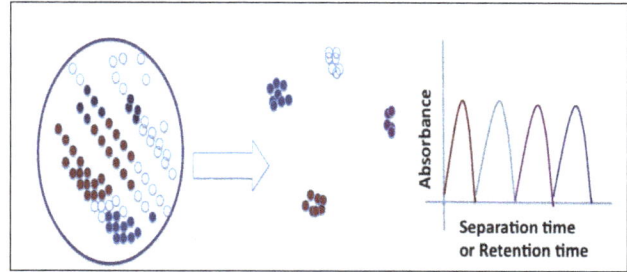

Figure shows separation of different coloured balls. Our hand can simply separate different coloured balls from a mixture of balls visually. Here, the detector is our eye, the separator is our hand and the physical property used is colour. Measured coloured index on the right hand side indicates the chromatogram in scientific jargon.

There are many different variations on what is shown on a chromatogram — depending on the settings used in each laboratory and any regulatory requirements. As an example, the minimum shown on a GC run of an in-process sample might be:

- Sample identification (Product, batch number, stage number).

- Sample information (weight or concentration of sample).

- Date and time the injection was made.

- Analyst's name or identification.

- Instrument identification and name of analytical method used.

- Filename and location of raw data generated during the run.

- Chart recording showing the peaks generated and the baseline, known as a trace.

- Results table (containing raw data and calculated data).

Chromatography Basic Requirements

As illustrated in figure, the basic requirements of chromatography are the sample to be separated, solvent in which the compounds to be separated (mobile phase), instruments based on the separation property of the sample constituents (a stationary phase and related instruments), a detector to quantify or qualify the separated molecule (e.g., spectrophotometer) and finally the collector to hold the separated molecules individually (e.g., fraction collector). These requirements can easily be understood while making particle free fruit juice. Similar requirements are illustrated for classical column based chromatography in figure.

Figure shows simulation of chromatography requirements with particle free fruit juice making. The kettle holding water is like a reservoir for mobile phase, water is a mobile phase, the spoon regulates the speed of water flow, sieve is a stationary phase to separate small and big particles in fruit juice and the mug is a collector of particle free fruit juice. The separation property used here is particle size and water solubility.

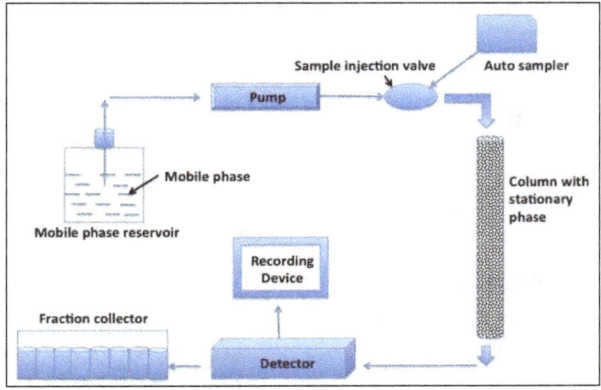

Figure: Parts of classical column chromatography.

Properties of Stationary Phase

Stationary phase in chromatography is generally made up of solid particles or solid particles coated with liquid. The solid particles should offer good flow characteristics, posses' sufficient mechanical strength and provide chemical inertness for biomolecules to be separated. Sometimes, the chemical groups are covalently attached to solid particles in bonded-phase chromatography (e.g., cellulose or agarose). In the stationary phase containing solid particles coated with liquid, the liquid is attached by non-covalent attraction (liquid-liquid chromatography). e.g., silica coated with non-polar hydrocarbon.

Properties of Mobile Phase

Mobile phase in chromatography could generally be either liquid or gas. The liquid phase could be aqueous or organic buffers. Aqueous buffers are generally used to retain the native structure of biomolecules as the biomolecules are evolved to function in an aqueous environment. Organic solvents are used as mobile phase if the native structure of biomolecules is not required. Inert gases are used as mobile phase for the separation of volatile compounds.

Basic Principle of Chromatography

Partition coefficient is the main principle of chromatography. It is the ratio of solute concentration between mobile and stationary phase. It is represented as K_D, which depends on physico-chemical interactions among sample, mobile phase, and stationary phase, and also by experimental conditions, like temperature, solvent polarity etc. The common physico-chemical interactions are solubility (Partition chromatography), adsorption (adsorption chromatography), size (size exclusion chromatography), attraction between opposite charges (ion-exchange chromatography) and biospecific interactions with ligand (Affinity chromatography). Efficient separation of biomolecules depends on the exploitation of tiny differences in their partition coefficients.

Performance Parameters of Chromatography

Performance of chromatography is merely separating the individual or required compounds from a sample without overlap with other compounds in measured quantities. The following parameters determine the chromatography performance. These parameters are mainly useful to compare different columns in chromatography.

- Retention of solute,
- Retardation factor of solute,
- Resolution,
- Peak broadening,
- Theoretical plates in column,
- Capacity factor,
- Peak symmetry.

Retention of Solute

Retention of solute is measured either by retention time or retention volume. Retention time is simply how long a solute is present in a column during chromatography separation. Retention volume is the total volume of eluent (the mobile phase coming out of the stationary phase) that is passed out of the column while detecting a solute. The relationship between retention volume and retention time is as follows:

$$V_R = f . t_R$$

In the previous equation, V_R is the retention volume in ml, t_R is the retention time in minutes, and f is the flow rate (ml/minute). The following figure can exemplify the retention of solutes and the terms involved in it.

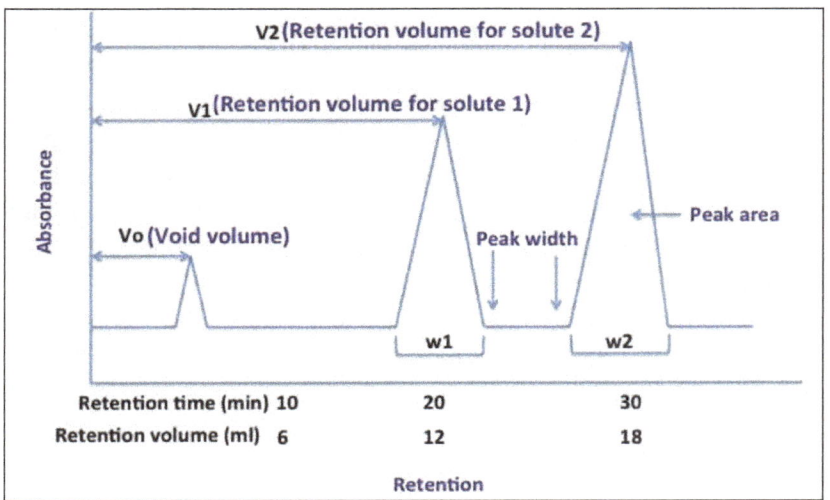

Figure shows schematic representation of the retention of two solutes in chromatography experiment. The V_o represents void volume, which is the volume required for the solvent or mobile phase to pass, unretained through the column and eluted. The V_o is equivalent to the volume of column plus injection valve and tubing, minus volume of packaging material in column. V_1 and V_2 indicate the retention volumes of solutes 1 and 2, respectively. W_1 and W_2 represent peak widths of the peaks of solutes 1 and 2, respectively. The peak area in each peak indicates the quantity of the solute in an analyzed sample.

Retardation Factor (RF) of a Solute

It is a ratio between the rates of movement of solute to the rate of movement of mobile phase. It is mainly useful in planar chromatography, like paper or thin layer chromatography, which are used for the separation of solutes visualized by specific color reactions.

$$RF = \frac{\text{Rate of movement of solute}}{\text{Rate of movement of mobile phase}}$$

Resolution (R)

Resolution is "how two peaks are resolved in a chromatogram". It is measured from the following formula based on the terms explained in figure.

$$R = \frac{V_2 - V_1}{(W_1 + W_2)/2},$$

Where, V_1 and V_2 are the retention volumes of solute 1 and 2, respectively. W1 and W_2 are the widths of peaks of solute 1 and 2, respectively. Narrow width (W) and well separated (large $V_2 - V_1$) indicate best resolution and best chromatography.

Peak Broadening

Elution of sample sometimes occurs in a much larger volume than that in which it was applied. Such situation arises due to a phenomenon called "band broadening". Primarily, a solute is eluted in a much larger volume of the eluent, so that its peak overlaps the peak of another solute eluent, resulting into poor resolution. That kind of the band broadening of the sample within a column could be either by the diffusion of the sample due to Eddy diffusion or by mass transfer phenomenon. The mass transfer phenomena may be due to mobile phase mass transfer or stagnant mobile phase mass transfer or stationary phase mass transfer. Specifically, Eddy's diffusion is mainly because of the variation in the size of stationary phase particles. When the stationary phase particles are not uniform in their size, there are narrow and broad channels for the movement of mobile phase and thus sample molecules. Such situation results into whirlpools or Eddies of mobile phase in narrow channels and the fast flow of mobile phase in wide channels (mobile phase mass transfer). Overall, lack of uniformity in the size of the stationary phase particles cause a wide distribution of sample between fast flowing wide and slow flowing narrow channels, resulting into broad band on chromatogram for an analyte. In addition to size, the orientation of stationary phase particles is also important. If the orientation is not uniform, the mobile phase will be stagnant at some places of the column, resulting into peak broadening. If the stationary phase particles are damaged, the mobile phase can be trapped into damaged stationary phase particles, resulting into stationary phase mass transfer.

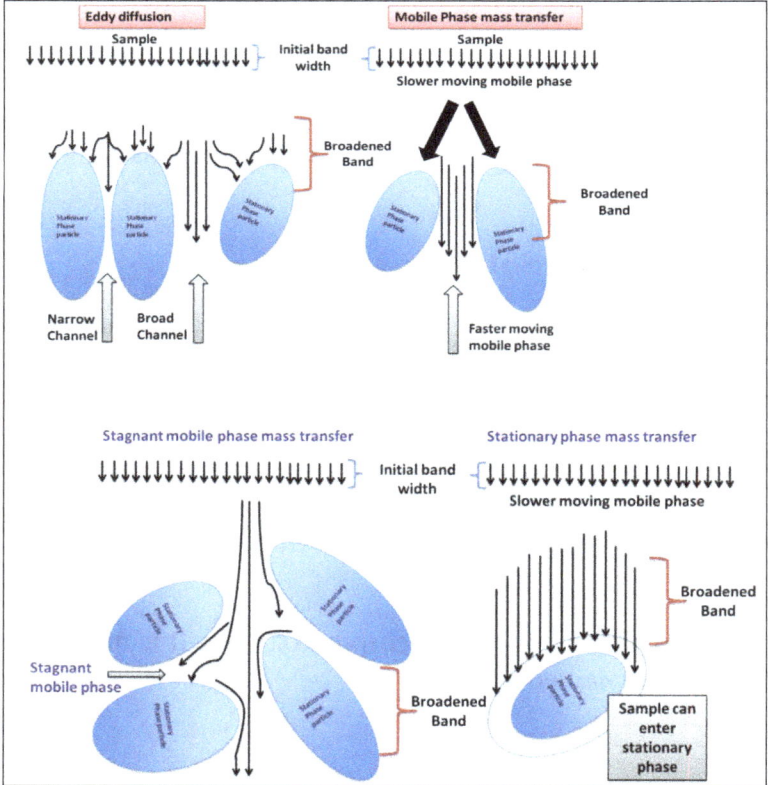

Figure: Peak broadening in chromatography by Eddy's diffusion and mass transfer phenomenon.

Theoretical Plates in Column

Theoretical plates are simply the number of equilibrations that a compound makes with the stationary phase. Chromatography columns are generally considered to have a number of adjacent plates or zones, where there is enough space for a compound to achieve complete equilibrium between the mobile and stationary phase. Each zone is called a theoretical plate and the length of the column is the addition of theoretical plate heights. Equilibration of a compound or analyte depends on its physiochemical interactions among itself, mobile and stationary phases. The equilibration or distribution of the compound between mobile and stationary phase is determined by its partition coefficient (K_d) based on its physiochemical interactions with these phases. For example, the distribution of two compounds having K_d of 1 and 0.1 is shown in the Figure. Typically, the compound with a K_d of 1 showed a bell shaped curve in the chromatogram, whereas the compound with a K_d of 0.1 could not come out of the column after theoretical plate 6. In other words, the second compound is more attached to the stationary phase.

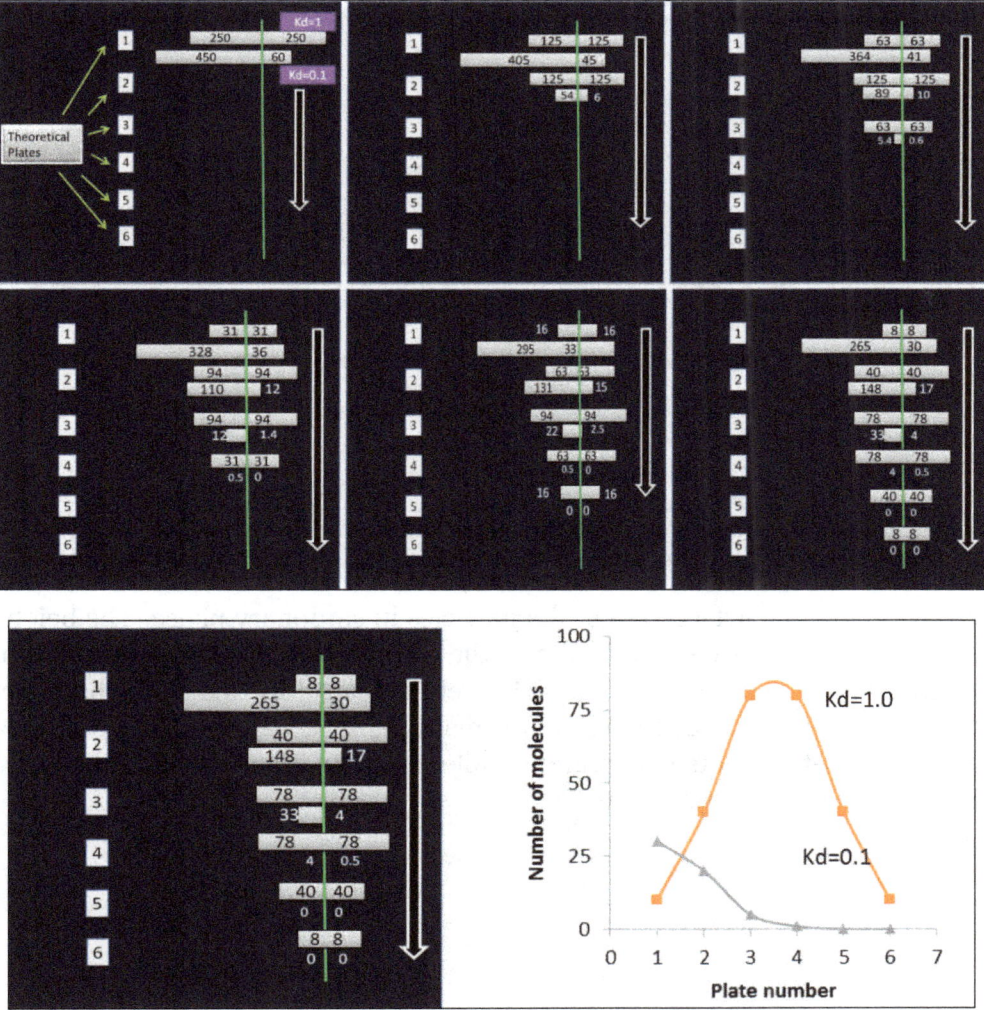

Figure shows distribution of two analytes in theoretical plates of chromatographic column. Two analytes with K_d values of 1 and 0.1 were equilibrated at the theoretical plates based on their physicochemical interactions with mobile and stationary phases. The graph indicates the chromatograms produced by these two analytes. The compound with a K_d value of 0.1 is more attached to stationary phase and hence did not come out of the column, resulting into truncated peak.

More number of theoretical plates or equilibration zones in a columns help for better resolution of the analytes due to narrow peaks in chromatogram. The number of theoretical plates can be estimated by the measures of the peaks in a chromatogram based on the mathematical relationship in figure. The width of a peak is inversely proportional to the number of theoretical plates.

$$N = 5.54 \cdot \left(\frac{t_R}{W_h} \right)^2 \qquad N = 16 \cdot \left(\frac{t_R}{W} \right)^2$$

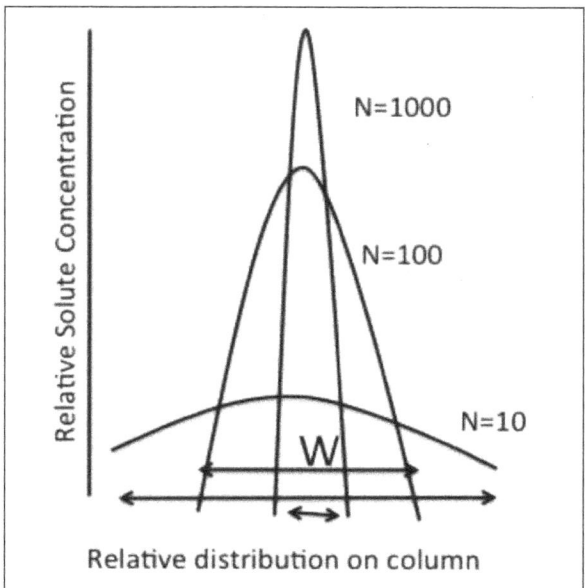

Figure: Estimation of theoretical plates from a chromatogram. N indicates the number of theoretical plates. W indicates width of a peak at base. W_h indicates width of half the width of a peak.

The N is directly related to surface area of the particles in stationary phase. The height of a theoretical plate (H) is a ratio between the length of the column to N. Different length columns (e.g., 15cm and 30cm) may have same number of theoretical plates. In this case, shorter column has better performance. Small H leads to better chromatography separation. The empirical relationship between the theoretical plate height and mobile phase (v) is as follows. This relationship can be appreciated from Van Deemter Curve.

$$H = A + \frac{B}{v} + C.v$$

Where,

- H = Theoretical plate height.

- A = Eddy's diffusion.

- B = Longitudinal diffusion (Diffusion along with directional flow).

- C = Mass transfer effects.

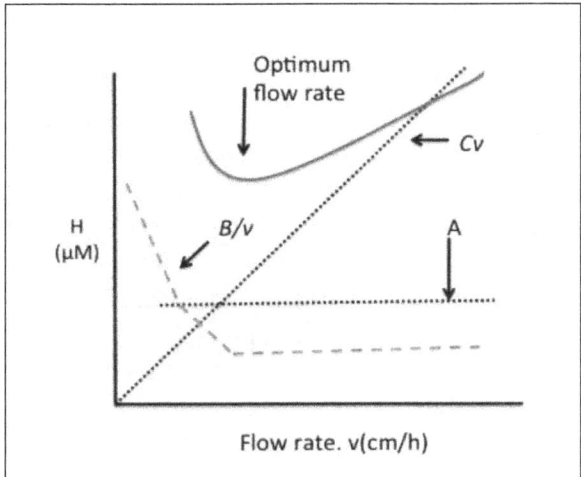

Figure: Van Deemter Curve.

Based on Van Deemter Curve, a decrease in C.v gives smaller H, which results in more N and better chromatography; extremely slow rates increases H due to increase in B/v (longitudinal flow), decrease in N and so poor chromatography; a balance between C.v and B/v is the optimum rate (v); low flow rate results long analyses times and higher flow rate than optimum leads shorter analyses times with inefficient separation.

Capacity Factor (k) and Separation Factor (α)

Capacity factor (k) can be calculated from the retention volumes of analytes. For instance, the relationship between retention volumes of two analytes and capacity factor is as follows:

- $V_1 = V_0 (1+ k_1)$
- $V_2 = V_0 (1+ k_2)$

Where, V_1 and V_2 are the retention volumes of analytes 1 and 2, respectively. K_1 and K_2 are the capacity factors of analytes 1 and 2. The ratio of K_1 and K_2 indicates separation factor α. High separation factor denotes good chromatography.

Peak Symmetry

An ideal condition of a peak is Guassian curve or normal curve without any overlaps. However, deviations are common in practical chromatography due to non-linear flow rates and gradient elution methods.

Significance of Performance Parameters in Chromatography

Performance parameters depend primarily on physical interactions between sample components and stationary phase. Hence, these can be used to compare different chromatography systems (e.g., HPLC and ion-exchange chromatography) as well as different sample components (e.g., peptides and sterol derivatives). A good chromatography needs high resolution (R), a large number of theoretical plates (N) in a column, a low plate height (H) and high values for separation factor (α).

High-performance Liquid Chromatography

A good chromatography needs to meet certain performance parameters, like high resolution between two peaks in a chromatogram, a large number of theoretical plates in a column, a low theoretical plate height and high values for the separation factor between two analytes. One possibility for improving the performance of a chromatography is decreasing the particle size of stationary phase, which increases the surface area, thereby increase in the number of theoretical plates or equilibrations among analyte, stationary and mobile phases, resulting into high performance. However, decreasing the particle size of stationary phase increases the resistance to mobile phase, backpressure in the column, which leads to a damage to the stationary phase and finally decreasing the eluent flow and resolution. The main solution to such problems due to small particle size is to select or prepare small particles (5-10µm) that can withstand the pressures of 40MPa, and the requirement of high pressure to increase the flow rate. Such chromatography with small particle size of stationary phase and the use of high pressure is called high performance or high-pressure liquid chromatography (or) HPLC is simply a separation of biomolecules with high performance parameters under high pressure by using appropriate instrumentation.

HPLC Instrumentation or Components

Different parts of HPLC are mobile phase reservoirs, stainless steel tubing, pumps, gradient mixer, injector, guard column, main column, detectors and fraction collector. There components illustrated in the figure.

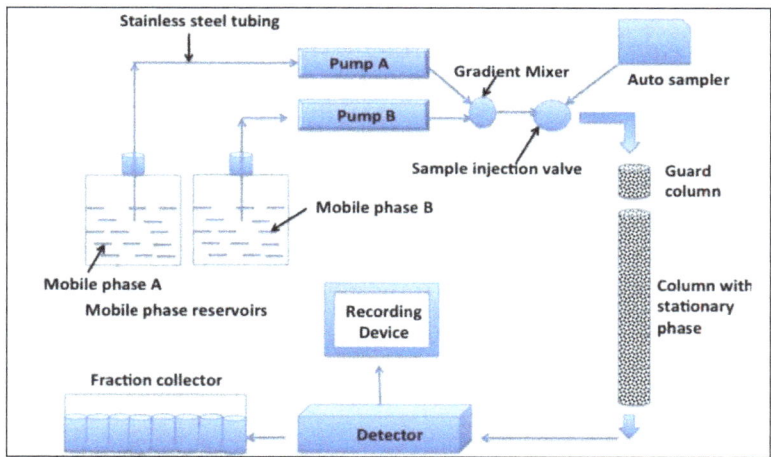

Figure: Components of HPLC

HPLC Mobile Phase

The HPLC mobile phase usually comprises of two or more solvents (e.g., water and acetonitrile) for the separation of biomolecules. The solvents should be highly purified, called HPLC grade, and they should be chemically unreactive. If these solvents are not pure, the impurities may interfere in detection system, especially at a wavelength of less than 200 nm. In addition, the solvents should be degassed before using them in HPLC. It is imperative, specifically, for those eluents with aqueous ethanol and methanol, because high pressure results in bubbling of gas or air present in solvents and those damage stationary phase. Several degassing methods, like warming, vigorous stirring with magnetic stirrer, vacuuming, ultrasonification and bubbling helium gas through eluent reservoir, can be used. However, the most common one is vacuuming.

Selection of HPLC mobile phase depends on the type of separation and the components in the sample. Usually, polar and non-polar solvents are used in certain ratio and they should not interfere in the detection mechanism. Based on the composition of mobile phase throughout HPLC experiment, HPLC elution can be classified as isocratic or gradient elution. In isocratic elution, the mobile phase composition is constant throughout the HPLC separation. The advantages of isocratic elution are, it can be used for simple mixtures, and the system and column are equilibrated all the time and does not suffer from fast chemical changes. In gradient elution, mobile phase composition is gradually changing in the ratio of polar to non-polar compounds during the sample run. Gradient elution is useful for the separation of complex mixtures.

HPLC Stationary Phase

Based on the interactions among sample analytes, mobile and stationary phases, the HPLC technique can be classified as normal phase HPLC and reverse phase HPLC. In normal phase HPLC, the stationary phase is more polar (e.g., Alumina or silica) than mobile phase. Hence, the first compound to be eluted is less polar. In reverse phase HPLC, the stationary phase is less polar (e.g., Alkyl, aliphatic or phenyl bonded phase) than mobile phase. The first compound to be eluted is more polar in nature and the retention time of a compound increases with decreasing polarity of the particular analyte.

The stationary phase particles in HPLC are rigid rather than soft gel as in open column chromatography. The particles are spherical and in uniform size to reduce space for diffusion. Mainly, there are three types of particles; micro porous particles, pellicular particles and bonded phase particles.

Micro Porous Particles

These are of 5-10µm in diameter composed of silica or aluminum and contain microscopic pores running through the particle to increase the surface area for more the interaction with analytes and mobile phase. E.g., Porous poly vinyl chloride (Fractogel TSK-HW-55) gel filtration of proteins.

Pellicular Particles

These are also called as superficially porous particles. They are of 40 µm in diameter. In has an inert core, on which the pores are coated. Inert core is made up of glass or some other hard material. The porous material could be made up of polymethacrylates or poly(styrenedivinylbenzene). These particles display excellent pH stability (pH 2-12) with high mechanical stability.

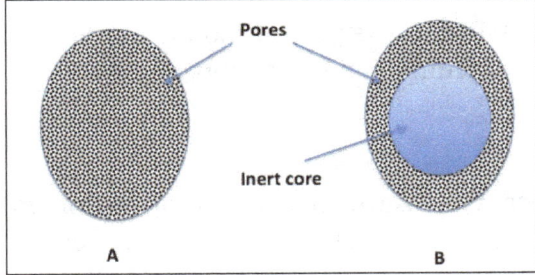

Figure: (A) Microporous (B) pellicular stationary phase particles.

Bonded Phase Particles

These particles also contain an inert core made up of silica. The stationary phase is chemically bonded on that inert core by different linkages. E.g., Silica C-18 for reverse phase chromatography of peptides/proteins.

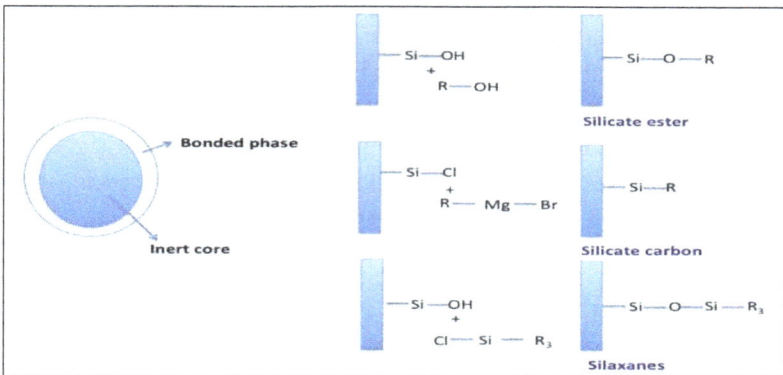

Figure: Bonded phase particles and their chemical linkages.

HPLC Columns to Hold Stationary Phase

- Conventional HPLC columns: These are made of stainless steel to withstand 50MPa pressure. They are generally 3-25 cm in length and 2-5mm in diameter. The flow rate of mobile phase in this column is 1-10ml/min.

- Capillary columns: The inner diameter of these columns is 100μm-1mm and the flow rate is 0.4-200μl/min.

- Nanobore columns: The inner diameter of these columns is 25-100μm and the flow rate is 25-4000nl/min. These columns are helpful to use at the interface with mass spectrometry. These columns are important in situations to analyze minute quantities of proteins –proteomics.

- Stationary phase packing: The stationary phase is packed into a HPLC column by high pressure slurring procedure. In this procedure, the stationary phase and solvent for slurry need to be in equal density and the slurry can be pumped into a column by high pressure. By this method, a dense, continuous bed of stationary phase without any cracks and other imperfections will be formed.

HPLC Pumps

An important requirement in HPLC is the flow of mobile phase should be very stable (i.e., pulse free). The following two types of pumps can reasonably attain this.

Constant Pressure Pumps

These pumps maintain constant pressure irrespective of stationary phase resistance to mobile phase flow. If the resistance increases in stationary phase, the flow rate decreases automatically to maintain constant pressure.

Constant Volume Pumps

These pumps maintain constant flow rate through stationary phase. The pumps increase the pressure if there is an increase in stationary phase resistance. However, increasing pressure is within the pre-set limits of the pump. If the pressure is required more than the preset limits, the pumps are inactivated automatically by safety-cut off mechanism.

These pumps are more popular in HPLC. An example for common constant volume pump is reciprocating pump, which uses a piston to deliver a fixed volume of solvent onto the column in repeated cycles of filling and emptying (pulses). In order to avoid pulses while the flow of mobile phase pulse dampners are built into the pumps.

Sample Application

Most common sample injector in HPLC is loop injector, which is operated manually. Some advanced systems have automatic sample injectors.

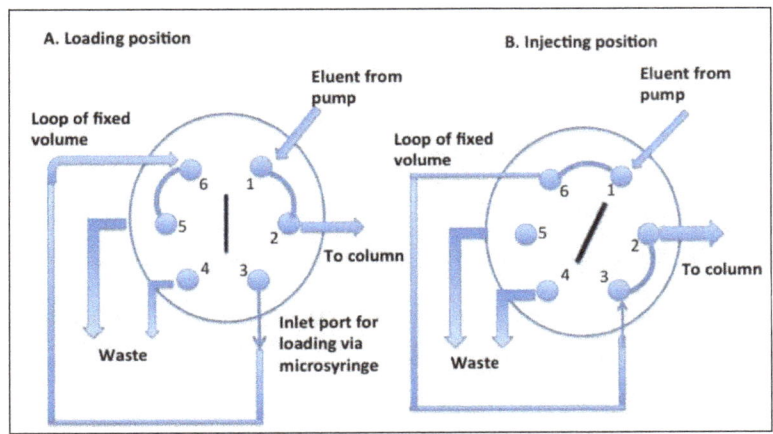

Figure shows HPLC loop injector. In the loading position of injector (A), the sample can be loaded into the loop of fixed volume through port 3, and the excess sample is lost though port 5. At this position, the eluent is directly passed to the column through ports 1 and 2. In the injecting position (B), a connection will be established between ports 1 and 6, and then the eluent passes through loop, ports 3 and 2 to the column.

Guard Column

It is a small column between injector and main HPLC column. This column is mainly used to protect the main column from impurities.

HPLC Detectors

Two major detection mechanisms, such as optical and electrical, are broadly involved in HPLC detectors. Optical detection based detectors are variable wavelength detectors (UV-Vis), scanning wavelength detectors (DAD), fluorescence detectors (FL), refractive index detectors (RI) and evaporative light-scattering detectors (EL). Electrical based detectors are electrochemical detectors (ECD), conductivity detector (CD). Additionally, mass spectrometer detectors and NMR spectrophotometer detectors can be integrated with HPLC. A criterion for selection of a specific detector for specific analysis is illustrated is the following figure.

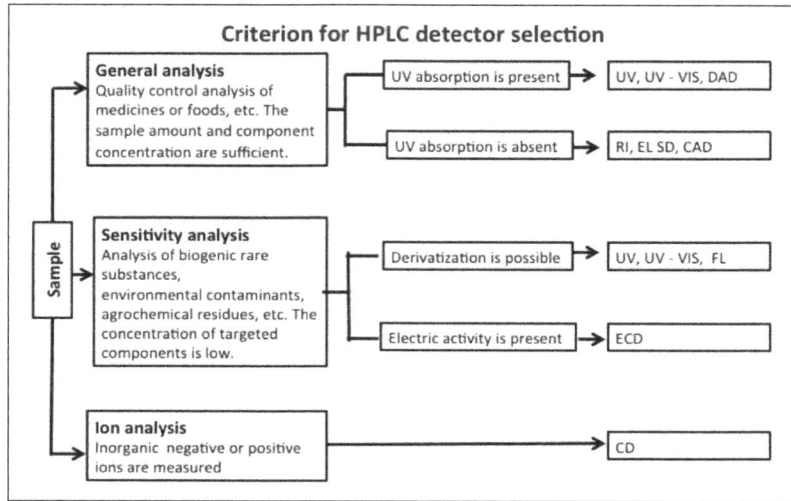

HPLC Limitations

- High cost of generation of high pressure.

- High pressure imposes some constraints on instrumentation.

- It has relatively low flow rates. Hence, the experiment is time taking.

- HPLC is especially useful in the separation and analysis of low molecular weight biomolecules. But it does not usually useful for the purification of complex biopolymers (e.g., proteins) in active form.

  - Reason 1: Some HPLC formats like reverse phase chromatography needs organic solvents which denatures proteins.

  - Reason 2: Although large capacity HPLC columns are available, they are expensive. Small capacity columns are quite common.

Probable solution: Development of another format for biopolymers- FPLC.

Probable solutions for above limitations:

- Decreasing the pressure and increasing the flow rate for comparable resolution of analytes by HPLC. It is possible with perfusion chromatography.

- Increasing the instrumentation that can withstand high pressure by considering cost economics for better resolution than HPLC. It is possible with UPLC.

HPLC Supplements

Perfusion Chromatography

With small particle size, high resolutions can be possible in short times with high flow rates in perfusion chromatography. In this chromatography, the stationary phase particles contain not only longitudinal pores throughout the particle, but also diffusive pores connecting the longitudinal pores (e.g., POROS particles). According to VanDeemter's curve, the height of theoretical plate inversely proportional to the diffusion coefficient of mobile phase. At high values of flow rate, the connective term is much important than diffusion term of mobile phase between the particles. Hence, theoretical plate height becomes largely independent of flow rate, but dependent on stationary particle diameter when the stationary phase particles have both longitudinal and diffusive pores. Therefore, inter-particle diffusion becomes unimportant relative to intra-particle convection in perfusion chromatography. These kinds of stationary phase particles have large surface area for interacting with mobile phase and analytes.

Ultra Performance Liquid Chromatography (UPLC)

Decreasing the particle size and increasing the capacity of instruments by Waters Corporation developed this kind of chromatography. It has stationary phase particle size of 1.7 μm diameter. The stationary phase particles are made up of Bridged Ethylsilixane Silica Hybrid (BEH). This chromatography is 10 times faster than conventional HPLC.

Applications of High-performance Liquid Chromatography (HPLC)

The HPLC has developed into a universally applicable method so that it finds its use in almost all areas of chemistry, biochemistry, and pharmacy.

- Analysis of drugs.

- Analysis of synthetic polymers.

- Analysis of pollutants in environmental analytics.

- Determination of drugs in biological matrices.

- Isolation of valuable products.

- Product purity and quality control of industrial products and fine chemicals.

- Separation and purification of biopolymers such as enzymes or nucleic acids.

- Water purification.

- Pre-concentration of trace components.

- Ligand-exchange chromatography.

- Ion-exchange chromatography of proteins.

- High-pH anion-exchange chromatography of carbohydrates and oligosaccharides.

Advantages of High-performance Liquid Chromatography (HPLC)

- Speed,

- Efficiency,

- Accuracy,

- Versatile and extremely precise when it comes to identifying and quantifying chemical components.

Limitations

- Cost: Despite its advantages, HPLC can be costly, requiring large quantities of expensive organics.

- Complexity.

- HPLC does have low sensitivity for certain compounds, and some cannot be detected as they are irreversibly adsorbed.

- Volatile substances are better separated by gas chromatography.

Paper Chromatography

Paper chromatography (PC) is a type of a planar chromatography whereby chromatography procedures are run on a specialized paper. PC is considered to be the simplest and most widely used of the chromatographic techniques because of its applicability to isolation, identification and quantitative determination of organic and inorganic compounds. It was first introduced by German scientist Christian Friedrich Schonbein.

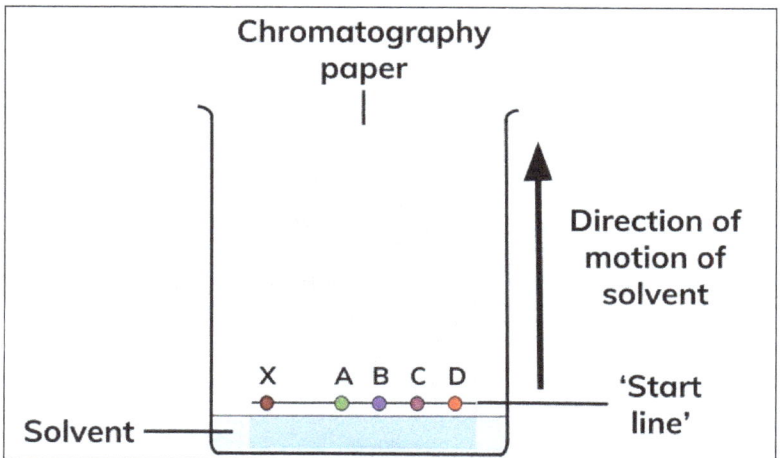

Types of Paper Chromatography

- Paper Adsorption Chromatography: Paper impregnated with silica or alumina acts as adsorbent (stationary phase) and solvent as mobile phase.

- Paper Partition Chromatography: Moisture/Water present in the pores of cellulose fibers present in filter paper acts as stationary phase & another mobile phase is used as solvent In general paper chromatography mostly refers to paper partition chromatography.

Principle of Paper Chromatography

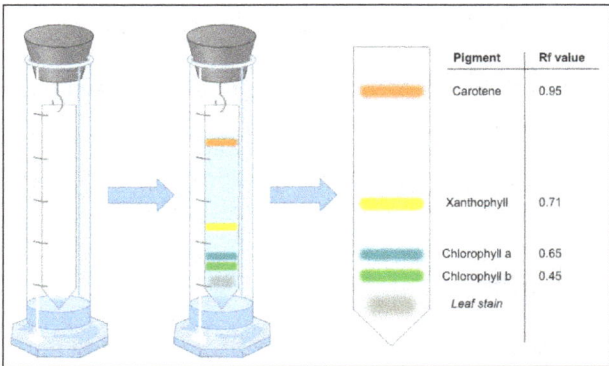

The principle of separation is mainly partition rather than adsorption. Substances are distributed between a stationary phase and mobile phase. Cellulose layers in filter paper contain moisture which acts as stationary phase. Organic solvents/buffers are used as mobile phase. The developing solution travels up the stationary phase carrying the sample with it. Components of the sample will separate readily according to how strongly they adsorb onto the stationary phase versus how readily they dissolve in the mobile phase.

Instrumentation of Paper Chromatography

Stationary Phase and Papers

Whatman filter papers of different grades like No.1, No.2, No.3, No.4, No.20, No.40, No.42 etc. In general the paper contains 98-99% of α-cellulose, 0.3 – 1% β - cellulose.

Other modified papers:

- Acid or base washed filter paper.

- Glass fiber type paper.

- Hydrophilic Papers – Papers modified with methanol, formamide, glycol, glycerol etc.

- Hydrophobic Papers – Acetylation of OH groups leads to hydrophobic nature, hence can be used for reverse phase chromatography.

- Impregnation of silica, alumna, or ion exchange resins can also be made.

Paper Chromatography Mobile Phase

Pure solvents, buffer solutions or mixture of solvents can be used. Examples:

- Hydrophilic mobile phase:

    ◦ Isopropanol: Ammonia:water 9:1:2.

    ◦ Methanol: Water 4:1.

    ◦ N-butanol: Glacial acetic acid: water 4:1:5.

- Hydrophobic mobile phases:

  ○ Dimethyl Ether: Cyclohexane kerosene: 70% isopropanol.

  ○ The commonly employed solvents are the polar solvents, but the choice depends on the nature of the substance to be separated.

  ○ If pure solvents do not give satisfactory separation, a mixture of solvents of suitable polarity may be applied.

Chromatographic Chamber

The chromatographic chambers are made up of many materials like glass, plastic or stainless steel. Glass tanks are preferred most. They are available invarious dimensional size depending upon paper length and development type. The chamber atmosphere should be saturated with solvent vapor.

Steps in Paper Chromatography

In paper chromatography, the sample mixture is applied to a piece of filter paper, the edge of the paper is immersed in a solvent, and the solvent moves up the paper by capillary action. The basic steps include:

- Selection of Solid Support: Fine quality cellulose paper with defined porosity, high resolution, negligible diffusion of sample and favouring good rate of movement of solvent.

- Selection of Mobile Phase: Different combinations of organic and inorganic solvents may be used depending on the analyte. Example. Butanol: Acetic acid: Water (12:3:5) is suitable solvent for separating amino-acids.

- Saturation of Tank: The inner wall of the tank is wrapped with the filter paper before solvent is placed in the tank to achieve better resolution.

- Sample Preparation and Loading: If solid sample is used, it is dissolved in a suitable solvent. Sample (2-20ul) is added on the base line as a spot using a micropipette and air dried to prevent the diffusion.

- Development of the Chromatogram: Sample loaded filter paper is dipped carefully into the solvent not more than a height of 1 cm and waited until the solvent front reaches near the edge of the paper. Different types of development techniques can be used:

  ○ Ascending development: Like conventional type, the solvent flows against gravity. The spots are kept at the bottom portion of paper and kept in a chamber with mobile phase solvent at the bottom.

  ○ Descending type: This is carried out in a special chamber where the solvent holder is at the top. The spot is kept at the top and the solvent flows down the paper. In this method solvent moves from top to bottom so it is called descending chromatography.

  ○ Ascending – descending development: A hybrid of above two techniques is called ascending-descending chromatography. Only length of separation increased, first ascending takes place followed by descending.

- ◦ Circular/radial development: Spot is kept at the centre of a circular paper. The solvent flows through a wick at the centre & spreads in all directions uniformly.

- Drying of Chromatogram: After the development, the solvent front is marked and the left to dry in a dry cabinet or oven.

- Detection: Colourless analytes detected by staining with reagents such as iodine vapour, ninhydrin etc. Radiolabeled and fluorescently labeled analytes detected by measuring radioactivity and florescence respectively.

R_f Values

Some compounds in a mixture travel almost as far as the solvent does; some stay much closer to the base line. The distance travelled relative to the solvent is a constant for a particular compound as long as other parameters such as the type of paper and the exact composition of the solvent are constant. The distance travelled relative to the solvent is called the R_f value.

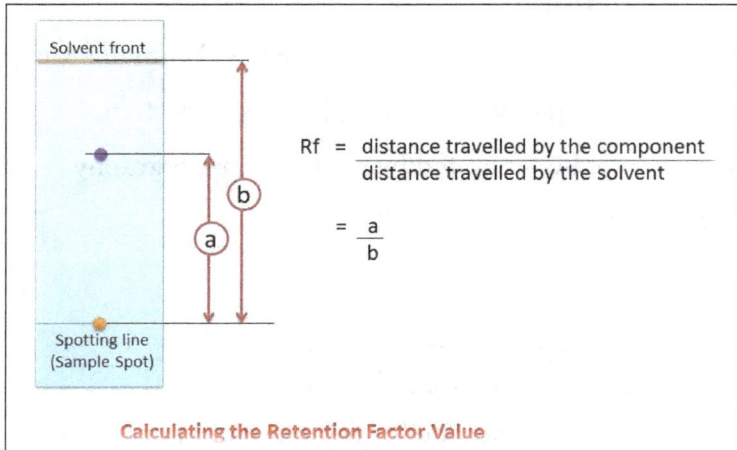

Calculating the Retention Factor Value

Thus, in order to obtain a measure of the extent of movement of a component in a paper chromatography experiment, "R_f value" is calculated for each separated component in the developed chromatogram. An R_f value is a number that is defined as distance traveled by the component from application point.

Applications of Paper Chromatography

- To check the control of purity of pharmaceuticals.

- For detection of adulterants.

- Detect the contaminants in foods and drinks.

- In the study of ripening and fermentation.

- For the detection of drugs and dopes in animals & humans.

- In analysis of cosmetics.

- Analysis of the reaction mixtures in biochemical labs.

Advantages of Paper Chromatography

- Simple.

- Rapid.

- Paper Chromatography requires very less quantitative material.

- Paper Chromatography is cheaper compared to other chromatography methods.

- Both unknown inorganic as well as organic compounds can be identified by paper chromatography method.

- Paper chromatography does not occupy much space compared to other analytical methods or equipment.

- Excellent resolving power.

Limitations of Paper Chromatography

- Large quantity of sample cannot be applied on paper chromatography.

- In quantitative analysis paper chromatography is not effective.

- Complex mixture cannot be separated by paper chromatography.

- Less Accurate compared to HPLC or HPTLC.

Ion-exchange Chromatography

Ion exchange chromatography is a technique for the separation of different ionised biomolecules based on the differences in their net charge. It involves the separation of analytes based on the attraction between oppositely charged stationary phase, called ion- exchanger, and the analytes. This technique is a fast, economical and versatile technique for an effective separation of amino acids, peptides, nucleotides and nucleic acids etc. These molecules have ionisable groups on their surface and hence they carry either a positive or negative net charge depending on their pKa and pH of the solution. This net charge can be utilized for separating the mixture of such biomolecules on ion exchange chromatography. This technique is widely used in the prefractionation or purification of a target protein from crude biological samples.

The basic principle of ion exchange chromatography is the reversible exchange of ionic analytes bound to solid support or stationary phase with similar ions generated from salt in liquid mobile phase. The analyte molecules are retained in the chromatographic column based on their ionic interactions with the stationary phase. The surface of the stationary phase contains ionic functional groups that interact with ionised analytes. The elution of desired analyte is done by increasing salt gradient. The salt ions act as counter ions in ion exchange chromatography. These salt ions compete with the analytes to bind with stationary phase. Increasing the salt concentration causes the elution of desires analyte. Most commonly used salt is NaCl, which exists in equilibrium with cations (Na^+) and anions (Cl^-), in aqueous solution.

Basic Steps in Ion Exchange Chromatography

In general, Ion exchange chromatography offers effective separation of contaminants from the substances of interest. One can choose whether to bind the substances of interest to the ion exchnager and allow the contaminants to pass through the column or to bind the contaminants and allow the substance of interest to pass through column. Generally, earlier method is more useful as it promotes a higher degree of fractionation besides concentrating the substances of interest. Most ion exchange experiments are taken place in five main stages. These steps are schematically illustrated below:

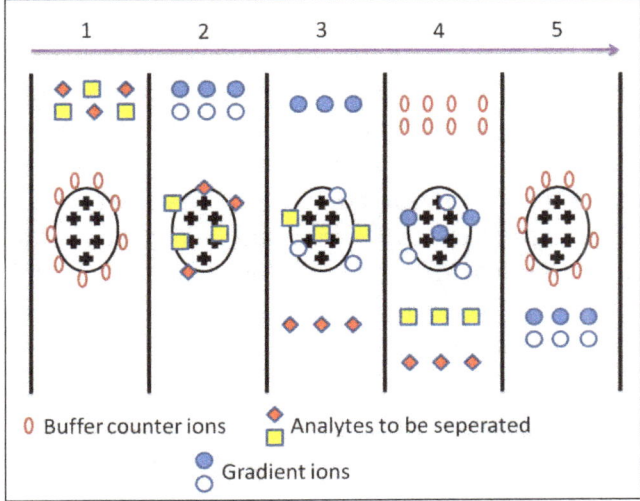

Stage 1: Equilibration

In this step, the ion exchanger or stationary phase is equilibrated with counter ions in a suitable buffer with specific pH and ionic strength. The counter ions could be either sodium or chloride based on the type of ion exchanger. If it is a cation exchanger, positive sodium ions interact with the negative charge on stationary phase. If it is an anion exchanger, the chloride ions interact with the stationary phase. The equilibration step is essential to allow the binding of the desired solute molecules to the exchanger as well as the separation of sample analytes based on their exchange ability with the similar changed counter ions on the stationary phase.

Stage 2: Sample Application and Washing

In this step, the sample is applied on the stationary phase in similar buffer used for equilibration of the ion exchanger. The analyte molecules carrying the appropriate charge displace the counter-ions and bind reversibly to the ion exchanger. Unbound analytes can be washed out from the exchanger by using the starting buffer.

Stage 3: Changing the Elution Conditions

In this stage, the bound analytes are eluted from the column by changing the buffer conditions in such a way that the ionic bound analytes will be replaced by the counter buffer ions. Increasing the ionic strength of the eluting buffer or changing its pH can achieve such a replacement. In the above figure, desorption of the analytes is achieved by the introduction of an increasing salt concentration gradient, which promote the release of analytes from the column in the order of their strengths of binding. The weeklies bound substances are eluted first.

Stages 4 and 5: Changing the Elution Conditions and Column Equilibration

These steps involve the removal or elution of analytes from the column which are not eluted under the previous experimental conditions. Then the column is subjected to re-equilibration by giving the starting conditions (Stage 1) for the next purification experiments.

Ion Exchanger

An ion exchanger is an insoluble matrix to which charged groups have been covalently attached. The charged groups on the matrix are associated with mobile counter ions. These counter ions can be exchanged with other ions of the same charge without altering the matrix. The matrix may be made up of inorganic compounds, synthetic resins or polysaccharides. The characteristics of the matrix determine its chromatographic properties such as capacity, efficiency, chemical stability and recovery as well as its mechanical strength and flow properties. On the basis of the exchanger types used for separation, this chromatography is further subdivided into anion exchange chromatography and cation exchange chromatography.

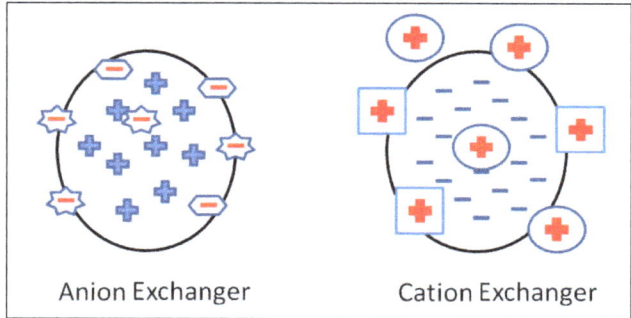

Anion Exchanger   Cation Exchanger

Anion Exchangers

- These exchangers possess positive charged groups on the stationary phase. Hence they can attract the negatively charged buffer counter ions (anions), which are available for exchange with sample analytes with negative charge. Therefore, they are called anion exchangers.

- These are also called as basic ion exchangers because of the positive charge on the stationary phase which is a resultant of the association of protons with basic groups in the matrix.

Cation Exchangers

- These exchangers possess negatively charged groups on stationary phase and so attract positively charged analytes in the buffer and sample.

- These are also called acidic ion exchanger because the negative charges are due to ionization of acidic groups on the stationary phase.

Example Structures of Anion Exchanger and Cation Exchanger

Anion Exchanger | Cation Exchanger

Ionised Groups of Ion Exchanger Matrix

The very presence of ionized groups is a basic and fundamental property of all ion exchangers. The type of charged group confers the type and capacity of the ion exchanger. The total number of ionic groups determines its capacity to bind with exchangeable ions. There is a variety of groups which have been chosen for use in ion exchangers.

Examples:

Anion Exchanger	Functional group	Cation Exchanger	Functional Group
Diethylaminoethyl (DEAE)	$-O-CH_2-CH_2-N+H(CH_2CH_3)$	Carboxymethyl (CM)	$-O-CH_2-COO$
Quaternary aminoethyl (QAE)	$-O-CH_2-CH_2-N+(C_2H_5)_2-$ $CH_2-CHOH-CH$	Sulphopropyl (SP)	$-O-CH_2-CHOH-CH_2-OCH_2-$ $CH_2-CH_2SO_3^-$
Quaternary ammonium (Q)	$-O-CH_2-CHOH-CH_2-O-CH_2-$ $CHOH-CH_2-N+(CH_3)_3$	Methyl sulphonate (S)	$-O-CH_2-CHOH-CH_2-OCH_2-$ $CHOH-CH_2SO_3^-$

Strong Ionic Exchangers

- Ion exchangers with functional ionic groups that are totally ionized at all working pH values are called as strong ionic exchangers. Ex: Sulphonate ($-SO_3^-$) and Quaternary ammonium ($-N^+R_3$).

Weak Ionic Exchangers

- Ion exchangers with functional ionic groups that are ionized only a narrow pH ranges. Ex: Carboxylate (-COO-) and Diethylammonium ($-NH^+ (CH_2CH_3)_2$).

Choice of Exchangers

The ion exchangers are basically selected based on stability of test analytes in the sample. Many biological molecules are stable within a small pH range. For separation of such molecules, the type of exchanger selected must operate within that narrow range. For example, if protein is highly stable below its isoelectric point, there will be net positive charge on the protein surface. Hence, for separation of such a protein, a cation exchanger should be used. The experimental pH value should be at lowest pH where the protein is stable. If protein is most stable above its isoelectric point, there will be net negative charge on the surface of the protein. So, an anion exchanger should be used to separate that protein. The experimental pH value in such cases should be between highest pH where protein is stable. If protein is stable over a wide range of pH, it can be separated by either type of ion exchanger. In such cases, the experimental pH value may be decided considering lowest and highest pH value stability of the protein. Weak electrolytes generally require a very high or very low pH for ionisation. So, it can only be separated by using a strong ion exchanger matrix, because they only operate over a wide pH range. For the separation of strong electrolytes, weak exchangers are generally preferred.

Different analytes have different degrees of interaction with the ion exchanger due to the differences in their charges, distribution of charge and charge densities on their surfaces. This interaction can be controlled by varying ionic strength and pH of medium. The small differences in such charge properties of biological compounds are often considerable as ion exchange chromatography is capable of separating different molecular species with very minor differences. Example: two proteins differing by only one charged amino acid can be separated using this technique.

Elution Buffer pH and Ionic Strength

Initial buffer pH and ionic strength should allow the binding of analytes to the exchanger. Many chromatographic procedures will get affected by applying changes in pH to affect a separation. For example, increasing the pH of the mobile phase buffer will lead the molecule to lose the proteon groups and the analyte become less protonated. This result in the generation of new ionic species with less positively charge on their surface. In cation exchange chromatography, such protein can no longer form an ionic interaction with the negatively charged solid support, which ultimately results in the molecule to elute from the column. In anion exchange chromatography, decreasing the pH of the mobile phase buffer will cause the molecule to become highly protonated. The resultant analyte is more positively (therefore less negatively) charged. This prevents the protein to form an ionic interaction with the positively charged solid support which causes the molecule to elute from the column.

It is important to keep the pH of the mobile phase buffer one unit above or below the isoelectric point (pI) of the analytes. In other words, the pH of the buffer should be between the acid dissociation constant (pKa) of the charged analyte molecule and the pKa of the charged group on the solid support. In cation exchange chromatography, for example, if a functional group on the solid support with a pKa of 1.2 is using, a sample molecule with a pI of 8.2 may be run in a mobile phase buffer of pH 6.0. In case of anion exchange chromatography, a molecule with a pI of 6.8 may be run in a mobile phase buffer at pH 8.0 when the pKa of the solid support is 10.3. Generally, anionic buffers are used for cation exchange chromatography; e.g., Tris, pyridine and alkalmine. On the contrary, cationic buffers are used for anion exchange chromatography; e.g., Acetate, barbituare and phosphate. Lowest ionic strength buffer should initially be used to allow minimum binding of contaminants. In gradient elution, initial conditions should be chosen that the exchanger binds all the test analytes at the top of column. Gradient elution is common than isocratic elution. For anion exchanger, the pH gradient decreases and ionic strength increases during elution. For cation exchanger, the pH gradient and ionic strength increase during elution.

Flow Rate of Buffer

Similar to other chromatography methods, the resolution in ion exchange chromatography is affected by flow rate. Flow rate means how quick the buffer is being passed through resin or stationary phase. Therefore, the flow rate affects interaction time of analytes with the column resin or matrix. This interaction time is called the residence time of a specific column at a particular flow rate. Flow rate affects both resolution, which is the ability to separate two peaks well, and capacity factor of the chromatography. Longer residence times increase both the capacity factor and the better resolution of a resin. As flow rates increase, pressure on the resin increases. If the backpressure is too high, it can crush the column resin. Generally, the fastest flow rate that still renders the desired capacity and resolution. Although slower flow rates may provide even better resolution and capacity, this is often at the expense of protein activity, as many proteins lose activity with time under the conditions in the chromatography system.

Applications of Ion Exchange Chromatography

- It is extremely used in the analysis of amino acids. The amino acid "Autoanalyzer" is based on in exchange principle.

- To determine the base composition of nucleic acids. Chargaff used this technique for established the equivalence of Adenine and Thymine; Guanine and Cytosine.

- This is the most effective method for water purification. Complete deionization of water (or) a non-electrolyte solution is performed by exchanging solute cations for hydrogen ions and solute anions for hydroxyl ions. This is usually achieved by the method is used for the softening of drinking water.

- Proteins are also successfully separated by this technique.

- It is also used for the separation of many vitamins, other biological amines, and organic acids and bases.

Advantages and Disadvantages of Ion-Exchange Chromatography

The advantages of ion-exchange chromatography are as follows:

- Ion-Exchange chromatography is one of the most powerful methods of separating charged particles.

- Using this method the inorganic ions can also be separated.

- It can use more commonly for both analytical and preparative purposes.

- Almost all charged molecules such as small amino acids, nucleotides and large proteins can separate using this method.

- This is very effective and powerful water softening method.

- Resins have a long life.

- It has cheap maintenance.

The disadvantages of ion-exchange chromatography are as follows:

- The buffer requirement is the major disadvantage of ion-exchange chromatography.

- It has a high working cost since the buffer is used for the separation of components.

- This method can only be used to isolate charged molecules.

- Sodium ions entering the soft water will increase the acidity level in the water.

Gel Filtration Chromatography

Gel Filtration chromatography is a separation technique of biomolecules based on the differences in their size or mass, when these biomolecules are passed through a column filled with the gel beads containing pores. The gel beads with pores are also called molecular sieve. Hence, another name for gel-filtration chromatography is molecular sieving. Like household sieve, small molecules that can enter into the pores of beads, and the large molecules cannot enter the pores of the beads. As the column is generally cylindrical in shape, the large molecules not entering into the bead will move faster between the beads and come out of the column soon. Small molecules travel through the pores and take more time to come out of the column, and hence elute later.

Exclusion Limit

In gel filtration chromatography, the pores in a stationary phase allow biomolecules below a particular mass or size. That particular mass or size is called exclusion limit. The molecules over and above that particular mass are excluded from entering into the pores of beads. Therefore, gel filtration chromatography is also called size exclusion chromatography. Different stationary phases of gel filtration chromatography have different exclusion limit. A few examples of different stationary phases along with their exclusion limit are shown in table.

Table: Different Gel filtration gels.

Gel filtration gel	Fractionation range in KDa	Application
Sephadex (Dextron) G-25	1-5	Desalting
Sephadex G-200	5-600	Protein fractionation
Sepharose 2B	70-40000	Fractionation of nucleic acids, particles and viruses
Sephacryl S-200 HR	5-250	Protein fractionation
Biogel P10	1.5-20	Fractionation of small proteins
Superdex-200	10-600	FPLC gel filtration
FRACTOGEL	1-700	HPLC gel filtration (proteins)
TSK HW-55		

Gel Filtration Chromatogram

Small molecules retain longest in the column than large molecules (or) large molecules elute first outside of the column. Therefore, the first peak in Gel Filtration Chromatogram represents larger size molecules that are excluded to enter into the pores of stationary phase. The distribution co-efficient (K_d) is low (minimum 0) for large analytes, high for small analytes (maximum 1) and within a range of 0-1 for intermediate analytes. That means, K_d represents the function of analyte molecular size. The distribution co-efficient (K_d) and retention volume of a particular analyte are related as follows:

$$V_s = (Kd_1 - Kd_2) V_i$$

Where,

- V_s: Difference between elution volumes between two analytes.

- Kd_1: Distribution co-efficient of analyte 1.

- Kd_2: Distribution co-efficient of analyte 2.

- V_i: Inner volume of mobile phase within the particle.

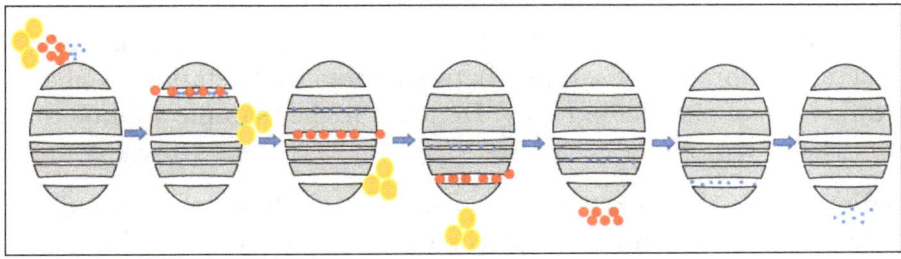

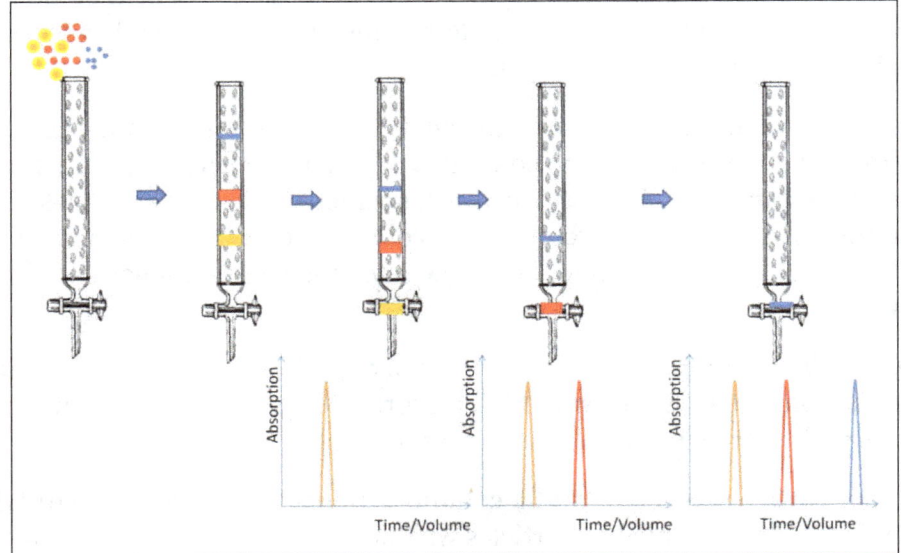

Figure shows gel filtration chromatography. Panel A indicates separation of three molecules with different size over the gel-filtration beads with different pores. The larger yellow molecules were excluded out of beads and hence they come out of the column first. These molecules are followed by red colour molecules and then blue colour molecules. This kind of separation is seen as bands in panel B and as peaks in gel-filtration chromatogram as panel C.

Factors Affecting Gel Filtration Chromatogram

The following factors determine the separation of molecules and thus the resolution in gel filtration chromatography:

- Sample volume: Large sample volumes allow the overlap of closely spaced peaks in chromatogram. Therefore, smaller samples are usually preferred. Sample volumes are expressed as a percent of total column volume. Most of the applications require a maximum of 2% sample volume out of the column volume to achieve maximum resolution. But depending upon the nature of the sample and the interest of application, larger sample volumes are also used, especially if the desired peak of interest is fully resolved and has no nearby overlapping peaks. For high resolution fractionation, 0.5% to 4% sample volumes are preferred. Sample volumes of up to 30% of total column volumes are also applied for group separation of molecules. Sample volumes of less than 0.5% are usually did not improve the resolution of complex molecules. Hence, it is preferred to use 0.5% of sample volume for complex samples and analytical applications.

- Ratio of sample volume to column volume: Column volumes need to be selected based on the sample volumes to be applied. Larger sample volumes require larger columns to allow the complete equilibration of samples between mobile and stationary phase for preventing the multiple runs of chromatography. Hence, a higher sample volume to column volume lowers the resolution.

- Column dimension: The height of column affects both the resolution and elution time. Resolution increases as the column height (bed-height) increases. Doubling the bed-height increases the resolution with a factor equivalent to the square root of 2. Sufficient bed heights along with low flow rates allow the complete diffusion of sample molecules between mobile and stationary phases for better resolution.

- Particle size: Gel filtration column efficiency, especially resolution, can be increased by decreasing the particle size of the column as smaller particles allow the maximum equilibration of the sample molecules. However, smaller particles create back pressure within the column which further can damage the column. So while using smaller particles, it is advised to decrease the flow rate. Low flow rates increase the length of chromatographic run time.

- Particle size distribution: Uniform distribution of particle size allows the uniform flow rate through the column. Columns with high uniformity (ex: small size) facilitate the elution of molecules with narrow peaks in chromatogram.

- Pore size of particles: Selectivity of a column depends on the pore size distribution of the particles in gel filtration media. Particles with smaller pore size can efficiently separate the small particles from large particles. Pore size of particles defines the fractionization range and exclusion limit of sample molecules.

- Flow rate: Flow rates determine the equilibration of sample between mobile and stationary phases. Low flow rates allow the complete diffusion of sample molecules between two phases and lead to better resolution.

- Viscosity of sample and buffer: Viscosity of samples limits the concentration of sample applied for separation in chromatography. Highly viscous samples cause poor separation of molecules with irregular flow patterns. Viscous samples create back pressure in the column and give broad elution peaks.

Sample Preparation

Sample preparation is one of the crucial steps in gel filtration chromatography. Samples to be subjected to chromatographic purifications should be clear and free of any particulate matter. The sample clarification will avoid blockage in the column, reduce the need for rigorous washing of column and also extends the life of chromatography column. Therefore, the samples should be properly clarified before applying to the column to avoid such risks and for better performance of chromatography. For small volumes of samples, syringe tip filter made of PVDF or cellulose acetate can be used for clarifying the samples.

Composition of Sample Buffer

Sample buffer should be carefully selected in such a way that it should not affect the stability and activity of the proteins, which are required to be separated by gel filtration chromatography. Keeping this point in mind, sample buffer with optimum pH, ionic strength and composition within in the range of stability of gel filtration medium will be selected. It is not necessary to use the buffer which is needed for equilibration of column as a sample buffer.

Concentration and Viscosity of Sample

The ability of gel filtration chromatography to resolute the mixture is independent of the sample concentration. It is possible to get the high resolution rates even for the concentrated samples. But, viscosity of the sample should be optimum as high viscosity can cause unstable separation and irregular flow rates. This can also result in increased back pressure and skewness in the peaks. Generally, protein samples of concentration up to 70mg/ml are subjected for chromatography separation.

Sample Volume

Volume of the sample should be optimum and within the range 0.5-4% of the column volume. It is one of the crucial factors that can affect the resolution of samples in gel filtration chromatography. Generally, 2 % sample volume out of the column volume results in better resolution.

Composition and Preparation of Buffer

Buffer should be prepared in such a way that it should not affect the size, stability and biological activity of the samples to be separated. However, it does not directly affect the resolution pattern in gel filtration. pH and ionic strength are the two important properties of the buffer that must be taken care as they can affect the conformation of molecules by affecting the intra and inter sub unit bonding of proteins, which further affect the interaction of hormone - carrier protein, enzymes - cofactors. Hence, buffer should be selected in such a way that it should be compatible with conformation as well as activity of the protein sample. Ionic interactions with the matrix are undesirable in gel filtration. Usually 0.15M NaCl can be used to avoid such interactions. If the separated products of gel filtration are required to be lyophilized, volatile buffer like ammonium acetate, ammonium bicarbonate or ethylene diamine acetate should be used. Buffer should be subjected to filtration through 0.45 µm or 0.22 µm filters and degassed in vacuum as it can affect the performance of this technique.

Denaturing Agents and Detergents

Solubility of sample can be increased by using detergents like guanidine hydrochloride or urea. These two agents are recommended in cases where sample denaturation is required. In cases of working under harsh conditions (dissociating or denaturing) or extreme pH, modern media like Superdex, Sephacryl or Superose is recommended. Hence, suitable concentrations of detergents are advised in running buffer in order to avoid the sample precipitation. Moreover, use of denaturing agents in the running buffer helps to maintain the extended configuration and thus, will be useful to determine the accurate molecular weight.

Applications of Gel Filtration Chromatography

One of the principal advantages of gel filtration chromatography is that separation can be performed under conditions specifically designed to maintain the stability and activity of the molecule of interest without compromising resolution. Absence of a molecule-matrix binding step also prevents unnecessary damage to fragile molecules, ensuring that gel-filtration separations generally give high recoveries of activity.

This technique, however, is not without its disadvantages. When separating proteins by gel filtration chromatography, for example, proteolysis becomes an increasing problem, since the target protein frequently becomes the abundant substrate for proteases also present in the mixture, consequently reducing recovery of activity. Because of the large size of gel filtration columns, large volumes of eluent are usually required for their operation, often creating excessive running costs. Gel filtration also has an inherent low resolution compared to other chromatographic techniques because none of the molecules are retained by the column and non-ideal flow occurs around the beads. In addition, this technique has a low sample-handling capacity dictated by the need to optimize resolution.

Despite these disadvantages, gel filtration chromatography still occupies a key position in the field of biomolecule separation because of its simplicity, reliability, versatility, and ease of scale-up.

Separation of Proteins and Peptides

Because of its unique mode of separation, gel filtration chromatography has been used successfully in the purification of literally thousands of proteins and peptides from various sources. These range from therapeutic proteins and peptides, which together constitute a multibillion euro worldwide market, to enzymes and proteins for industrial applications.

Recombinant human granulocyte colony stimulating factor (rhG-CSF) was refolded from inclusion bodies in high yield, with great suppression of aggregates formation, by urea-gradient size-exclusion chromatography on a Superdex 75 column. A similar technique was used to purify human interferon-gamma, solubilized from inclusion bodies by 8 M urea, to a specific activity of 12,000,000 IU/mg with protein recovery of 67 %. Luteinizing hormone (LH) was purified 46-fold from crude pituitary extract by gel filtration on two Sephacryl S-200 columns. The method exploited differential binding of LH (in the crude extract) to blue dextran for the first chromatography step. Before the second step, addition of high salt released LH from the blue dextran, enabling effective purification. Fusion ferritin (heavy-chain ferritin plus light-chain ferritin) has also been purified by urea-gradient gel filtration. In this case, fusion ferritin solubilized from inclusion bodies with 4 M urea was applied to the column. Refolding enhancers were included in the urea-diluent buffer subsequently applied to the column to produce properly folded fusion ferritin multimers.

A continuous rotating annular size-exclusion chromatography system permitted the purification of crude porcine lipase with productivity of approximately 3 mg lipase per mg gel per hour and an activity recovery of almost 99 %.

Among food-use proteins, hen egg lysozyme has been successfully refolded using both acrylamide- and dextran-based gel columns (Sephacryl S-100 and Superdex 75, respectively). Gel filtration has also proven useful for the purification of the whey proteins alpha-lactalbumin and beta-lactoglobulin from aqueous two-phase systems.

Protein engineering techniques enable the design of self-assembling multimeric protein cages for applications in nanotechnology. Grove et al. describe a gel-filtration method to examine the metal ion-mediated assembly of protein cages.

Size-exclusion Reaction Chromatography: Protein PEGylation

Covalent attachment of PEG (polyethylene glycol; "PEGylation") to a protein can attenuate its antigenicity and/or extend its biological half-life or shelf life. Size-exclusion reaction chromatography (SERC) permits one to control the extent of a reaction (such as PEGylation) that alters molecular size and to separate reactants and products. In SERC, injection of reactants onto a size-exclusion chromatography column forms a moving reaction zone. Reactants and products partition differently within the mobile phase leading to different flow rates through the column. Thus, products are removed selectively from the reaction zone, shortening their residence time in the reaction zone and separating them into the downstream section of the column. In PEGylation, addition of PEG groups to the protein significantly increases molecular size, allowing the use of SERC to obtain a dominant final PEGylated protein size in high yield. The principle was successfully demonstrated using two model proteins, alpha-lactalbumin and beta-lactoglobulin.

Separation of Nucleic Acids and Nucleotides

Gel filtration chromatography has for many years been used to separate various nucleic acid species such as DNA, RNA, and tRNA as well as their constituent bases, adenine, guanine, thymine, cytosine, and uracil. Linear phage lambda DNA and circular double stranded phage M13 DNA, for example, can be completely separated from chromosomal DNA and RNA by gel filtration on Sephacryl S-1000 Superfine. Plasmid DNA can also be purified by gel filtration, although modern commercial kits often use a centrifugal spin column format for greater convenience. Limonta et al. describe the novel use of two gel-filtration steps, one before and one after a reverse-phase operation, to purify plasmid DNA from a clarified alkaline E. cell lysate.

Virus Particles

Krober et al. devised an open loop simulated moving bed (SMB) for the continuous size-exclusion chromatographic separation of influenza virus (derived from cell culture) from contaminating proteins. Overall productivity of the SMB process was estimated to be up to 3.8-fold greater than that of an optimized batch process.

Size-exclusion chromatography was used downstream of expanded bed adsorption chromatography to recover active recombinant hepatitis B core antigen (HbcAg) in 45 % yield with a purification factor of 4.5. A Sephacryl S-1000 SF proved to be effective and economical in the purification of recombinant Bombyx mori nucleopolyhedrosis virus displaying human pro renin receptor. Sephacryl S-1000 gel-filtration chromatography gave more effective purification of turkey coronavirus from infected turkey embryos than did use of a sucrose gradient.

Endotoxin Removal

The presence of bacterial endotoxin is unacceptable in injectable recombinant biologicals, since endotoxin in the bloodstream can induce a pyrogenic response. Good manufacturing practice (GMP) will effectively remove endotoxin, but preclinical biologics may be produced under non-GMP conditions. London et al. investigated various means of endotoxin removal from preparations of a recombinant human protein. Endotoxins typically form aggregates, which may be quite large. A Superdex 200 size-exclusion column (1.75 L bed volume) removed most of the "spiked" endotoxin from an applied sample of monomeric monoclonal antibody, which was obtained in good yield.

Absolute Size-exclusion Chromatography (ASEC)

Absolute size-exclusion chromatography (ASEC) is a technique that couples a dynamic light scattering (DLS) instrument to a size-exclusion chromatography system for absolute size measurements of proteins and other macromolecules as they elute from the chromatographic system. Dynamic light scattering (DLS; also known as photon correlation spectroscopy or quasi-elastic light scattering) is a technique that uses light scattering patterns (usually from a laser source) to determine the size distribution profile of small particles in suspension, or of polymers (such as proteins) in solution. DLS can also be used to probe the behavior of complex fluids such as concentrated polymer solutions.

The sizes of the macromolecules are measured as they elute into the flow cell of the DLS instrument from the size-exclusion column. It should be noted that the technique measures the hydrodynamic size of the molecules or particles and not their molecular weights. For proteins, a Mark–Houwink type of calculation can be used to estimate the molecular weight from the hydrodynamic size.

A big advantage of DLS coupled with SEC is the ability to obtain enhanced DLS resolution. Batch DLS is quick and simple to perform. Using SEC, the proteins and protein oligomers are separated, allowing oligomeric resolution. ASEC can also be used for aggregation studies: although the aggregate concentration may not be calculated, the size of the aggregate will be measured, being limited only by the maximum size eluting from the SEC columns. Limitations of ASEC include flow rate, concentration, and precision. Because a correlation function requires anywhere from 3 to 7 s to properly build; only a limited number of data points can be collected across the peak.

Molecular Mass Estimation

Gel filtration chromatography is an excellent alternative to SDS-PAGE for the determination of relative molecular masses of proteins, since the elution volume of a globular protein is linearly related to the logarithm of its molecular weight. One can prepare a calibration curve for a given column by individually applying and eluting at least five suitable standard proteins (in the correct fractionation range for the matrix) over the column, determining the elution volume for each protein standard, and plotting the logarithm of molecular weight versus V_e / V_o. When a protein of unknown molecular weight is applied to the same column and eluted under the same conditions, one can use the elution volume of the protein to determine its molecular weight from the calibration curve.

Group Separations

By selecting a matrix pore size which completely excludes all of the larger molecules in a sample from the internal bead volume, but which allows very small molecules to enter this volume easily, one can effect a group separation in a single, rapid gel-filtration step which would traditionally require dialysis for up to 24 h to achieve. Group separation can be used, for example, to effect buffer exchanges within samples, for desalting of labile samples prior to concentration and lyophilization, to remove phenol from nucleic acid preparations and to remove inhibitors from enzymes.

Affinity Chromatography

Affinity chromatography is a separation technique which makes use of affinity of ligand towards molecule to be separated. Ligand is usually attached to insoluble support. Ligand forms complex with target molecule in reversible way. Ligand can only form complex with target molecule since it lacks affinity for other molecules. Thus, affinity chromatography has potential to separate target molecule in presence of large number of other molecules. This also means that in this technique, higher level of purification can be achieved in a single step. Since binding is reversible, target molecule from complex can be released by selecting appropriate conditions which favours breakdown of ligand-target complex. This can be achieved by altering pH, ionic strength, polarity, temperature of eluting solution.

$$\text{Molecule} + \text{Ligand} \rightleftharpoons \text{Molecule} - \text{Ligand Complex}$$

Affinity chromatography essentially requires:

- Ligand immobilization to inert matrix.
- Packing of immobilized ligand in chromatographic column.
- Equilibration of column with binding buffer.
- Binding of target molecules to ligand.
- Removal of unbound molecules.

- Release of bound molecules.

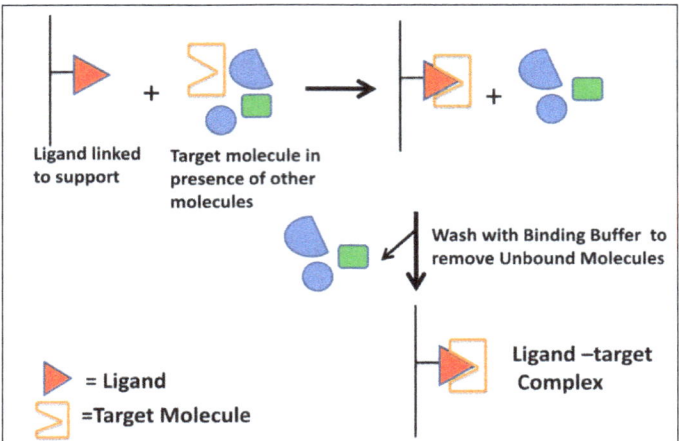

Figure: Binding of target molecules in presence of other molecules to affinity matrix.

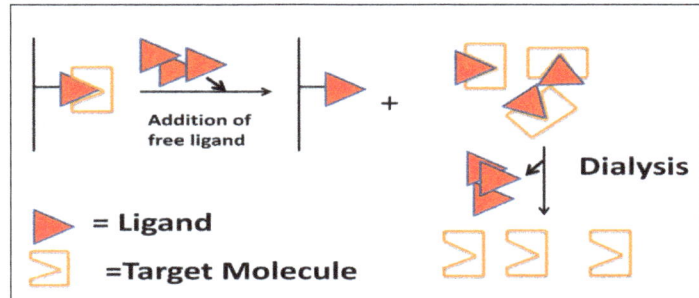

Figure: Release of Bound Molecule by addition of Free Ligand.

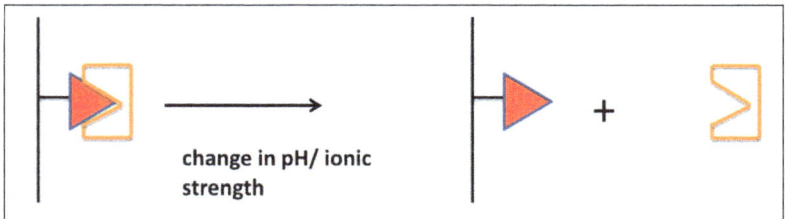

Figure: Release of bound target by change in pH/ionic strength.

Unlike many other separation techniques, affinity chromatography does not depend on physical properties of molecules. Specific interaction between ligand and target molecule is exploited for separating molecules from others. There are numerous examples where two different molecules interact with each other and provide resolution. Important interactions include the following:

- Interaction of substrate or competitive inhibitor with enzyme.

- Interaction of antigen with antibody.

- Interaction of lectin with glycoprotein.

- Interaction of borate with glycoprotein.

- Interaction of receptor with protein.

- Interaction between complementary strands of DNA.

- Interaction of cibacron blue dye with nucleotide requiring enzymes.

- Interaction of avidin with biotin.

- Interaction of polyA with mRNA.

- Interaction of protein A or G with immunoglobulins.

- Interaction of 5'-AMP with NAD^+-dependent dehydrogenase.

- Interaction of heparin with lipase.

- Interaction of metal ions with imidazole, thiol and indole groups.

- Interaction of aptamer with target molecule.

Affinity chromatography is useful step in overall purification strategy of protein. Only requirement is availability of ligand which has specificity for target protein. Several inert supports with different chemistry are available for linking ligand. Spacer is attached towards one end of ligand. This enables ligand to acquire flexibility. Also, stearic hindrance between support and target is reduced. This helps in efficient interaction between ligand and target. Binding capacity of matrix is high and thus affinity columns are of low bed volume.

Thus, affinity chromatography can be completed in short duration. Because of these several advantages, affinity chromatography is widely used for purification of proteins. Interaction between ligand and protein is non-covalent. Ionic interaction, hydrogen bond, van der Wall interaction and hydrophobic interaction may participate in interaction between ligand and target molecule. These non-covalent interactions can be disrupted by selecting conditions which disrupts ionic interaction, hydrogen bond, van der Wall interaction and hydrophobic interaction. Ionic interaction can be disrupted by adding 0.5 NaCl or 1.0 M NaCl in elution buffer. Alternately, this interaction can be broken by changing pH of elution buffer. Hydrogen bond between ligand and target can be broken by incorporating urea (~ 6 M) in elution buffer. The objective in elution is also to recover protein in biological active form. Thus, elution is avoided with buffers of extreme pH or hydrophobic solvents. Several thousand fold purification can be achieved. Recovery is also good and recovered molecules retain biological activity. If extreme of pH is used for elution of protein, eluted protein may lose activity.

In this situation, eluted fraction should be neutralized immediately after elution. Elution can also be achieved by using higher concentration of ligand or competitive inhibitor or substrate in eluent. However, these materials are required to be removed from eluted fraction for obtaining purified protein. Small molecular weight ligand, substrate and inhibitor can be removed by dialysis or gel filtration. The technique is useful for isolating substances in pure form from large volume sample containing low concentration of sample. Affinity chromatography is only technique which results in purification of biomolecule on the basis of biological function. The technique can be used for separating native protein from denatured protein. Availability of pre-packed column, affinity matrix and ready to reconstitute binding buffer & elution buffer has made affinity chromatography user friendly. The power of technique is clear from the fact that many persons prefer to express recombinant protein as fusion protein having 'His' tag for allowing its separation by metal chelate affinity chromatography.

Components of an Affinity Medium

The Matrix

The matrix is an inert support to which a ligand can be directly or indirectly coupled. Matrix should have following properties:

- It should not interact with molecules which are to be separated. Thus it should be inert.

- It should have functional group to which ligand can be attached. Hydroxyl groups can be easily derivatized for covalent attachment.

- It should have open structure which ensures availability interior of matrix for ligand coupling.

- It should have good flow property.

- It should have good mechanical strength. Cross-linked polymers have high mechanical strength.

- It should be stable under experimental conditions such as high and low pH, detergents and dissociating agents.

Sepharose is widely used matrix. It is a tradename for a crosslinked, beaded-form of agarose, a polysaccharide polymer material extracted from seaweed. Its brand name is derived from Separation-Pharmacia-Agarose. Sepharose has been modified to suit the particular requirements for each application. In affinity chromatography the particle size and porosity are designed to maximize the surface area available for coupling a ligand and binding the target molecule. A small mean particle size with high porosity increases the surface area. Increasing the degree of crosslinking of the matrix improves the chemical and mechanical stability. Sepharose matrices with different degree of crosslinking, and bead size are commercially available and named as:

- Sepharose High Performance 6% highly cross-linked agarose 34 μm.

- Sepharose 6 Fast Flow 6% highly cross-linked agarose 90 μm.

- Sepharose 4 Fast Flow 4% highly cross-linked agarose 90 μm.

- Sepharose CL-6B 6% cross-linked agarose 90 μm.

- Sepharose CL-4B 4% cross-linked agarose 90 μm.

- Sepharose 6B 6% agarose 90 μm.

- Sepharose 4B 4% agarose 90 μm.

Matrix in ready to use format are also commercially available. These are NHS-activated sepharose and CNBr- activated Sepharose.

NHS-activated agarose- Amino hexanoic acid is linked to hydroxyl group of agarose by epichlorohydrin. Then, carbodiimide is allowed to react and this result in formation of O-acylurea .This ester is unstable. N-hydroxysuccinimde then reacts with O-acylurea ester and semi-stable NHC ester is formed. Protein containing primary amino group can be linked to semi-stable ester and in the process n-hydroxysuccinimide is released. The matrix is stable at high pH.

CNBr-activated Agarose

Cyanogen bromide reacts with hydroxyl groups on Sepharose to form reactive cyanate ester groups. Proteins through primary amino group can react with cyanate ester bond to form isourea bond. Cyanate activated agarose is an alternative to NHS-activated agarose. The method is an alternate method to NHS- activated agarose. Cyanogen bromide is hazardous chemical and is handled in safety hood.

Ligand

Ligand must contain functional groups which can be modified to allow coupling of ligand to matrix. Ligand must bind reversibly to target. The dissociation constant (kD) for the ligand - target complex should ideally be in the range 10^{-4} to 10^{-8} M. Dissociation constants greater than 10^{-4} M, are likely to be too weak for successful affinity chromatography. Dissociation constant lower than 10^{-8} M will require harsh elution conditions and these may inactivate eluted molecules. If several functional groups are available, coupling of ligand via the group least likely to be involved in the specific affinity interaction is to be preferred.

Spacer Arm

In purification of large size molecules such as protein, ligands may bind deep into molecule. This is likely that ligand is unable to access the binding site of the target molecule. It will result in low binding capacity due to steric interference. In these circumstances a "spacer arm" is interposed between the matrix and the ligand to facilitate effective binding. Spacer arms must be designed to maximize binding, but to avoid non-specific binding effects. The length of spacer should be optimum. Ligand with short arm will fail to bind target molecules. If spacer arm is too long, proteins may bind non-specifically to the spacer arm. Spacer arm is not generally needed for larger molecules used as ligand.

Immunoaffinity Chromatography

Immunoaffinity chromatography (IAC) is a type of LC in which the stationary phase consists of an antibody or antibody-related reagent. This technique represents a special sub category of affinity chromatography, in which a biologically related binding agent is used for the selective purification or analysis of a target compound. The selective and strong binding of antibodies for their given targets has made these agents of great interest for many years as immobilized ligands in affinity chromatography. Other methods sometimes included under the heading of IAC are those that use immobilized targets for antibody purification.

The earliest use of IAC and related methods was in the selective purification of compounds from biological samples. For instance, target compounds immobilized onto kaolin and charcoal were used as early as 1935 to isolate antibodies associated with syphilis and TB. A total of 1 year later, Karl Landsteiner and co-workers isolated antibodies by using targets immobilized onto chicken erthyrocyte stroma by a diazo coupling method. The first modern use of IAC is generally attributed to Campbell et al., who immobilized serum albumin to p-aminobenzylcellulose in 1951 for use in antibody purification. Since that time, there have been thousands of reports using IAC for both the isolation of chemicals or biochemicals and analytical applications.

Basic Components of IAC

Structure and Properties of Antibodies

The basis for IAC relies on the selective binding of antibodies. This binding is a result of a large variety of non-covalent interactions that can occur between an antibody and an antigen and can result in association equilibrium constants in the range of $10^5–10^{12}$ M^{-1}. It has been estimated that the human body is able to produce between 10^7 and 10^8 different types of antibodies, with each capable of binding to a separate antigen. The typical structure of an antibody, using IgG as an example, consists of four polypeptide chains. These four chains consist of two identical heavy chains and two identical light chains that are linked by disulfide bonds to from a Y-shaped structure. The lower stem region of an antibody is referred to as the F_c region and is highly conserved from one antibody class to the next. The upper arms of the antibody are called the F_{ab} regions. The amino acid sequences in the F_{ab} regions are identical within a single type of antibody but are highly variable between different antibodies. It is this variability that allows antibodies to bind a wide range of antigens.

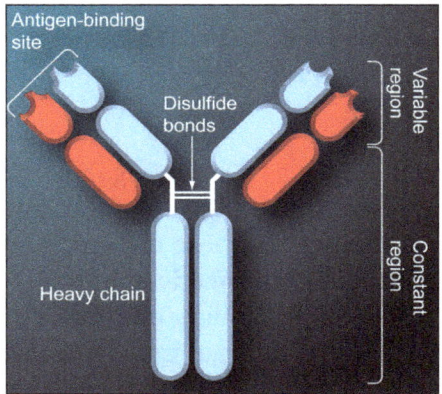

Figure: Typical structure of an IgG-class antibody.

IgG-class antibodies consist of four polypeptide chains that are linked by disulfide bonds to create a 'Y'-shaped structure with two identical binding sites.

A foreign agent that is capable of initiating antibody production is called an antigen. Common antigens include viruses, bacteria and foreign proteins from animals and plants that are capable of producing an immune response. Due to the large size of naturally occurring antigens, antibodies that bind to several different regions of the antigen with a range of binding affinities are often generated. Each individual location on an antigen that can bind to a given antibody is called an epitope. In order for a substance to be recognized by the body's immune system and to lead to the production of antibodies, this substance must have a size that corresponds to a mass of several thousand Daltons. Antibodies can also be produced against smaller substances, but these substances first must be coupled to a larger species (e.g., a carrier protein) before antibody production can occur. A small substance that is used to produce antibodies after being linked to a carrier agent is known as a hapten.

The two main types of antibodies that are used in IAC are polyclonal antibodies and monoclonal antibodies. Polyclonal antibodies are produced from multiple cell lines within the body and as a population can bind a variety of epitopes on a single antigen with a range of binding strengths. Monoclonal antibodies are produced by the fusion of a myeloma cell line with spleen cells obtained from an animal that has been immunized with the desired antigen. Because monoclonal antibodies are generated from a single cell line, they bind to a single epitope with identical binding affinities. Two other types of antibodies that can be used in IAC are autoantibodies, which are polyclonal antibodies obtained from patients with auto immune diseases, and anti-idiotypic antibodies, which are antibodies that can mimic the interactions of antigens, hormones or substrates for cell receptors.

Recent advances have led to the production of artificial antibodies that have high binding affinities. These artificial antibodies can be produced by combining two or more ligands with moderate affinity on a synthetic tether or polymer. One method of producing these so-called synbodies utilizes a small library of short, unstructured polypeptides that are capable of binding independent sites on a protein target. These polypeptides can easily be linked to bivalent reagents at different distances and orientations.

Production of Antibodies

Polyclonal antibodies can be produced against a given target by injecting the antigen or a hapten–carrier conjugate into a laboratory animal (e.g., a mouse, rabbit, goat or sheep). Often this solution of the antigen or hapten–carrier conjugate contains an enhancing agent called an adjuvant. After this initial injection, blood samples from the animal are collected after approximately a month (e.g., 3–4 weeks, although the exact times used in this sequence may vary) and tested for the presence of antibodies that are specific for the desired target. Another injection of the antigen or hapten conjugate (called a 'booster') is then made into the animal. The animal's blood is then retested later (e.g., after 10 days) for the presence of antibodies. Based on the antibody levels that are detected, the animal can be allowed to rest for a period of time (e.g., a few weeks) before being administered another booster injection. This booster/bleed routine can be repeated several times until the antibody concentrations for the required antigen reach the desired level (i.e., as determined by an assay of the blood). At this point, antibody-containing serum can be collected from the animal and stored for later use.

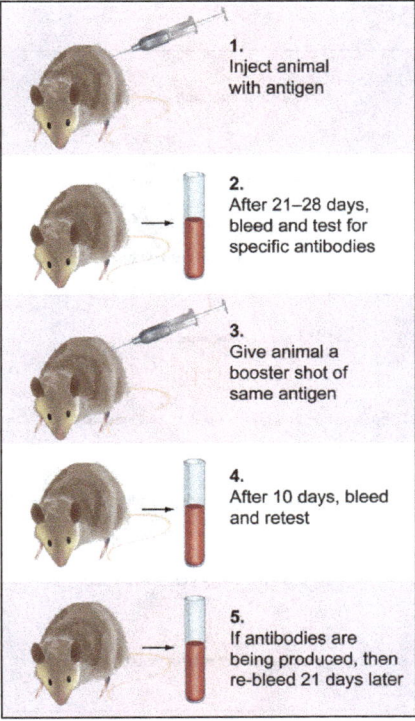

1.
Inject animal
with antigen

2.
After 21–28 days,
bleed and test for
specific antibodies

3.
Give animal a
booster shot of
same antigen

4.
After 10 days, bleed
and retest

5.
If antibodies are
being produced, then
re-bleed 21 days later

Figure: Process for polyclonal antibody production.

The antibodies that are produced upon the first exposure of an animal to a foreign agent are typically IgM class antibodies. After repeated exposure, IgG class antibodies will also be produced. It is this secondary immune response that produces antibodies that are best suited for IAC applications. Antibodies that are produced through the normal immune response are polyclonal antibodies that can bind with various strengths and to a variety of epitopes on the antigen. Before these antibodies can be used in IAC, some further purification is often required. This purification may involve the use of ion-exchange chromatography, precipitation with ammonium or dextran sulfate or isolation on a protein A or protein G column. The isolated products from each of these techniques will contain some antibodies that do not bind to the desired antigen, but these other antibodies can later be removed by using an immobilized antigen column.

Monoclonal antibodies can be produced by isolating a single antibody-producing cell and combining this cell with a carcinoma or myeloma cell. The resulting hybrid cell line, called a hybridoma, is relatively easy to culture and grow for long-term antibody production. This approach for monoclonal antibody production was first reported by Köhler and Milstein in 1975. In this method, a solution of the antigen or the hapten–carrier conjugate is mixed with adjuvant and injected subcutaneously into an animal. The animal is later given a booster shot, killed and the spleen harvested. The lymphocyte cells are mixed with myeloma cells in the presence of polyethylene glycol, which is added to promote cell fusion. The cells are then grown in the presence of drugs that kill myeloma cells and unfused lymphocytes, but permit the growth of hybridoma cells. Individual cultures of hybridomas are examined for the production of specific antibodies and those that make the desired antibody are cloned to produce a homogenous culture of cells making a monoclonal antibody. Although this process can be tedious and difficult, it has the advantage of producing antibodies in relatively large quantities that have well-defined specificity.

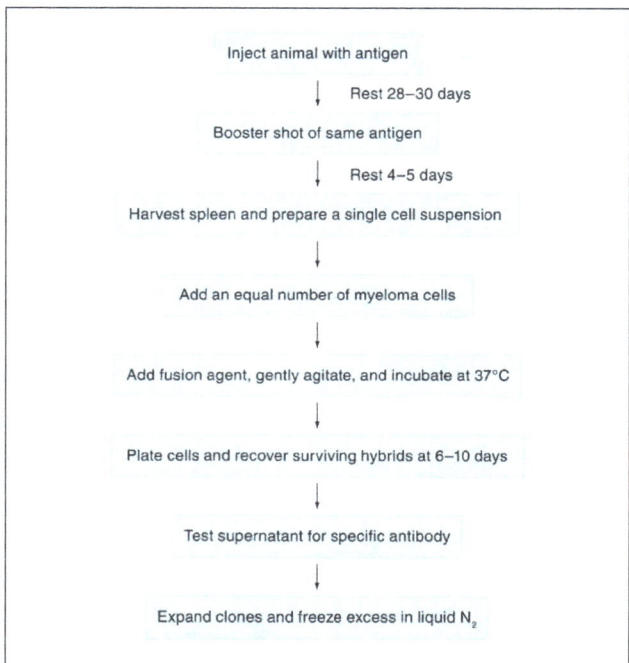

Figure: Process for monoclonal antibody production.

Supports for IAC

There are several types of supports that can be utilized to place antibodies within columns for use in IAC. Traditional immunoaffinity supports have been based on low-performance materials such as carbohydrate-related media (e.g., agarose and cellulose) or synthetic organic supports (e.g., acrylamide polymers, copolymers or derivatives, polymethacrylate derivatives and polyethersulfone matrixes). Table lists several commercial supports in these categories that can be used to immobilize antibodies for IAC. The low cost of these materials has made these supports popular for IAC applications involving target purification or offline immunoextraction. However, most of these materials can be used at only relatively low back pressures and are best suited for work under gravity flow or with a peristaltic pump. These supports also can have slow mass-transfer properties.

Table: Commercially available supports that can be used for immunoaffinity chromatography.

Support type	Supplier
Low or medium performance supports	
Affi-Gel	BioRad
Affinica Agarose/Polymeric Supports	Schleicher and Schuell
AvidGel	BioProbe
Bio-Gel	BioRad
Fractogel	EM Separations
HEMA-AFC	Alltech
Reacti-Gel	Pierce
Sephacryl	Pharmacia
Sepharose	Pharmacia
Superose	Pharmacia
Trisacryl	IBF
TSK Gel Toyopearl	TosoHaas
Ultragel	IBF
High performance supports	
AvidGel CPG	BioProbe
HiPAC	ChromatoChem
Protein-Pak Affinity Packing	Waters
Ultraafinity-EP	Bodman
Emphaze	3M Corp./Pierce
POROS	ABI/PerSeptive Biosystems

These disadvantages have limited the use of the low-performance immunoaffinity supports in applications requiring their direct use with HPLC. When antibodies are used within an HPLC column, the resulting method is referred to as HPIAC. Immunoaffinity supports that are used for HPIAC must be rigid and have higher efficiency than typical low-performance supports. Examples of materials that have been used in HPIAC include derivatized silica, glass, azalactone beads, methacrylate polymeric supports and polystyrene-based perfusion media.

In addition to good efficiency and mechanical stability, the ideal support for IAC should have low nonspecific binding and be easily modified for antibody attachment. Another item to consider for a porous material is the size of its pores for immobilization of antibodies and binding by antibodies to the target. Supports with small pores have the largest amount of total surface area, but much of this area may not be accessible for antibody immobilization. By contrast, supports with larger pores do not have accessibility problems, but their lower surface area can result in small amounts of immobilized antibodies. As a result, the maximum coverage for antibodies is often observed for supports with pore sizes of 300–500 Å, which is approximately three- to five-times the diameter of an antibody. This size range is also suitable for the binding of immobilized antibodies to many small- or medium-sized targets (i.e., agents with sizes less than about 100–150 kDa), although larger pore sizes may be needed for larger targets. Other support materials that have recently been used in IAC are disks, fibers and monolithic rods. Unique features offered by these newer support materials include their good mass-transfer and flow properties. A general procedure for immobilizing antibodies onto monoliths can be seen in figure.

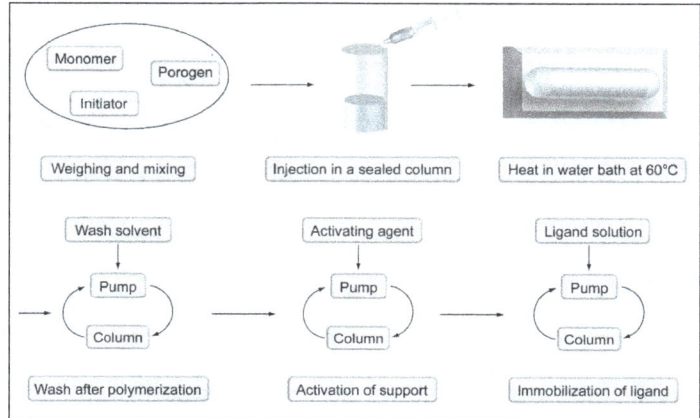

Figure: General procedure for immobilizing antibodies within a monolithic column.

Antibody-immobilization Methods

Antibodies can be immobilized onto supports by using a variety of techniques that range from covalent attachment to adsorption-based methods. Of these techniques, those that make use of covalent attachment are the most common, but even these methods can range from the use of random attachment via amino or carboxyl groups to more site-selective immobilization approaches that make use of modified carbohydrate residues or thiol groups. The ideal situation in any of these immobilization methods is to have antibodies attached to the support in a way that does not affect the activity of the binding sites or the accessibility of these sites to their target compound.

Antibodies can be immobilized through free amine groups by using supports that have been activated with agents such as N,N'-carbonyldiimidazole, cyanogen bromide, N-hydroxysuccinimide or tresyl chloride/tosyl chloride. Antibodies can also be immobilized through amine groups using a support that has been treated to produce reactive epoxy or aldehyde groups on its surface. The use of amine groups is one of the easiest ways to immobilize antibodies but can cause a decrease in activity if the antibodies have some of these amine groups in their antigen-binding sites. In addition, this approach can cause the antibodies to be immobilized in a random orientation, leading to steric hindrance and a decrease in binding efficiency.

Antibodies or F_{ab} fragments can be covalently linked to IAC supports through more site-selective methods. This can be achieved by utilizing the free sulfhydryl groups that are created when F_{ab} fragments are generated. These groups can be used for immobilization by using techniques such as the divinylsulfone, epoxy, iodoacetyl/bromoacetyl, maleimide or tresyl chloride/tosyl chloride methods. In addition, site-selective immobilization can be accomplished by coupling antibodies through the carbohydrate residues that are located in their F_c regions. This process is carried out by first oxidizing the carbohydrate residues under mild conditions to generate aldehyde residues. These aldehyde groups are then reacted with a hydrazide- or amine-containing support.

A third way antibodies can be immobilized onto supports is by using a secondary ligand to adsorb these antibodies. This can be accomplished by using antibodies that have been reacted with biotin or biotinylated, and then adsorbed to a support that contains immobilized avidin or streptavidin. The most popular biotinylation technique is to incubate antibodies with N-hydroxysuccinimide-D-biotin at pH 9. The strong non-covalent linkage of biotin to strepavidin or avidin can then be used to immobilize these antibodies. These linkages have association equilibrium constants in the range of 10^{13}–10^{15} M^{-1} and can resist many types of elution conditions without dissociation. In addition, a modified version of avidin called neutravidin (from Pierce) can be used for the immobilization of biotinylated antibodies. If amine-based coupling of biotin to the antibodies is employed, the biotin can attach at or near the antibody's binding sites and cause a decrease in binding. There will also be random orientiation of the resulting biotinylated antibodies on the streptavidin support. These problems can instead be minimized by using hydrazide-biotin, which is reacted with the carbohydrate residues of antibodies after these regions have been oxidized under mild conditions to produce some aldehyde groups.

Another approach for antibody immobilization utilizes the strong binding of protein A or protein G to the F_c regions of many antibodies. Protein A and protein G are bacterial cell wall proteins that have strong binding for many types of antibodies under physiological conditions. However, if the pH of the surrounding solution is decreased to approximately pH 2–3, this binding will be weakened and the retained antibodies can be eluted. Protein A and protein G supports are useful when high antibody activity is needed and it is desirable to have frequent antibody replacement in an IAC column. Long-term reproducibility of the IAC binding capacity is possible when using protein A or protein G supports, but a much larger amount of antibodies will be needed than is required for traditional immobilization methods if the antibodies are eluted and replaced on a regular basis. If desired, protein A or protein G supports can be prepared with a permanent coating of antibodies by cross-linking the antibodies to the immobilized protein A or protein G by using carbodiimide or dimethyl pimelimidate.

Application and Elution Conditions

The application and elution conditions are another important set of factors to consider in the design and use of an IAC method. The application buffer used in IAC is generally chosen for its ability to promote fast and efficient binding of the desired analyte or target compound to the immobilized antibodies. Optimum binding for antibodies typically occurs under physiological conditions, so IAC generally makes use of a neutral pH buffer (i.e., pH 7.0–7.4). Under these conditions, the equilibrium constants for antibody binding are usually in the range of 10^6–10^{12} M^{-1}. Due to this strong binding between the antibody and its target, isocratic elution is often not feasible unless the

IAC method is using low-affinity antibodies (i.e., those with association equilibrium constants of less than 10^6 M^{-1}).

The elution conditions for IAC need to allow for fast elution of the analyte while still allowing later regeneration of the immobilized antibodies. The need for fast but reversible binding and regeneration is especially important when an IAC column is to be used for a large number of samples. Elution is often carried out by temporarily lowering the effective strength of antibody binding to the target. The most-common approaches for elution in IAC include changing the mobile phase pH or adding a chaotropic agent to the mobile phase. Other, less-common IAC elution methods include adding a competing agent, organic modifier or denaturing agent to the mobile phase, or changing the temperature of the column during elution. Usually the elution buffer is applied in a step gradient, but gradual or nonlinear gradients can be used as well. Figure shows a common scheme by which an IAC column can be used to selectively bind and elute analytes from a sample.

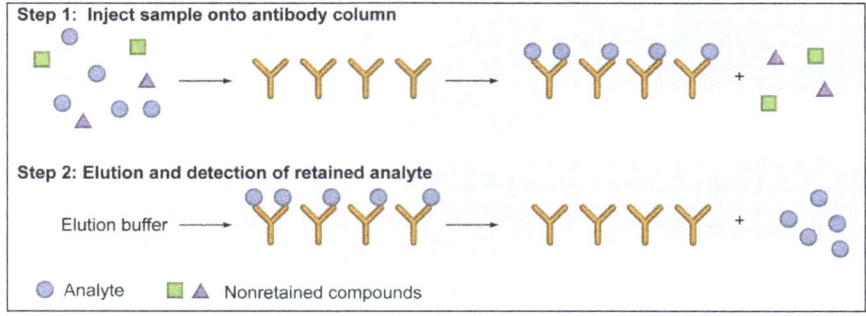

Figure shows typical format in which a sample containing the analyte is applied to an immunoaffinity column and nonretained sample components are allowed to pass through. The analyte is later eluted by disrupting the antibody–antigen interactions with an appropriate elution buffer. This on/off mode can be used for direct detection and/or purification of the analyte.

Changing the mobile phase pH is the most popular method for eluting retained compounds from IAC columns. This approach is usually conducted by applying an acidic buffer (pH 1–3) to the column. Alkaline elution conditions have been used in conjunction with low-performance IAC supports, but cannot be used with common HPIAC supports, such as silica or glass beads, due to the instability of these supports at a pH greater than 8.0. One difficulty with changing the pH of the mobile phase is the possibility of denaturing the immobilized antibodies or any retained compounds that are susceptible to variations in pH. However, many IAC columns have been shown to be quite stable when moderate pH changes in the range of 7.0–7.5 to 2.5–3.0 are used for elution.

To avoid denaturing effects that are caused by lowering the pH, the elution of retained compounds in IAC can alternatively be performed by adding chaotropic agents such as thiocyanate (SCN^-), trifluoroacetate (CF_3COO^-), perchlorate (ClO_4^-), iodide (I^-) or chloride (Cl^-) to the elution buffer. The elution strength of these agents follows the approximate order $SCN^- > CF_3COO^- > ClO_4^- > I^- > Cl^-$. These agents are typically used at concentrations of 1.5–8 M and have been shown to be effective in dissociating high-affinity antibody–antigen complexes. When an organic modifier is used in the mobile phase, care must be taken to ensure that the concentration of the organic additive does not permanently denature the antibodies. When methanol or other organic modifiers are used, the capacity of the immunoaffinity column has been shown to decrease. This denaturation is not irreversible, but kinetic regeneration can take a few days. In one study, an anticlenbuterol immunoaffinity column was regenerated 20-times after offline elution with 2 ml of ethanol 80% in water. However, this column was shown to lose half its activity over these 20 regenerations. Therefore, if harsh elution conditions are required, the capacity of the immunoaffinity column should be much larger than the actual amounts of analyte that are to be measured or isolated.

Formats and Applications of IAC

Immunoaffinity chromatography is a powerful technique that can selectively isolate a given compound from complex samples. As a result, many formats utilizing IAC have involved preparative applications or selective analyses. IAC has also been used with both direct and indirect detection methods and has been coupled with other methods such as HPLC, GC, MS and CE.

On/Off Elution and Direct Detection Methods

Preparative applications generally use the on/off mode of IAC, as shown in figure. This format can also be used with direct detection for chemical analysis. In the on/off mode, a sample is applied to the IAC column in the presence of an application buffer. As the sample is applied to the column, only the analyte and closely-related compounds are retained by the column, while other sample components pass through non-retained. After the analyte has been bound to the column and other materials have been washed away, an elution buffer is applied and the analyte is eluted. After this elution step, the application buffer is reapplied to the column to allow for regeneration of the antibodies prior to another sample application.

The on/off mode of IAC is commonly used in biochemistry and other fields for the selective purification of target compounds from complex samples. Compounds that have been isolated by this approach include proteins, glycoproteins, carbohydrates, lipids, bacteria, viral particles, drugs and environmental agents. This mode can also be used for the direct detection of an analyte by placing a suitable detector after the IAC column. For this type of application, the analyte must be present at relatively high concentration and be eluted in a sharp, well-defined peak that allows a good detection limit. Depending on the desired level of detection, UV/visible absorbance, fluorescence and MS have all been used for detecting analytes in the on/off mode of IAC. Specific examples of analytes that have been measured by this approach include human serum albumin, recombinant tissue-type plasminogen activator, recombinant antithrombin III, IgG, Escherichia coli , isoproturon, phenylurea herbicides, benzidine, dichlorobenzidine, aminoazobenzene, azo dyes, triazine, diethylstilbestrol, acetylcholinesterase, transferrin and insulin.

Immunoextraction and Immunodepletion

When IAC is used to remove a specific analyte or group of analytes from a sample prior to analysis by a second analytical method, this approach is referred to as immunoextraction. immunoextraction is coupled with a second analytical method such as LC. Immunoextraction can be carried out either offline or online with the second analytical method. In the offline mode, antibodies are typically immobilized onto a low-performance support and packed into a small disposable syringe or SPE cartridge. Samples are then applied through the affinity support, which binds the analytes of interest while other sample components are washed away. An elution buffer is then passed through the affinity support to elute the extracted analytes. In addition to removing undesirable sample components, immunoextraction can allow for analyte concentration. In fact, by applying a larger sample volume to the immunoaffinity support (as long as column capacity is not exceeded), more analyte can be made available for detection by the second analytical method due to the essentially irreversible binding of the analyte to the antibodies under typical application conditions. Just as in traditional SPE methods, in offline IAC the eluted fraction can be collected, dried down, and dissolved in a solvent more suitable for analysis. If necessary, the sample can also be derivatized prior to analysis. Offline immunoextraction coupled with other methods has been used in the analysis of urine, food, water and soil extracts. Examples of analytes that have been examined by this approach include α_1-anti-trypsin, atrazine, benzylpenicilloyl-peptides, bovine serum albumin, carbendazim, chloramphenicol, cortisol, clenbuterol and phenytoin, among others.

A related method that uses IAC is immunodepletion. In immunodepletion, an antibody column is used to remove abundant analytes from a complex sample prior to using a second method of analysis for the minor sample components. Typically, this method is used to remove high- and mid-abundance proteins from serum samples prior to the analysis of low-abundance proteins, as is often required in proteomics. In contrast to other methods that can be used to remove high- and mid-abundance proteins (e.g., precipitation, SPE, ultracentrifugation, molecular-weight separation and pI separation), immunodepletion can provide highly selective depletion of multiple high-abundance proteins simultaneously.

Hybrid IAC Methods

The on/off mode of IAC can also be used for online analytical applications. In this format, IAC is directly coupled to a second analytical technique for analysis. By directly combining immunoextraction with a second analytical method, such as HPLC, sample pretreatment can be automated. IAC is often coupled with reversed-phase (RP)LC, but methods in which IAC has been coupled with size exclusion, ion-exchange chromatography, CE, MS, GC and microfluidic devices have also been reported.

One reason immunoextraction is commonly coupled with RPLC is due to the widespread use of RPLC in chemical separations. In addition, the elution buffer used in IAC is an aqueous solution, which acts as a weak mobile phase for RP columns. Online immunoextraction coupled with RPLC has been used to measure compounds in food extracts, bodily fluids, cell extracts and environmental samples. An example of an HPLC system which allows for the combination of immunoextraction with RPLC is shown in figure.

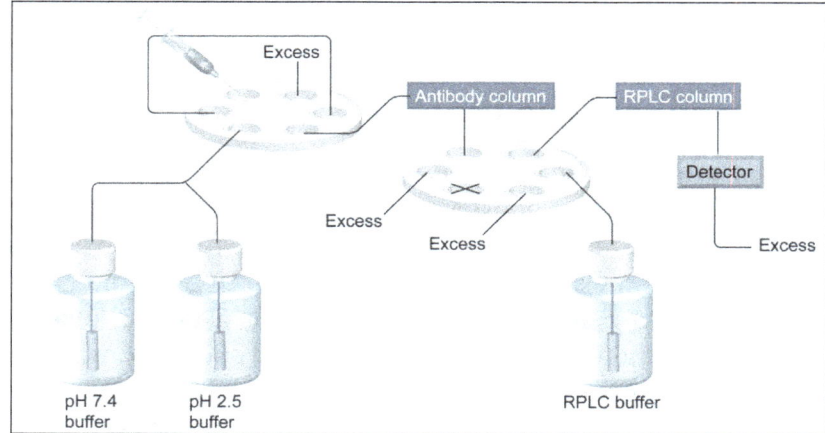

Figure: System for combining immunoaffinity chromatography with reversed-phase chromatography.

Figure shows an antibody column is used to extract or preconcentrate analytes from a sample prior to separation of these analytes using a reversed-phase column. This particular system has been used to measure virginiamycin in water samples with a LOD of 1 ppb.

Competitive Binding Assays

Another way in which IAC can be used as an analytical tool is in an immunoassay format. This approach is known as a chromatographic immunoassay or flow-injection immunoassay. One general type of chromatographic immunoassay is a competitive binding assay, in which a signal is generated as the analyte competes with some labeled species for antibody binding sites.

The most common type of competitive binding immunoassay in IAC is the simultaneous injection immunoassay. A schematic diagram of this format is shown in Figure. In this method, a sample is mixed with a labeled analog of the analyte and applied to a column that contains a limited amount of immobilized antibodies. The limited amount of antibodies causes the labeled analog (A*) and the analyte (A) to compete for binding sites. Due to the presence of this competition, the amount of labeled analog that is detected in the bound and/or retained fractions is affected by the presence of analyte. Typical calibration curves are prepared by plotting the relative response of the labeled analog, B/B_0, versus the concentration of analyte in the sample, where B is the amount of labeled analog bound in the presence of a given amount of analyte and B_0 is the amount of labeled analog bound in the absence of any analyte. This type of calibration curve will have a maximum value of 1 when no analyte is present and should approach 0 at high analyte concentration, assuming there is no non-specific binding between the labeled analog and the column. A large variety of analytes have been measured by using simultaneous injection competitive binding immunoassays. Factors that affect the response of these methods include the relative amount of analyte applied to the column, the flow rate and the amount of labeled analog.

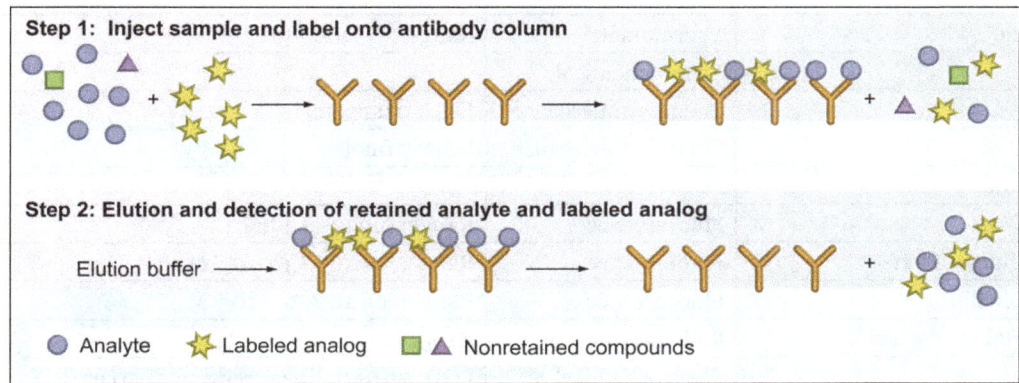

Figure: Simultaneous injection immunoassay format.

Figure shows that this type of competitive immunoassay, the sample and a labeled analog of the analyte are mixed and injected onto an immunoaffinity column. The labeled analog and analyte can bind to this column while other sample components pass through non-retained. The analyte and labeled analog are then eluted with an appropriate mobile phase/elution buffer. In this method the analyte concentration in the sample is inversely related to the amount of retained labeled analog that is detected.

Table: Examples of simultaneous injection immunoassays.

Analyte	Detection method	Assay characteristics
Methods based on antibodies adsorbed on protein A or protein G		
Human transferrin	Fluorescence	Range: <500 μg/ml
Transferrin	Fluorescence	LOD: 25 μg/ml
Adrenocorticotropic hormone	Fluorescence	Range: 0.2–10 mg/l
Testosterone	Fluorescence	LOD: 0.5 μg/ml
Theophylline	Fluorescence	LOD: 0.3 ng/ml; range: <500 μg/l
Atrazine	Fluorescence	LOD: 2.1 μg/l; range: 2.1–50 μg/l
Cephalexin	Electrochemical	LOD: 1 μg/l
IgG	Absorbance	LOD: 333 zmol
Theophylline	Electrochemical	LOD: 25 ng/ml
Anti-BSA	Fluorescence	LOD: 0.2 nM; linear range: 0.4–0.8 nM
Methods based on covalently immobilized antibodies		
HSA	Thermometric	LOD: 10^{-10} M
Theophylline	Fluorescence	LOD: 3 μg/l; range: 3–75 μg/l
Pullulanase, IgG, antithrombin	Fluorescence	Range: μg/ml
IgG	Fluorescence	LOD: 155 ng/ml
IgG	Fluorescence	LOD: 4×10^{-9} M
Gentamicin	Fluorescence	LOD: 200 ng/ml; working range: 250–5000 ng/ml
IgG	Fluorescence	Linear range: 1–5 μg/ml
Insulin	Thermometric	LOD: 0.025 μg/ml; working range: 0.05–2 μg/ml
Carbaryl	Fluorescence	LOD: 20 ng/l
Atrazine	Fluorescence	LOD: 75 ng/l
Isoproturon	Absorbance	LOD: 0.09 μg/l
Gentamicin	Thermometric	Range: 10–400 μg/l
IgG	Electrochemical	
Digoxin	Chemiluminescence	LOD: 0.2 ng/ml
IgG	Chemiluminescence	LOD: 7 fmol
Theophylline	Fluorescence	
Theophylline	Fluorescence	Range: 0.025–0.4 mg/l
Theophylline, caffeine	Fluorescence	Range: 3×10^{-5}–3×10^{-8} M
Antitheophylline	Fluorescence	Range: 4×10^{-7}–6×10^{-9} M
Methotrexate	Radioactivity	Range: 1–100 μg/l
Thyroxine	Electrochemical	LOD: 25 μg/l; linear range: 25–50 μg/l
Cortisol	Fluorescence	Range: 1–60 μg/dl

A second type of competitive binding immunoassay is the sequential injection immunoassay. As shown in Figure, this method differs from the simultaneous injection immunoassay format in that the sample is injected onto the column followed by a later injection of label analog. A calibration curve is again generated by plotting the relative response (B/B_o) versus the analyte concentration. The sequential injection binding assay has one important advantage over the simultaneous injection format in that the labeled analog never comes into contact with the analyte, which means sample matrix effects during the detection of the labeled analog can be eliminated. This feature allows for improved reproducibility, a wider range of detection formats and lower background signals. Several applications of sequential injection immunoassay applications can be seen in Table. The sequential injection format tends to give lower LOD than the simultaneous format, because the analyte has a better chance of binding to the column in this format. However, the simultaneous injection format has higher upper LOD and a wider dynamic range; this wider dynamic range is a result of the increased ability of the analyte to compete with the labeled analog when they are applied to the immunoaffinity column at the same time.

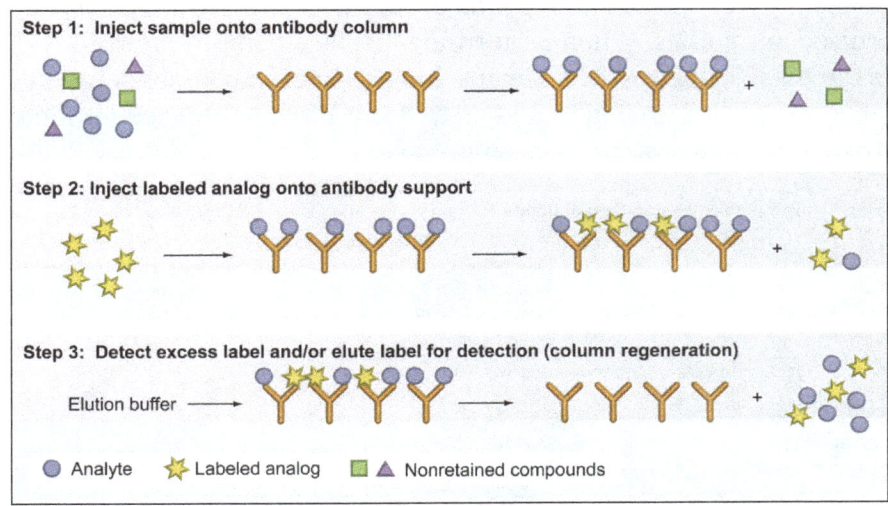

Figure: Sequential injection immunoassay format.

Figure shows the sample is applied onto an immunoaffinity column and the analyte is allowed to bind. A labeled analog of the analyte is then injected onto the same column and also allowed to bind to any remaining free antibody sites. An elution buffer is used to later remove both the retained analyte and labeled analog for the column. The amount of retained labeled analog will be inversely related to the amount of analyte that was in the original sample.

Table: Examples of sequential injection immunoassays.

Analyte	Detection method	Assay characteristics
Imazethapyr	Fluorescence	LOD: 500 ppb
Imazethapyr	Fluorescence	LOD: 0.5 ppb
IgG	Thermometric	Range: 10–400 µg/ml
IgG	Thermometric	LOD: 33 pmol
Digoxin	Electrochemical	LOD: 10 pg/ml; range: 10–1000 pg/ml
Atrazine	Fluorescence	Range: 0.02–0.3 µg/l
Atrazine	Fluorescence	Range: 0.03–0.5 µg/l
HSA	Electrochemical	
α-amylase	Absorbance	
Anti-IgG	Chemiluminescence	LOD: 1 fmol
Imazethapyr	Fluorescence	LOD: 0.1 ppb
Imazethapyr	Fluorescence	Range: 0.1–100 ng/ml

A third type of competitive binding immunoassay is the displacement immunoassay, as illustrated in Figure. In this method, an IAC column is first saturated with a labeled analog of the desired analyte. Sample is then injected onto this column and displaces any labeled analog that is momentarily free in solution. This displaced analog is then eluted from the column and gives a response that is proportional to the amount of analyte in the sample. As long as enough labeled analog remains bound to the column to give a consistent and measurable signal, several samples can be injected onto the IAC column before this column must be regenerated. To help ensure enough labeled analog is present, IAC columns with large capacities are often employed for this format. Additionally, the stability of the signal depends on the rate of dissociation of the labeled analog from the immobilized antibodies. When performing displacement immunoassays, slow flow rates tend to increase the displacement effect because longer times of contact between the sample and column allow for more labeled analog to dissociate from the immobilized antibodies. Several applications for displacement assays can be seen in table.

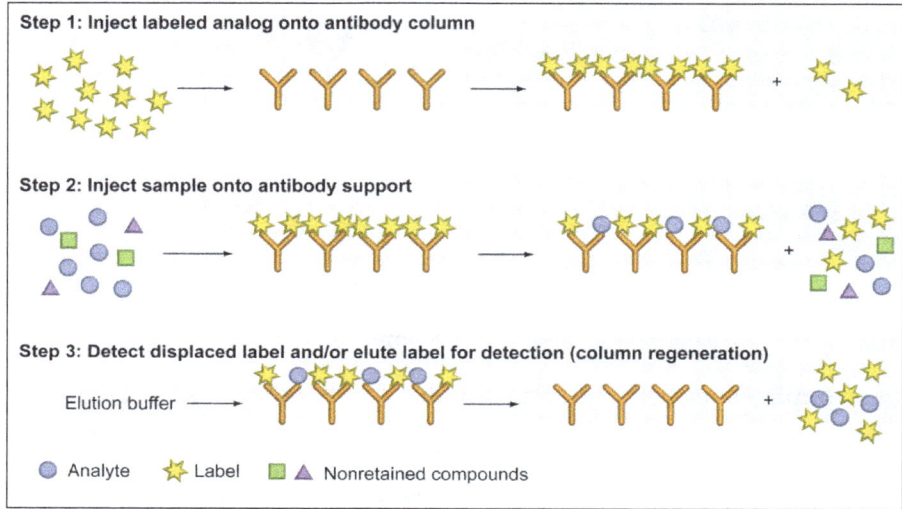

Figure: Displacement immunoassay format.

Figure shows that this type of competitive binding immunoassay, a labeled analog of the analyte is injected onto an immunoaffinity column. A sample is then injected onto the column and the analyte is allowed to displace some of the labeled analog. The size of the peak for the displaced label is directly related to the amount of analyte that was in the sample.

Table: Examples of displacement immunoassays.

Analyte	Detection method	Assay characteristics
Cocaine, benzoylecgonine	Fluorescence	
TNT, DNT	Fluorescence	LOD: 2.5 ng/ml; range: 20–1200 ng/ml
Cocaine, benzoylecgonine	Fluorescence	
2,4-dinitrophenol	Radioactivity	LOD: 140 nM; linear range: 570–4600 nM
2,4-dinitrophenol	Fluorescence	Linear range: 290–2300 nM
Cortisol	Fluorescence	Dynamic range: 12.5–1250 pmol
Polychlorinated biphenyls	Fluorescence	LOD: 4 ppm; linear range: 4–20 µg/ml
Transferrin, HSA	Absorbance	

Noncompetitive Immunoassays

Noncompetitive immunoassays are another group of immunoassays that use indirect detection. These assays are often referred to as immunometric methods. In these methods, there is no competition between the analyte and other substances. Two types of non-competitive immunoassays have been used in IAC: sandwich immunoassays and one-site immunometric assays.

Sandwich immunoassays utilize two different antibodies that bind the same analyte. One of the antibodies is immobilized onto a solid support and is used to extract the analyte from samples. A second, labeled, antibody is either mixed with the sample prior to application or applied to the column directly after the sample is applied to the IAC column. Once the analyte is 'sandwiched' between the two antibodies, an elution buffer is applied to elute the analyte and labeled antibodies. The labeled antibodies can then be detected and give a response that is directly proportional to the amount of retained analyte. A typical scheme for this type of assay is given in figure. When the sample and labeled antibodies are mixed prior to application onto the column, better detection limits can be obtained than when sequential injection of the sample and labeled antibodies is employed, because there is more effective binding between the labeled antibody and analyte. A calibration curve for sandwich immunoassays is constructed by plotting the relative response of the eluted labeled antibody against the amount of analyte in a sample and can give a linear response over a broad range of analyte concentrations.

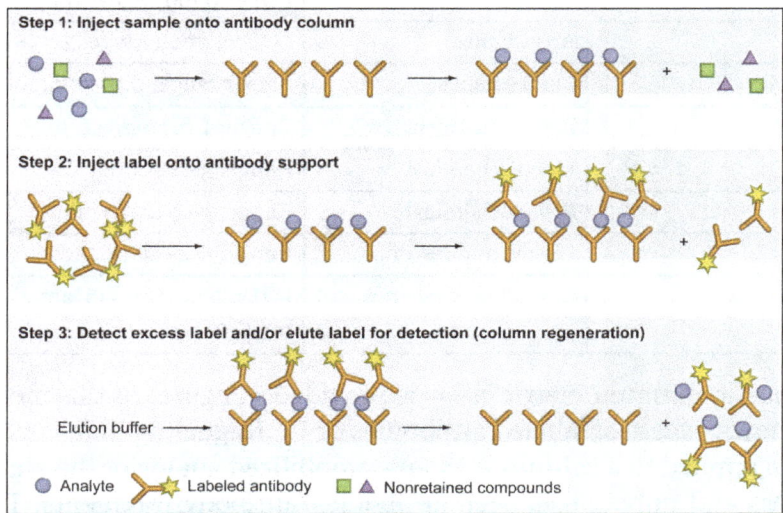

Figure: Sandwich immunoassay format.

Figure shows that this type of noncompetitive binding immunoassay, the sample is injected onto an immunoaffinity column and the analyte allowed binding to the immobilized antibodies. A labeled antibody that is specific for the analyte is then injected onto the same column and also allowed to bind, creating a sandwich immune complex for the analyte. An elution buffer is applied to disrupt the antigen–antibody binding and regenerate the column. The amount of retained, labeled antibody that is eluted during this step is directly proportional to the amount of analyte that was present in the original sample.

Due to the direct relationship between the amount of analyte and the response in a sandwich immunoassay, plus the ability to use an excess of labeled antibodies to help promote good detection, this approach allows for better signal-to-noise ratios and lower limits of detection than competitive binding immunoassays. Sandwich immunoassays also tend to be more selective than competitive binding immunoassays, because two types of antibodies are used for analyte binding and detection instead of one. In general, low flow rates are desirable to allow sufficient time for analyte–antibody complexes to bind to an IAC column in this format. In addition, columns that can effectively capture the antibody–analyte complex with low nonspecific binding are best for sandwich immunoassay development. The main disadvantage of the sandwich immunoassay is that only large analytes (e.g., large peptides, proteins and biomacromolecules) can be quantified by this approach because the analyte must be able to bind two antibodies simultaneously. One other disadvantage of this approach is the added cost of performing a sandwich immunoassay due to the need for two different antibodies per analyte. Examples of sandwich immunoassays that have been carried out by using IAC can be found in table.

Table: Examples of sandwich immunoassays.

Analyte	Detection method	Assay characteristics
Thyroid-stimulating hormone	Absorbance	Range: 0–0.29 nM
hCG	Fluorescence	Range: 0–66.6 ng/ml
HSA	Fluorescence	LOD: 0.001 mg/ml
PTH, interleukin-5	Fluorescence	LOD: 10 nM; linear range: <250 μM
IgG	Absorbance	LOD: 3 fmol; range: 3.33–130 fmol
IgG	Fluorescence	Linear range: 0.5–50 pmol/l
HSA	Electrochemical	Range: 1–10 mg/ml
Anti-IgG	Electrochemical	Range: 3–225 fmol
IgG	Electrochemical	Range: 5–400 ng/ml
IgG	Chemiluminescence	Range: 0.2–20 fmol
Parathyroid hormone	Chemiluminescence	LOD: 0.24 pM (16 amol); linear range: 0.24–67 pM

The format for a one-site immunometric assay is provided in figure. In this method, the sample is incubated with a known excess of labeled antibodies or F_{ab} fragments that can bind to the analyte. This mixture is then applied to a column with an immobilized analog of the analyte, which is used to remove the excess and unbound labeled antibodies/antibody fragments. The bound analyte-labeled antibodies elute in the nonretained fraction with the analyte and give a signal that is proportional to the amount of analyte in the original sample. Typically, calibration curves for one-site immunometric assays are created by plotting the relative response of the labeled antibody–analyte complex against the amount of analyte in the sample. To obtain a maximum signal in a one-site immunometric assay, the binding between the labeled antibodies and the analyte must reach equilibrium and the flow rate must allow for the capture of all the excess binding agents on the IAC column. The column should also have a binding capacity that exceeds the amount of binding agent that is applied between regeneration steps. Some advantages of one-site immunometric assays are that they can detect both large and small analytes, their signal is directly proportional to the amount of analyte and a multitude of elution conditions can be used with the immobilized analog columns. The main disadvantage of one-site immunometric assays is obtaining affinity ligands with the required activity and purity to create an assay in which all the excess agent binds to the affinity column and gives a low background signal. Examples of reports that have utilized the one-site immunometric format in IAC are listed in table.

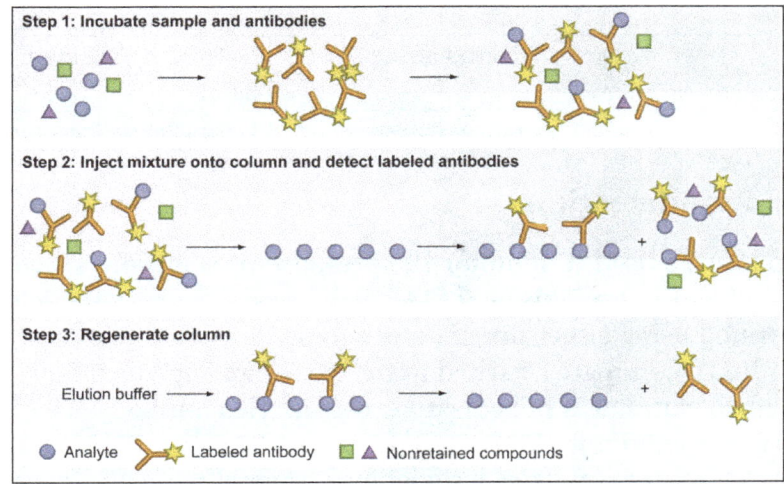

Figure: One-site immunometric assay.

Figure shows that this noncompetitive immunoassay format, the analyte and labeled antibodies are mixed and incubated prior to injection onto an immobilized analyte column. The analyte column binds any excess antibodies and the analyte-bound antibodies are eluted in the nonretained fraction, providing a signal that is directly related to the analyte's original concentration. The column is later regenerated by eluting off the excess labeled antibodies.

Table: Examples of one-site immunometric assays.

Analyte	Detection method	Assay characteristics
Granulocyte colony- stimulating factor	Fluorescence	LOD: 1.5 ng/120 µl
Digoxin and metabolites	Fluorescence	LOD: 2×10^{-10} M
Digoxin and metabolites	Fluorescence	LOD: 160 pg/ml; linear range: 0.2–2 nmol/l
Digoxin	Fluorescence	LOD: 200 fmol
Digoxigenin	Fluorescence	LOD: 50 fmol; linear range 50–1000 fmol
Digoxin	Fluorescence	LOD: 0.025 nM
Digoxigenin	Fluorescence	LOD: 0.01 nM
Interleukin-10	Fluorescence	LOD: 40 fmol
Digoxin	Absorbance	LOD: 0.2 µg/l
2,4-D	Electrochemical	LOD: 0.25 µg/l
Digoxigenin	Electrochemical	LOD: 0.5 amol; linear range: 0.38–7.7 fmol
α-(difuoromethyl)-ornithine	Fluorescence	LOD: 200 amol; linear range: 5×10^{-11}–2.5×10^{-9} M
Fatty acid-binding protein	Absorbance	Range: 1–12 µg/l and 12–2000 µg/l

Thyroxine	Chemiluminescence	LOD: 10^{-11} M
4-amino-l- and d-phenylalanine	Chemiluminescence	LOD: 1.76 pmol/ml
17-estradiol	Fluorescence	
α-fetoprotein	Fluorescence	LOD: 0.1 ng/ml; linear range: 0.5–60 ng/ml
Terbutryn	Grating coupler	LOD: 15 μg/l; linear range: 20–200 μg/l

Postcolumn Immunodetection

An IAC column can also be used to monitor the presence of a specific analyte as it elutes from another chromatographic column. This use of IAC is referred to as immunodetection. Immunodetection can be performed using either direct detection or indirect detection. Typically, a postcolumn reactor and an IAC column are attached to the exit of an analytical HPLC column as shown in Figure. As the analyte elutes, it is mixed with excess labeled antibody. The excess labeled antibodies are removed using an immobilized analog of the analyte column. Bound, labeled antibodies pass through the analog column and give a signal proportional to analyte concentration. Reviews of this technique can be found elsewhere. When performing postcolumn immunodetection, it is especially important to ensure the eluant from the HPLC column is properly adjusted to allow for maximum antibody binding in the postcolumn reactor (pH, ionic strength and removal/minimization of organic modifiers).

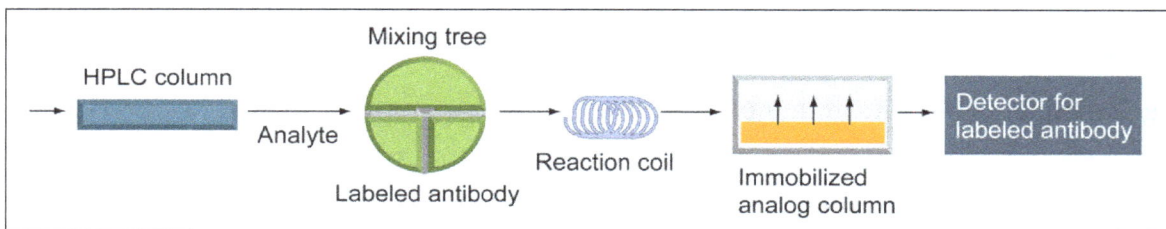

Figure: General scheme for postcolumn immunodetection.

Figure shows that this detection method, eluent from an HPLC column is directed into a reaction coil along with an excess of labeled antibodies that are able to bind the desired analyte. The excess antibodies bind to an immobilized analog column and the labeled analyte–antibody complexes are monitored as they elute from the analog column, providing a signal that is directly related to the original amount of eluting analyte.

Recent Developments in IAC

The development of IAC is ongoing and continues to be integrated with other analytical techniques, including CE and MS. Other new developments include ultrafast immunoaffinity CE and the use of antibodies in microanalytical systems.

When immobilized ligands are used in CE, the method is referred to as affinity electrophoresis. Affinity ligands, such as antibodies, can be immobilized in CE capillaries by several methods, including physical entanglement in gels and covalently binding the ligand to the capillary wall, or polymers, frits or beads inside the capillary. CE can also be used with antibodies/antibody fragments to quantitatively measure analytes by allowing CE to separate free analytes from analyte–antibody complexes. These formats can be either competitive or noncompetitive, with noncompetitive formats giving better LOD. CE immunoassays are utilized due to their ease of automation and their relatively fast separation of antibodies, analytes and/or antibody–analyte complexes. Another advantage of CE immunoassays is that only small amounts of sample and reagents are used while still maintaining good LOD. The best LOD in CE immunoassays are generally achieved when using laser-induced fluorescence detection or MS.

By combining IAC with MS (IAC–MS), a technique is produced that utilizes the selectively of antigen–antibody interactions and the sensitivity of MS. When performing online IAC–MS, the IAC elution buffer should contain only volatile buffer salts (e.g., ammonium acetate and ammonium formate) to avoid lowering ionization efficiency. In addition, methanol or acetonitrile is often added to the eluant prior to ionization to increase sensitivity. MALDI can use immobilized antibodies on the target to help extract desired compounds from a sample prior to analysis by MS.

Due to the speed and specificity of antibody–antigen interactions, IAC can also be used for very fast immunoextractions that often take less than a second to perform. Ultrafast immunoextraction has been used to quantify the free fraction for drugs and hormones in clinically related samples. Measuring the free fractions of drugs and hormones in serum is often difficult for other methods, because any removal of the free fraction can perturb protein binding and cause additional drug or hormone to dissociate in the sample. Ultrafast immunoextraction conducted in less than a few hundred milliseconds has been used with both direct detection based on fluorescence and chromatographic immunoassays using a displacement format for detection of warfarin, phenytoin and thyroxine in protein and serum samples.

Owing to recent advances in microfabrication technologies, micro and ultramicro (nano) analytical systems can also be developed to utilize the specificity of antibodies. These micrototal analysis systems can be engineered in a variety of formats, including the construction of microarrays which utilize several channels that can each separate multiple analytes all at once. Much like CE immunoassays, these systems need to use quite sensitive detectors, due to the small amounts of analyte that must be measured. One example of a micrototal analysis systems utilizing laser-induced fluorescence detection detected α-fetoprotein.

Metal Chelate Chromatography

Meal ions such as Zn^{++}, Cu^{++}, Cd^{++}, Hg^{++} or transition metals ion such as Ni^{++}, Co^{++}, Mn^{++} can be attached to iminodiacetate or tris (carboxymethyl)ethylelnediamine substituted agarose. Thiol group of cysteine, imidazole group of histidine and indole group of tryptophan in protein can form coordinate bonds with meals ions. The number of coordinate bonds will determine the strength of interaction and retention of protein. Protein with low number of coordinate bonds will remain unbound and will be removed in washings. Bound proteins can be eluted by lowering pH or addition of chelating agents such as EDTA in elution buffer.

Dye-ligand Chromatography

Cibacron blue F3G-A is triazine dye having conjugated rings and ionic groups. The dye has affinity for some protein. Hydrophobic and ionic interaction play role in providing affinity of dye for some protein. It has not possible to point out which protein will bind to dye while which other protein will not. Dye is cheap, can be coupled to matrices and is stable. Since which protein will bind to dye cannot be predicted, interaction between dye and protein is ascertained with trial and error methods. Neutral to slightly alkaline pH (7.0 to 8.5) is kept for binding of protein to immobilized dye. The dye can be immobilized on Sepharose 4B. Bound protein can be eluted by salt gradient.

Lectin Affinity Chromatography

Lectins are proteins that have recognition for carbohydrate. These are produced by animal, plants and moulds. Since these bind carbohydrate, these have affinity for glycoproteins. Different lectins recognize different carbohydrates and thus these will be specific for particular glycoprotein. Lectins exist in polymeric form, most of them being tetrameric. Sub unit of lectin may or may not be identical. Each identical sub unit will recognize only one particular carbohydrate. Lectin comprising of two different sub units will recognize two different carbohydrates. Lectins of plant origin are commonly used because of their abundance in pea, soybean. These can be immobilized on CNBr-activated or NHS-activated agarose. Immobilized lectins are also commercially available. Interaction between lectin and glycoprotein is not ionic and therefore, binding of glycoprotein to lectin column can be achieved without removal of salt after salt fractionation. Bound glycoprotein can be eluted by change in pH, use of carbohydrate in elution buffer, use of borate buffer or by addition of ethylene glycol in buffer to reduce hydrophobic interaction.

Elution Methods

After binding of target molecule to column and removal of unbound material, bound material is to removed. This step is called elution. Elution conditions used for collecting bound materials is called elution method. It is preferred to leave column in an environment of elution conditions (buffer, temp., polarity) for certain period before continuing elution. This gives more time for dissociation of ligand-target complex and this also helps in improving recoveries of bound substances.

pH Elution

pH brings changes in degree of ionization and thus can change ionic interaction between ligand molecule and target. Change in pH can also bring change in conformation of ligand and target molecule. This will disrupt interaction between ligand and target resulting in release of target molecule. Matrix, ligand and target should be stable at change pH. If low pH is used, fractions are collected in into neutralization buffer such as 1 M Tris-HCl, pH 9. This brings pH of fraction to a neutral range. The column should also be re-equilibrated to neutral pH immediately.

Ionic Strength Elution

Ionic interaction between ligand and target molecule can be easily broken by using buffer containing sodium chloride. A linear gradient of NaCl or fixed concentration of NaCl is used for elution. Since buffers of neutral pH range are used during elution, eluted molecules will retain biological activity.

Competitive Elution

In competitive elution, ligand or similar molecule is added to elution buffer. Bound target molecules compete for ligand molecules linked to matrix and free ligand molecules in elution buffer. Usually competitive molecules are small sized molecule and these are removed by dialysis or gel filtration for obtaining target molecules in pure form.

Reduced Polarity of Eluent

Conditions are used to lower the polarity of the eluent promote elution without inactivating eluted substances. Dioxane (up to 10%) and ethylene glycol (up to 50%) are typical of this type of eluents.

Chaotropic Eluents

Chaotropic agents can disrupt the hydrogen bonding network between water molecules and reduce the stability of the native state of proteins by weakening the hydrophobic effect. Chaotropic agents include guanidinium hydrochloride, gtuanidinium thiocyanate (GITC), urea, thiourea, lithium perchlorate, lithium acetate, magnesium chloride, butanol, propanol, ethanolphenol, sodium dodecyl sulfate. Chaotropic agents can be added to elution buffer to elute bound target molecules.

Table: Problems in Affinity Chromatography, their Cause and Solutions.

Sr. No.	Cause	Solution
1. Sample does not bind to affinity column	(i) Sample has particulate matter.	• Filter or centrifuge sample. • Clean the column by extensively washing with binding buffer. • Repeated suspension of matrix after removal from column in binding buffer and decantation may help in solubilization and removal of particulate matter.
	(ii) Sample (protein) has altered structure on storage.	• Use fresh sample.
	(iii) pH and ionic strength of binding buffer is incorrect.	• Prepare fresh buffer. • Protein sample can be dialyzed against. • Binding buffer. • Protein sample can be passed through. • Desalting column equilibrated with fresh. • Buffer.
	(iv) Column is not equilibrated with binding buffer.	• Equilibrate column with at least three bed volume of binding buffer.
	(v) More sample in comparison to column capacity is loaded.	• Decrease sample amount.

	(vi) Microbial growth has occurred in column.	• Discard the column. • Never leave column in buffer without preservative. Use 20% ethanol or buffer containing 0.05% azide. However, column is to be equilibrated with buffer before use. • Microbial growth during run of column is usually not noted.
2. Sample does not elute or elution is incomplete	(i) Protein is degraded during run.	• Use protease inhibitors during extraction of protein and also equilibrate column with binding buffer containing protease inhibitors. Avoid using protease inhibitors during elution.
	(ii) Protein is precipitated during run.	• Use chaotropic agents.
	(iii) Activator in sample extract is removed as unbound fraction.	• Measure enzyme activity in under optimum conditions.
	(iv) Elution occurs at low pH.	• Collect eluted fractions in neutralization buffer such as 1M Tris-HCl, pH 9.0.
3. Flow rate is retarded	(i) Filters are clogged.	• Replace filters.
	ii) Sample is precipitated or sample has particulate matter.	• Replace column. • Check stability of protein in equilibration buffer.
4. Air bubbles in column	(i) Buffers are not degassed.	• Degas buffer use.
	(ii) Column and buffer are at different temperature.	• Warm up buffer stored at low.
5. Bed is compressed	(i) Pressure used during run is higher than pressure used for packing.	• Use similar pressure during packing and run. • Repack the column.

Applications of Affinity Chromatography

Purification of Soluble Cytokine Receptors and Binding Proteins

The affinity chromatography is a powerful separation technique that is based on unique interaction between the target molecules and a ligand coupled covalently to a resin. It provides the target molecule in a reasonably pure state and enables its identification by partial sequencing, either by N-terminal micro-sequencing or by mass spectrometry. The biological activity of the purified proteins is retained in most cases allowing the assessment of their function. The method has been successfully applied to purify the receptors for IL-6, IL-1β, IL-2, IL-4, IFN-γ, TNF-α, IFN- α/β, IL-13, IL-18, IL-22, and IL-33. The ligand affinity chromatography enabled rapid and efficient isolation of seven soluble receptors corresponding to the cell-associated receptors. It is a rapid, simple, selective, and efficient purification procedure for proteins to yield thousand-fold purification in a single step.

Characterization of pH-dependent Protein Switches

Enzymatic reactions are regulated and controlled by conformational changes of the involved proteins and enzymes. Specific movements of side chains, secondary structures, or protein domains facilitate the regulation of substrate selection, binding, and catalysis. The study was conducted to assess the structural effects of engineered ionizable residues in glutathione-S-transferase to convert it into a pH-dependent allosteric protein. In the charged state, the residues invoke unfavorable interactions, which induce conformational changes affecting the function of the enzyme. To test this, the researchers engineered a single aspartate, cysteine, or histidine residue at a distance from the active site of the enzyme. To evaluate the pH-dependent behavior of the mutant proteins binding the glutathione, the mutants were subjected to GSH-affinity chromatography. It was found that these mutations exhibit a strong effect on the protein's affinity to bind the glutathione. Whereas, the aspartate or histidine mutations lead to permanent nonbinding or binding versions of the protein, respectively. The crystal structures of the mutant protein GST5oC under ionizing and nonionizing conditions showed that the water molecules are recruited into the hydrophobic core to produce conformational changes influencing the protein's active site. The affinity chromatography is an indispensable technique to understand the intimate relationship between the structure and dynamics of the enzymatic reactions.

Purification of Therapeutic Proteins

Recombinant molecules extracted from plants are potential biotherapeutics that could be used to cure several medical conditions. In the study, histamine (HIM), phenylamine (PHEM), tryptamine (TRM), and tyramine (TYRM) coupled to Sepharose CL-4B via a 1,4-butanediol diglycidyl ether spacer was used to purify human monoclonal anti-HIV antibody 2F5 (mAb 2F5) from maize (seed) and tobacco (leaf) extracts. The factors affecting the chromatographic behavior of mAb 2F5, and maize seed and tobacco leaf proteins were also determined. The adsorbents showed a reduced affinity to purify proteins from tobacco extract as compared to the maize extract. Under optimal conditions, histamine exhibited high selectivity for mAb 2F5 and presented a high extent of purification (A95% purity) and recovery (A90%) in a single step with salt elution from the maize seed extract. The antibody fractions, purified from affinity chromatography, were further analyzed on ELISA and Western blot. This showed that the antibody was completely active and free of degraded variants or modified forms. It was concluded that the affinity chromatography is a versatile technique for biotherapeutics and antibody purification.

Purification of Albumin and Macroglobulin Contamination

Affinity chromatography is a helpful tool for cleaning and removing excess albumin and α2- macroglobulin from samples as these components could interfere with the downstream process analysis. For this, Blue Sepharose affinity chromatography is the suitable type of affinity purification. In this method, the ligand is covalently coupled to Sepharose via a chlorotriazine ring. Albumin, to be removed, binds in a nonspecific manner by electrostatic or hydrophobic interactions with the aromatic anionic ligand. The most commonly used dye is Cibacron blue F-3-GA which can be adsorbed onto Sepharose to create an affinity column. The method has been found efficient for 90% albumin clearance.

Gas Chromatography

Gas chromatography differs from other forms of chromatography in that the mobile phase is a gas and the components are separated as vapors. It is thus used to separate and detect small molecular weight compounds in the gas phase. The sample is either a gas or a liquid that is vaporized in the injection port. The mobile phase for gas chromatography is a carrier gas, typically helium because of its low molecular weight and being chemically inert. The pressure is applied and the mobile phase moves the analyte through the column. The separation is accomplished using a column coated with a stationary phase.

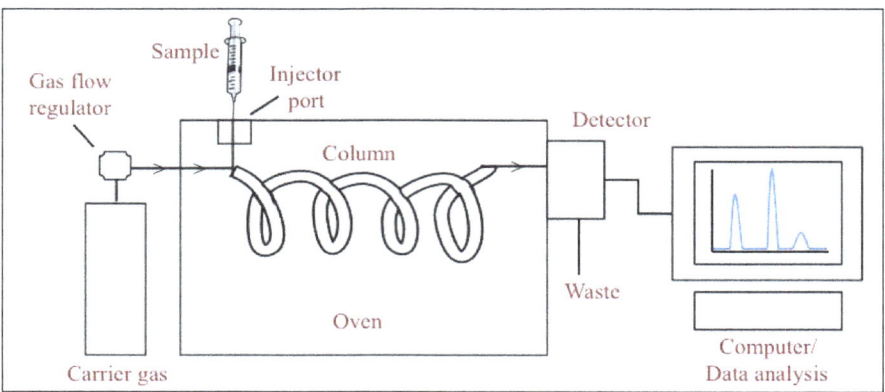

Principle of Gas Chromatography

The equilibrium for gas chromatography is partitioning, and the components of the sample will partition (i.e. distribute) between the two phases: the stationary phase and the mobile phase. Compounds that have a greater affinity for the stationary phase spend more time in the column and thus elute later and have a longer retention time (Rt) than samples that have a higher affinity for the mobile phase. Affinity for the stationary phase is driven mainly by intermolecular interactions and the polarity of the stationary phase can be chosen to maximize interactions and thus the separation. Ideal peaks are Gaussian distributions and symmetrical, because of the random nature of the analyte interactions with the column.

- The separation is hence accomplished by partitioning the sample between the gas and a thin layer of a nonvolatile liquid held on a solid support.

- A sample containing the solutes is injected into a heated block where it is immediately vaporized and swept as a plug of vapor by the carrier gas stream into the column inlet.

- The solutes are adsorbed by the stationary phase and then desorbed by a fresh carrier gas.

- The process is repeated in each plate as the sample is moved toward the outlet.

- Each solute will travel at its own rate through the column.

- Their bands will separate into distinct zones depending on the partition coefficients, and band spreading.

- The solutes are eluted one after another in the increasing order of their kd, and enter into a detector attached to the exit end of the column.

- Here they register a series of signals resulting from concentration changes and rates of elution on the recorder as a plot of time versus the composition of carrier gas stream.

- The appearance time, height, width, and area of these peaks can be measured to yield quantitative data.

Parts of Gas Chromatography

Gas chromatography is mainly composed of the following parts:

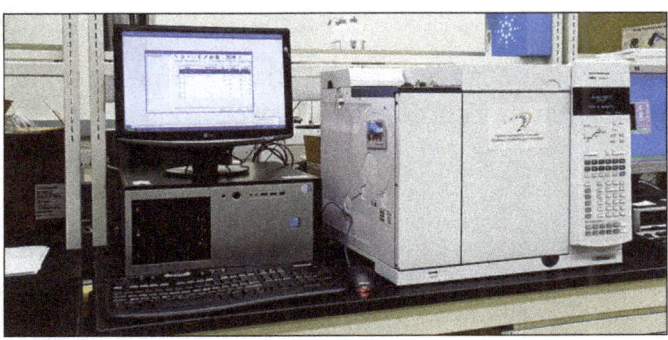

Carrier Gas in a High-Pressure Cylinder with Attendant Pressure Regulators and Flow Meters

Helium, N_2, H, Argon are used as carrier gases. Helium is preferred for thermal conductivity detectors because of its high thermal conductivity relative to that of most organic vapors. N_2 is preferable when a large consumption of carrier gas is employed. Carrier gas from the tank passes through a toggle valve, a flow meter, (1-1000 ml/min), capillary restrictors, and a pressure gauge (1-4 atm). Flow rate is adjusted by means of a needle valve mounted on the base of the flow meter and controlled by capillary restrictors. The operating efficiency of the gas chromatograph is directly dependant on the maintenance of constant gas flow.

Sample Injection System

Liquid samples are injected by a microsyringe with a needle inserted through a self-scaling, silicon-rubber septum into a heated metal block by a resistance heater. Gaseous samples are injected by a gas-tight syringe or through a by-pass loop and valves. Typical sample volumes range from 0.1 to 0.2 ml.

The Separation Column

The heart of the gas chromatography is the column which is made of metals bent in U shape or coiled into an open spiral or a flat pancake shape. Copper is useful up to 250°. Swege lock fittings make column insertion easy. Several sizes of columns are used depending upon the requirements.

Liquid Phases

An infinite variety of liquid phases are available limited only by their volatility, thermal stability and ability to wet the support. No single phase will serve for all separation problems at all temperatures.

- Non-Polar: Parafin, squalane, silicone greases, apiezon L, silicone gum rubber. These materials separate the components in order of their boiling points.

- Intermediate Polarity: These materials contain a polar or polarizable group on a long non-polar skeleton which can dissolve both polar and non-polar solutes. For example. di-ethyl hexyl phthalate is used for the separation of high boiling alcohols.

- Polar - Carbowaxes: Liquid phases with a large proportion of polar groups. Separation of polar and non-polar substances.

- Hydrogen bonding: Polar liquid phases with high hydrogen bonding e.g. Glycol.

- Specific purpose phases: Relying on a chemical reaction with solute to achieve separations. E.g. $AgNO_3$ in glycol separates unsaturated hydrocarbons.

Supports

The structure and surface characteristics of the support materials are important parameters, which determine the efficiency of the support and the degree of separation respectively. The support should be inert but capable of immobilizing a large volume of liquid phase as a thin film over its surface. The surface area should be large to ensure the rapid attainment of equilibrium between stationary and mobile phases. Support should be strong enough to resist breakdown in handling and be capable of packed into a uniform bed. Diatomaceous earth, kieselguhr treated with Na_2CO_3 for 900 °C causes the particle fusion into coarser aggregates. Glass beads with a low surface area and low porosity can be used to coat up to 3% stationary phases. Porous polymer beads differing in the degree of cross-linking of styrene with alkyl-vinyl benzene are also used which are stable up to 250 °C.

Detector

Detectors sense the arrival of the separated components and provide a signal. These are either concentration-dependent or mass dependent. The detector should be close to the column exit and the correct temperature to prevent decomposition.

Recorder

The recorder should be generally 10 mv (full scale) fitted with a fast response pen (1 sec or less). The recorder should be connected with a series of good quality resistances connected across the input to attenuate the large signals. An integrator may be a good addition.

The procedure of Gas Chromatography

Step 1: Sample Injection and Vapourization

A small amount of liquid sample to be analyzed is drawn up into a syringe. The syringe needle is positioned in the hot injection port of the gas chromatograph and the sample is injected quickly. The injection of the sample is considered to be a "point" in time, that is, it is assumed that the entire sample enters the gas chromatograph at the same time, so the sample must be injected quickly. The temperature is set to be higher than the boiling points of the components of the mixture so that the components will vaporize. The vaporized components then mix with the inert gas mobile phase to be carried to the gas chromatography column to be separated.

Step 2: Separation in the Column

Components in the mixture are separated based on their abilities to adsorb on or bind to, the stationary phase. A component that adsorbs most strongly to the stationary phase will spend the most time in the column (will be retained in the column for the longest time) and will, therefore, have the longest retention time (Rt). It will emerge from the gas chromatograph last. A component that adsorbs the least strongly to the stationary phase will spend the least time in the column (will be retained in the column for the shortest time) and will, therefore, have the shortest retention time (Rt). It will emerge from the gas chromatograph first. If we consider a 2 component mixture in which component A is more polar than component B then: 1. component A will have a longer retention time in a polar column than component B, 2. component A will have a shorter retention time in a non-polar column than component B.

Step 3: Detecting and Recording Results

The components of the mixture reach the detector at different times due to differences in the time they are retained in the column. The component that is retained the shortest time in the column is detected first. The component that is retained the longest time in the column is detected last. The detector sends a signal to the chart recorder which results in a peak on the chart paper. The component that is detected first is recorded first. The component that is detected last is recorded last.

Applications

- GC analysis is used to calculate the content of a chemical product, for example in assuring the quality of products in the chemical industry; or measuring toxic substances in soil, air or water.

- Gas chromatography is used in the analysis of:

  - Air-borne pollutants.

  - Performance-enhancing drugs in athlete's urine samples.

  - Oil spills.

  - Essential oils in perfume preparation.

- GC is very accurate if used properly and can measure picomoles of a substance in a 1 ml liquid sample, or parts-per-billion concentrations in gaseous samples.

- Gas Chromatography is used extensively in forensic science. Disciplines as diverse as solid drug dose (pre-consumption form) identification and quantification, arson investigation, paint chip analysis, and toxicology cases, employ GC to identify and quantify various bio-logical specimens and crime-scene evidence.

Advantages

- The use of longer columns and higher velocity of carrier gas permits the fast separation in a matter of a few minutes.

- Higher working temperatures up to 5000 °C and the possibility of converting any material into a volatile component make gas chromatography one of the most versatile techniques.

- GC is popular for environmental monitoring and industrial applications because it is very reliable and can be run nearly continuously.

- GC is typically used in applications where small, volatile molecules are detected and with non-aqueous solutions.

- GC is favored for non-polar molecules.

Limitations

- Compound to be analyzed should be stable under GC operation conditions.

- They should have a vapor pressure significantly greater than zero.

- Typically, the compounds analyzed are less than 1,000 Da, because it is difficult to vaporize larger compounds.

- The samples are also required to be salt-free; they should not contain ions.

- Very minute amounts of a substance can be measured, but it is often required that the sample must be measured in comparison to a sample containing the pure, suspected substance known as a reference standard.

References

- What-is-a-chromatogram-31537: chromatographytoday.com, Retrieved 27, August 2020

- High-performance-liquid-chromatography: microbenotes.com, Retrieved 05, May 2020

- Applications-of-Ion-exchange-chromatography, ion-exchange-chromatography: biochemden.com, Retrieved 17, Feb 2020

- Advantages-and-disadvantages-of-ion: chrominfo.blogspot.com, Retrieved 08, March 2020

- Affinity-chromatography-protocol: conductscience.com, Retrieved 25, June 2020

Spectroscopic Techniques

Spectroscopic techniques make use of light and its interaction with the matter to evaluate the consistency or structure of the sample. The different types of spectroscopic techniques include ultraviolet and visible light spectroscopy, fluorescence spectroscopy, and circular dichroism spectroscopy. All these types of spectroscopic techniques have been carefully analyzed in this chapter.

Spectroscopic techniques employ light to interact with matter and thus probe certain features of a sample to learn about its consistency or structure. Light is electromagnetic radiation, a phenomenon exhibiting different energies, and dependent on that energy, different molecular features can be probed.

Spectroscopic Tools

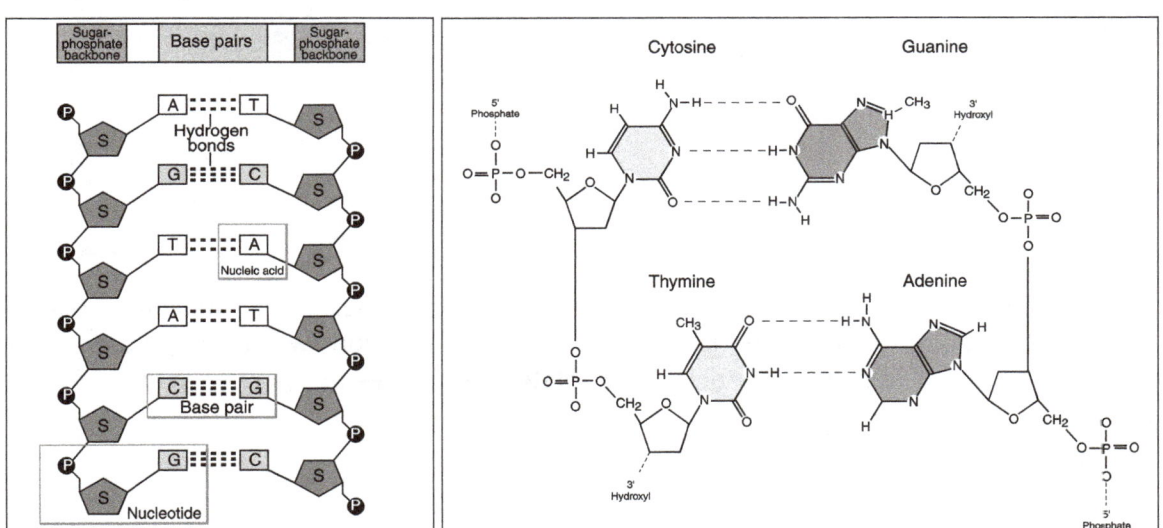

Figure: The 4 DNA bases form the letters of the genetic alphabet and specific hydrogen bonds allow for molecular recognition between the correct complementary bases.

To perform a measurement, we must observe the interaction of the compound of interest with another particle. So a basic spectroscopic measurement will consist of shooting particles with well defined properties at the sample and analyzing particles which are emitted by the sample as indicated in Figure. As a result, the measurement is due to the properties of the sample, the properties of the probing particle, and the physical laws governing the interaction between the two (in many cases called "selection rules").

In principle any particle can be used, but in most cases the secondary particle is a photon. Please be aware, that in some cases photons are conveniently described as particles with well defined energy/momentum, whereas in other cases a wave description is more useful. This does not reflect fundamental differences in the interaction, which always satisfies the same physical laws and fulfills conditions set by both, particle and wave nature of the photon. Other particles follow the same physical rules, but are usually described as particles only, reflecting the more classical nature of heavy particles.

Photons are special particles without any rest mass and without charge, therefore the only relevant particle properties are the photon energy, polarization, and coherence properties. According to deBroglie, a particle can be described as wave with a wavelength inversely proportional to its momentum. Equation below gives the relation between energy, momentum, wavelength and frequency relative to the Planck constant of $h = 6.62 \cdot 10^{-34} \, J \cdot s$ or $\hbar = h / (2 \cdot \pi)$.

$$\text{Wavelength}: \lambda = \frac{h}{p}$$

$$\text{Mass}: m_{ph} = \frac{hv}{c_0^2}$$

$$\text{Energy}: E_{ph} = \hbar v$$

Where,

- λ wavelength

- $h, \hbar$ Planck constant

- p momentum

- v frequency

- c_0 speed of light

- m_{ph} photon mass

If we restrict our discussion to photons, then the discussion of measurements can be tied to the linear scale of the photon energy/frequency. Most accessible frequency ranges are used for spectroscopy, but some particular frequencies and related properties are particularly useful for the analysis of biological systems.

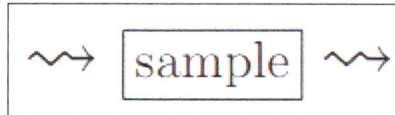

Figure: Spectroscopic measurement: Put the sample in a box, where it is isolated from the environment. Next, observe/characterize particles which enter or leave the box to learn about the sample properties.

In the radio wave regime, the very precise control and measurement of frequency and phase allows nuclear magnetic resonance spectroscopy in both, time and frequency domain. The resulting Fourier transformation (FT) methods allow the measurement of spin correlation ($\rightarrow$ "multidimensional spectroscopy") and therefore the structural characterization of macromolecular systems. The same properties are used for the tomographic imaging of tissue on a macroscopic scale (>mm). Optical spectroscopy in the visible can be directly combined with the spatial resolution of a microscope and has large historic and current importance. Diffraction methods are dominating the X-ray regime, because the short wavelength allows the spatial determination of atomic positions down to sub-$°A$ precision. In all cases, we can discuss the absorption, emission, or diffraction of photons. The following is a short, exemplary discussion of those three types of interactions to give an overview the fundamental possibilities and limitations of different spectroscopic methods.

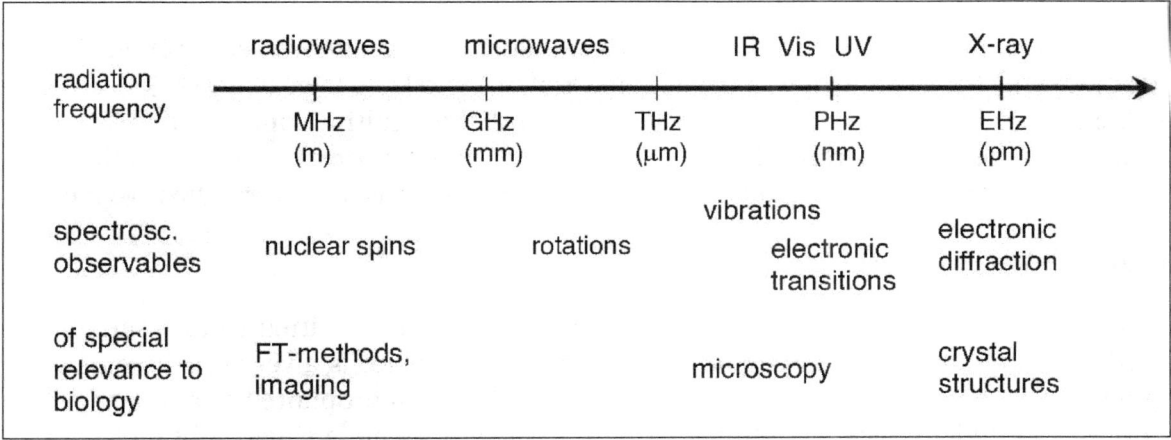

Figure: Qualitative frequency scale for electromagnetic radiation and the relevant molecular observables.

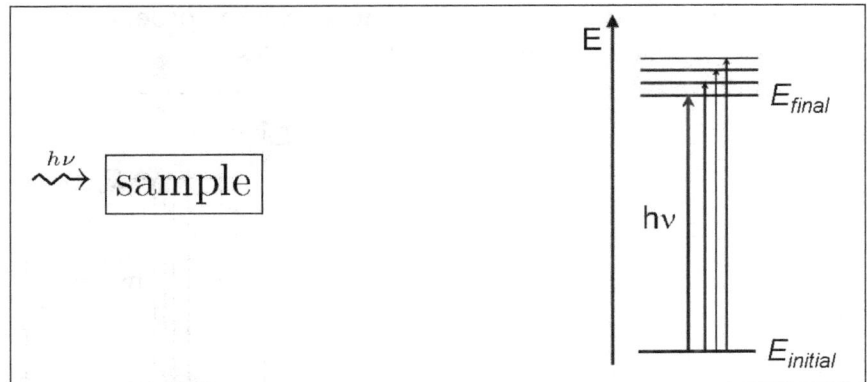

Figure: Absorption measurement: Irradiate the sample with photons of energy hν. The sample may absorb the photon according to the laws of energy conservation if hν = E_{final} − $E_{initial}$ and according to selection rules.

Optical Absorption

All chemical compounds absorb light in the infrared, visible or ultraviolet region. Hence, absorption spectroscopy is a very general method applicable to almost all samples. According to the laws of energy conservation, a photon ca only be absorbed, if the photon energy corresponds to the difference energy between an initial and final state of the sample: hν = E_{final} − $E_{initial}$. The absorbed frequencies therefore depend on the energy levels of the vibrational system (IR) or electronic system (Vis/UV). As a rule of thumb, delocalized electronic systems lead to stronger absorption in the visible and near UV, and aromatic compounds are therefore favorite chromophores for absorption spectroscopy. The strength of absorption is defined in Beer's absorption law:

Absorption: $I = I_o \cdot e^{-\varepsilon l}$

Where,

- I_o incident intensity

- I transmitted intensity

- ε extinction coefficient

- z sample length

We can estimate the absorption for a sample of the wild type green fluorescent protein (GFP) with $\varepsilon = 7000(M \cdot cm)^{-1}$ at 475 nm and a molecular mass of $M_R = 30$ kDa. If we dissolve 30 µg (= 1 µmol) in a cubic cuvette with 1 cm length and a volume of 1 cm^3, the resulting concentration is 3 µM. Previous Equation then gives a transmission of $I = I_0 \cdot e^{-0.007}$ or an absorption of $\approx 0.7\%$. Such a small absorption is difficult to detect, and it is evident that absorption is not a very sensitive method.

Optical Emission

Fluorescence is a widely used method in biochemistry. Photons emitted from a sample can be detected with a quantum yield approaching 1, hence fluorescence is a very sensitive method. To continue our example of GFP, we find that this extraordinary fluorophore has a fluorescence lifetime of $\tau \approx 1$ ns and a typical quantum yield of 0.5. If we irradiate a single GFP molecule with a focussed light source which saturates the absorption, we may collect up to 0.5 photons per ns, this corresponds to a rate of $5 \cdot 10^8$ photons/s. A realistic rate may be an order of magnitude lower, but the fluorescence rate is clearly enough to detect single molecules.

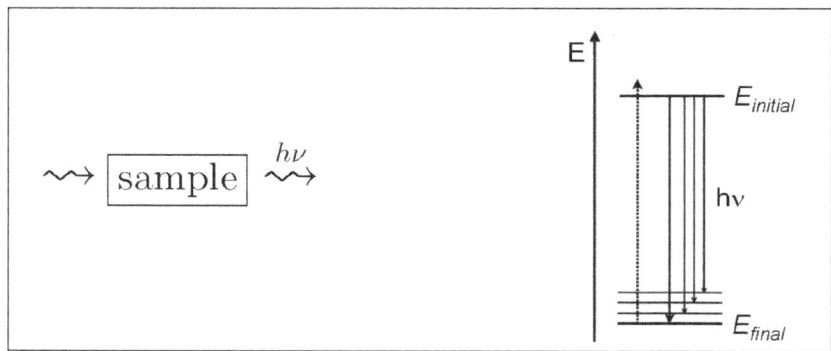

Figure: Fluorescence measurement: Irradiate the sample with photons and observe the emission of fluorescence photons. The sample may emit photons according to the laws of energy conservation and selection rules.

The downside of fluorescence is, that only very few molecules fluoresce with reasonable quantum yields. Fluorescence is therefore a very sensitive technique, but not applicable to a large set of samples. Nevertheless, many spectroscopic studies of biomolecules rely on the fluorescent labelling of compounds with either artificial fluorophores, or with natural fluorophores such as GFP. The location of the labelled compound can then be observed in a microscope. Fluorophores are also attached to antibodies, which bind only to specific molecular targets and thereby allow the very sensitive identification of the target molecule. Fluorophore markers are also used for sequencing, where they may replace other biological building blocks for easier detection. Finally, fluorophores found a particular application as probes of local structure by using donor-acceptor pairs, which can transfer excitation and fluorescent activity if they get spatially close to one another.

Diffraction

Due to the wave nature of light, scattered photons will lead to interference patterns which contain information about the relative position of the scattering centers as shown in figure.

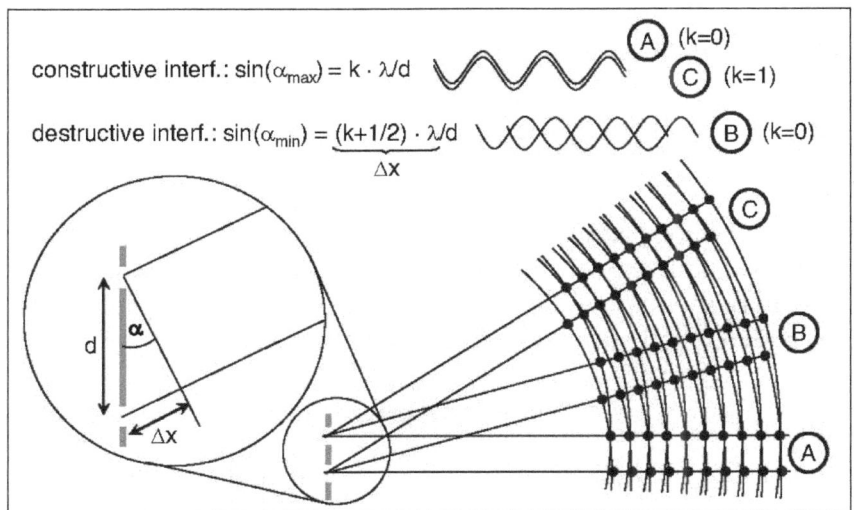

Figure shows interference from two point sources of light (e.g. pinholes, or scattering centers). Photons emitted from the two point sources at an angle α have a path difference Δx. The path difference can lead to constructive (A, C) or destructive (B) interference and determines the probability to observe photons at the corresponding angle.

The detected intensity is proportional to the square of the sum of wavefunctions from all waves, (e.g. two waves in Figure). The result can be as large as expected for the classical sum of two waves, but can also be zero if the waves interfere destructively. The latter occurs whenever two waves have opposite phase, which is the case if their path differs by k + 1/2 values of the wavelength (k=1, 2,...). Equations below calculate the expected amplitude for a diffraction angle α.

$$\text{Path}: \Delta x = d \cdot \sin(\alpha)$$

$$\text{Photon wave}: \Phi_n = \sin\left(2\pi\frac{t \cdot c_o}{\lambda_n} + 2\pi\frac{\Delta x_n}{\lambda_n}\right)$$

$$\text{Interference} \Rightarrow I = \int_t \left[\sum_{n=1}^{2} \Phi_n\right]^2$$

Where,

- Δx path difference

- α detection angle

- d emitter distance

- Φ_n photon wave function

- t time

- λ_n photon wavelength

- I detected intensity

$$\Delta x_{0,1} = 0, \Delta x: \quad I = \int_t \left[2 \cdot \sin\left(2\pi \frac{t \cdot c_0}{\lambda}\right) + \sin\left(2\pi \frac{t \cdot c_0}{\lambda} + 2\pi \frac{\Delta x}{\lambda}\right) \right]^2$$

$$\text{Trigonometry}: \quad I = \int_t \left[2 \cdot \sin\left(2\pi \frac{t \cdot c_0}{\lambda} + \pi \frac{\Delta x}{\lambda}\right) \cdot \cos\left(\pi \frac{\Delta x}{\lambda}\right) \right]^2$$

If more diffraction centers are present, the interference grows more complex. If the interfering centers are regularly spaced, however, then the pattern becomes much sharper because only the absolute interference maxima remain. The latter is the case in crystals and is the reason why X-ray diffraction works best for crystal structure determination.

Detection of Particles other than Photons

Spectroscopic or spectrometric information can be gained from the interaction of the sample with particles other than photons. Electrons have a much shorter deBroglie wavelength than X-Ray photons and can be used for absorption or diffraction experiments with very high resolution. The collision of electrons, ions, or photons with a sample can result in ionic species which may be detected in mass spectrometers. Radioactivity from labelled compounds yields information about chemical pathways or the spatial distribution of metabolites.

Physical Limits of Measurements

Sensitivity and Selectivity

The detection of photons and other particles can be highly efficient. Quantum efficiencies for photon detection with avalanche photodiodes exceed 50% and the detection of charged particles is possible with nearly 100% yield. Efficient detection, however, must be coupled to good contrast between the desired signal and undesired noise to allow sensitive measurements. Fluorescence can serve as an example for a very sensitive method if scattering and background fluorescence is low; in this case single molecule detection is possible. Absorption is also based on photon detection, but the limited absorption quantum yields of single molecules reduce the sensitivity of this method.

For selective measurements, signals from different species should be distinguishable. High resolution techniques (e.g. some kinds of NMR) can yield enough information to distinguish multiple compounds, whereas low resolution spectroscopy only offers the integrated signal of undistinguishable species (e.g. optical absorption and emission in a homogeneously broadened environment). The resolution of all spectroscopic techniques is inherently limited by the Heisenberg uncertainty principle.

Resolution and the Heisenberg Uncertainty Principle

The Heisenberg uncertainty principle can be interpreted as effect of the particle wave dualism. The cosine term in previous equation after trigonometric reformulation is responsible for the zero intensities at all values $\Delta x / \lambda = 1/2, 3/2, ...$ It is interesting to consider what happens if the distance d becomes similar to the wavelength: if $\Delta x / \lambda$ is always smaller than $1/2$, then the first minimum is no longer observed. In this case, the distance between the point sources cannot be determined by investigation of the diffraction pattern. Hence, there is an uncertainty limit for the determination of position which can be expressed as $d \cdot \sin(\alpha) \geq \lambda / 2$. As shown in figure, a substitution of the angle by the impulse components of the photon and the substitution of the de-Broglie wavelength $\lambda = h / p$ results in the famous Heisenberg uncertainty principle in its impulse formulation $\Delta x \cdot \Delta p \geq \hbar / 2$; $\hbar = 1.054 \cdot 10^{-34} J \cdot s$ with an additional factor of 2π. This factor is due to the simplistic assumption, that we must observe exactly one diffraction minimum within a 90° angle. Taking the uncertainty principle into account, it is obvious why X-rays with a wavelength in the picometer range are suited for molecular structure determination by diffraction. Photons in the optical domain with wavelength larger than interatomic distances ($\sim$ Å) on the other hand cannot be used to resolve molecular structure by diffraction.

The uncertainty principle can therefore be directly deduced from the wave nature of matter. This is also true for the other formulations of the uncertainty principle, e.g. if we consider $\Delta E \cdot \Delta t \gtrsim \hbar$. The term $E = h\nu$ is well-defined if the corresponding frequency ν is known. This is clearly the case for a constant, temporally infinite wave, where the wave can be observed for long times. If, on the other hand, the wave is truncated after a short time Δt, then the frequency is no longer well-defined and the corresponding energy becomes uncertain. Inverting this consideration, we note that the sum of multiple waves with different frequencies can result in a short pulse as illustrated in figure.

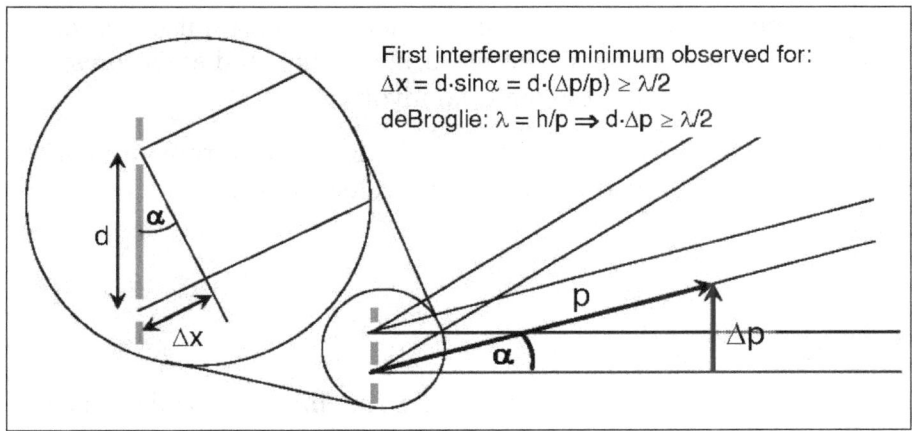

Figure shows direct deduction of the Heisenberg uncertainty principle from the requirement that a diffraction minimum must be resolved for the determination of a distance d. Substitution of the angle α with the corresponding impulse components p and Δp and substitution of the deBroglie wavelength does the trick!

The uncertainty principle limits the energy resolution of photon absorption/emission experiments: Rotational and vibrational states in solution have short lifetimes in the picosecond range due to collisions with the surrounding medium. We can estimate the energy uncertainty using $\Delta E = h\Delta\nu$ to get $\Delta E \geq \dfrac{1}{2\pi \cdot 1\text{ps}}$ and calculate a frequency uncertainty of ≈ 160 GHz (5 cm^{-1}). The natural line width is therefore reasonably narrow as compared to the spacing of vibrational lines, but not as compared to rotational lines. In addition to lifetime broadening, different particles in the condensed phase see different local environments (e.g. via dipole-charge or dipole-dipole interactions) which shift the observed transitions and lead to inhomogeneous broadening.

Ultraviolet and Visible Light Spectroscopy

Electromagnetic radiation or waves is composed of an electric vector and magnetic vector which oscillates in phase and 90 degrees (right angles) to each other and both the vectors are also right angles to the direction of propagation of wave. EMR radiates out from a source in all the directions and includes radiowaves, microwaves, infrared, visible light (400-700 nm), ultraviolet (200-400 nm), X-rays and Gamma rays depending on the source. Electromagnetic spectrum consists of radiations of different wavelength, each possessing different energy, frequency and intensity. High energy radiations are characterized by short wavelength and high frequency. Low energy radiations are characterized by long wavelength and low frequencies.

The interaction of electromagnetic radiation with a particular matter involves not only the properties of the radiation but also involves the structural characteristics of the matter. The electrons in a molecule are distributed at different energy levels, but mainly present in the lowest energy state (E_1) until promoted to the excited state (E_2). The difference in the energy, $\Delta E = E_1 - E_2$, in terms of quantum is required by the electron to reach the higher energy state (excited state). If this energy is derived from the EMR, it gives rise to the absorption spectrum (range of wavelengths which are absorbed by the electrons of the molecule to reach the excited state). Additionally, the source of radiation is also due to the energy transitions in the matter itself. This occurs when the electrons falls back from higher energy excited state to lower energy ground state. The range of wavelength released gives rise to emission spectrum. Therefore, absorption and emission spectrum is characteristic of the wavelength of EMR and the material involved.

During the absorption and emission which involves single electron transition, one quantum of the energy is absorbed or emitted respectively and is equivalent to ΔE.

$$\Delta E = h\nu$$

Where,

- h = Plank constant (6.63×10^{-34} Js).

- ν = Frequency of the electromagnetic radiation (number of oscillations made by the wave per second).

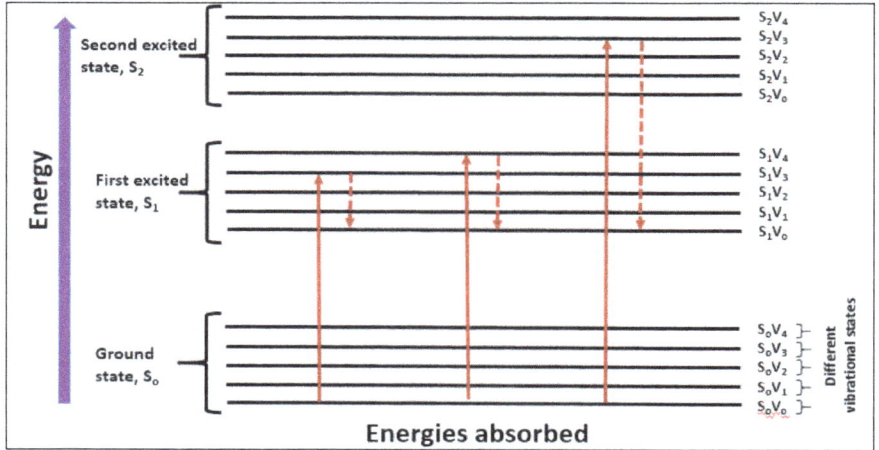

Figure: Energy levels and transitions of electrons.

Principle Ultraviolet-visible (UV/Visible) Spectroscopy

It is based on the principle of absorbance and works in the visible and UV region of the electromagnetic spectrum. This region of the electromagnetic spectrum when absorbed by the molecule it affects the electronic transition of the molecule and therefore, transition occurs from ground state to the excited state. The law which governs light absorption is referred as Lambert-Beer Law. The law is combination of two different laws and these are described here:

Lambert's Law

When a ray of monochromatic light of initial intensity (I_o) passes through a solution in a transparent vessel, the intensity of the transmitted light (I) is inversely proportional to the path length (l) of the absorbing medium. OR When a ray of monochromatic light passes through an absorbing medium, its intensity decreases exponentially as the length of the absorbing medium increases. OR The absorbance of a material sample is directly proportional to its thickness (path length).

$$I = I_o \, e - k_1 l$$

Beer's Law

When a ray of monochromatic light of initial intensity (I_o) passes through a solution in a transparent vessel, the intensity of the transmitted light (I) is inversely proportional to the concentration absorbing solution. OR When a ray of monochromatic light passes through an absorbing solution, its intensity decreases exponentially as the concentration of the absorbing solution increases. OR The absorbance of a material sample is directly proportional to concentration.

$$I = I_o \, e - k_2 c$$

Lambert-beer Law

When a ray of monochromatic light of initial intensity I_o passes through a solution in a transparent vessel, some of the light is absorbed so that the intensity of the transmitted light I is less than I_o and is inversely proportional to the path length (l) of absorbing medium and concentration (c) of absorbing solution.

The relationship between I and I_o depends on the path length of the absorbing medium, l and the concentration of the absorbing solution, c. These factors are related in the Beer and Lambert Law. On combining previous two equations:

$$I = I_o e - k_3 cl$$

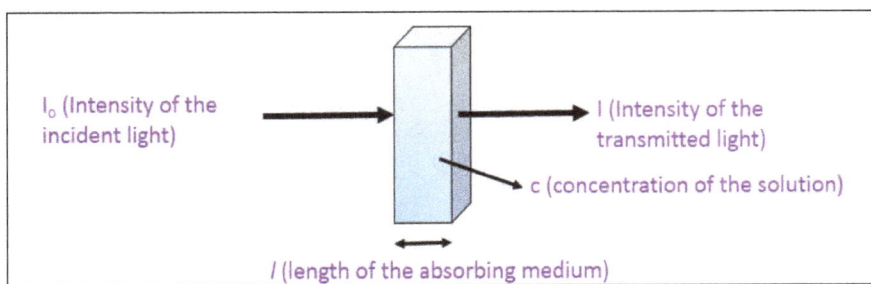

Intensity of the transmitted light (I_o) is always less than the intensity of the incident light (I_o). Decrease in the intensity of transmitted light is mainly because of the absorption of light by the analyte present in the solution. There is some loss of light intensity from scattering by particles present in the solution or by reflection at the interfaces.

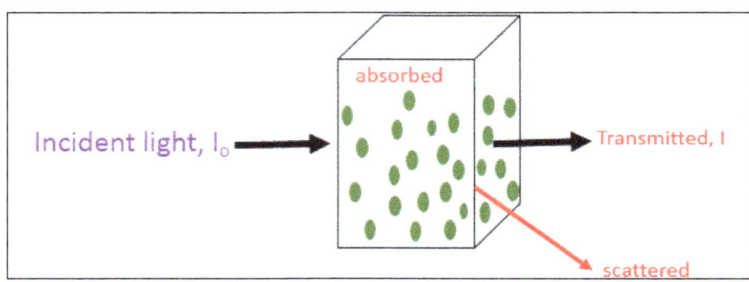

1. Transmittance is the ratio of intensity of the transmitted light (I) and intensity of the incident light (I_o).

2. Transmittance (T) = I/I_o:

   • 100% T represents a solution which does not absorb anything and 100% photons reaches the detector.

   • 60% T represents a solution which absorbs 40% of the photons and 60% of the photons reaches the detector.

   • 0% T represents a solution which does not transmits but absorbs all the photons (occurs in case of opaque solutions).

3. Absorbance (A) is the reciprocal of transmittance = (1/T) = I_o/I.

   • A = 0, represents no photons are absorbed by the solution (no analyte is present).

   • A = 1, represents 90% of the photons are absorbed by the analyte present in the solution and only 10% of the photons reaches the detector.

   • A = 2, represents 99% of the photons are absorbed by the analyte present in the solution and only 1% of the photons reaches the detector.

   $I = I_o\ e - k_1 l$ (Lambert Law)

   $I = I_o\ e - k_2 c$ (Beer Law)

   $I = I_o\ e - k_3 cl$ (Lambert-Beer Law)

   $I / I_o = e - k_3 cl$

   $I_o / I = 1 / e - k_3 cl$

   $\log(I_o / I) = \log(1 / e - k_3 cl)$

   $A = (k_3 cl) / 2.303$

   $A = \varepsilon \lambda cl$

Absorbance is unitless.

Where,

- ○   A = absorbance of the solution.

- ○ $\varepsilon\lambda$ = extinction coefficient ($\varepsilon\lambda$ varies with the wavelength of the light).

- ○   l = pathlength of the cuvette through which the light travels.

- ○   c = concentration of the solution.

4. Molar Extinction (Absorbance) Coefficient/Molar absorptivity, $\varepsilon\lambda = A$, when concentration of the solution is 1 mol/L and pathlength of the cuvette is 1 cm. Unit of molar extinction coefficient is L mol^{-1} cm^{-1} or M^{-1} cm^{-1}. In SI, unit of $\varepsilon\lambda$ is mol^{-1} m^2.

Instrumentation of UV-visible Spectrophotometerscopy

Spectrophotometer covers the range of wavelengths in the both UV and Visible region. Materials used in the spectrophotometer depend upon the wavelength used.

Light Source

Tungsten filament lamps are used in the visible region and hydrogen and deuterium lamps are used in the UV range. In UV-Visible spectrophotometer both the lamps are switched on while working, and it depends on the user which lamp to be used. For this, a mechanical switch is available for directing the light.

Monochromator

Monochromatic light refers to the light of single color or single wavelength or narrow band of wavelengths. Wavelength is selected by the monochromator (prisms and gratings). Monochromator is an optical instrument which selects the light of single or narrow range wavelengths (bandwidth) from wide range of wavelengths emitted from light source. Prism separates the beam of light into its components by the phenomenon of refraction or dispersion whereas gratings do this by the phenomenon of diffraction. The resolution is higher from gratings than from prisms. It is also to be noted that prisms have higher dispersion in UV range, therefore are best suited to work in UV region. Monochromators are also equipped with collimator which converts the diverging light to the collimated light (parallel rays of light).

Slits

Optical slit width affects the bandwidth. Light from a source is emitted in all directions. Slit allows only a portion of the light to reach the monochromator. Selected light is again allowed to pass through the slit. The reproducibility of the absorbance values increases as the slit width decreases. With the decrease in slit width, sensitivity also decreases as less radiation passes to the detector.

Cuvettes

Samples for the detection should be placed in a transparent cuvette, mostly rectangular in shape and commonly have internal width, 1 cm. Small test tubes are also used in some instruments.

Cuvettes are made of silica for the detection in the visible range. Borosilicate glass is used in visible region as well as in near UV region. Fused silica and quartz cuvette are required for the UV range but also covers the visible region.

Chromophores

It is the part/moiety of molecule which is responsible for imparting absorbance. This phenomenon occurs when a molecule absorbs certain wavelength in the visible range and promoted to the excited state. Biological molecules also cause conformation change when exposed to light. Examples include molecules or compounds that have unpaired electrons or unsaturated bonds and can absorb the radiation such as alkenes, ketones, aldehydes, phenyl and other aromatic species. Organic compounds absorbs in the UV or visible region. For the determination of organic compounds in the UV range, organic solvents are not suitable as these solvents absorbs in the UV region. Therefore, ethanol is best suited as a solvent for organic soluble compounds and water for water soluble compounds.

Single-beam and Double Beam Spectrophotometer

A spectrophotometer may be single beam or double beam. In single-beam spectrophotometer, beam of light passes through the cuvette present in the sample cell. Therefore, sample cell is first used to zero the instrument by using blank or reference solution in the cuvette and later by placing the sample cuvette in the same place and the absorbance is recorded. This is to minus the blank absorbance readings from the sample absorbance readings. In double-beam spectrophotometer, incident light is split into two beams by half mirror, one beam passes through the reference cuvette and the other beam passes through the sample cuvette. The transmitted light from both the cuvettes then reaches the detectors.

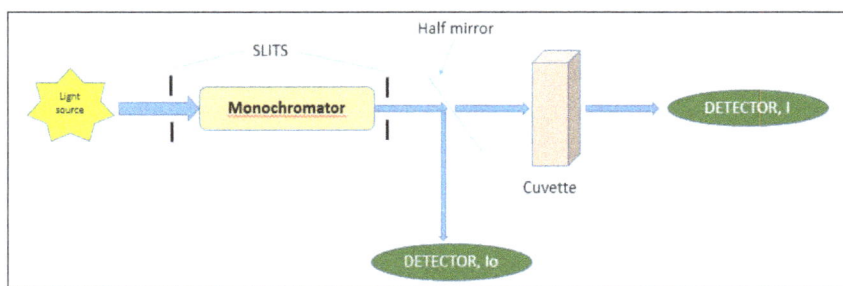

Figure: Single-beam spectrophotometer.

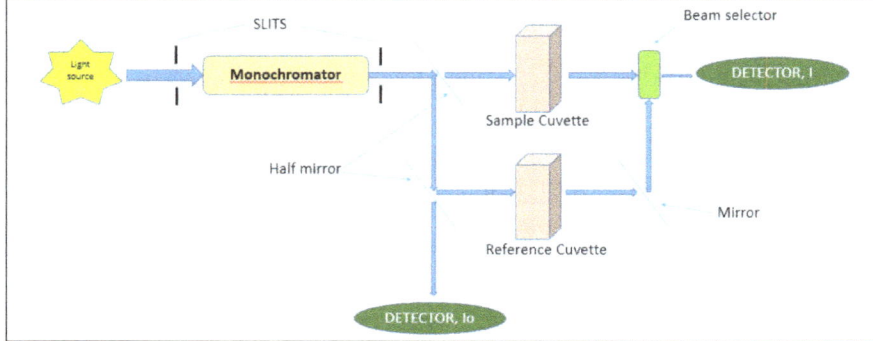

Figure: Double-beam spectrophotometer.

Applications Of UV-visible Spectroscopy

Detection of Impurities

UV absorption spectroscopy is one of the best methods for determination of impurities in organic molecules. Additional peaks can be observed due to impurities in the sample and it can be compared with that of standard raw material. By also measuring the absorbance at specific wavelength, the impurities can be detected. Benzene appears as a common impurity in cyclohexane. Its presence can be easily detected by its absorption at 255 nm.

Structure Elucidation of Organic Compounds

UV spectroscopy is useful in the structure elucidation of organic molecules, the presence or absence of unsaturation, the presence of hetero atoms. From the location of peaks and combination of peaks, it can be concluded that whether the compound is saturated or unsaturated, hetero atoms are present or not etc.

Quantitative Analysis

UV absorption spectroscopy can be used for the quantitative determination of compounds that absorb UV radiation. This determination is based on Beer's law which is as follows:

$$A = \log I_0 / I_t = \log 1/ T = - \log T = abc = \varepsilon bc$$

Where, ε is extinction co-efficient, c is concentration, and b is the length of the cell that is used in UV spectrophotometer. Other methods for quantitative analysis are as follows:

- Calibration curve method,

- Simultaneous multicomponent method,

- Difference spectrophotometric method,

- Derivative spectrophotometric method.

Qualitative Analysis

UV absorption spectroscopy can characterize those types of compounds which absorbs UV radiation. Identification is done by comparing the absorption spectrum with the spectra of known compounds. UV absorption spectroscopy is generally used for characterizing aromatic compounds and aromatic olefins.

Dissociation Constants of Acids and Bases

$$PH = PKa + \log [A^-] / [HA]$$

From the above equation, the PKa value can be calculated if the ratio of $[A^-] / [HA]$ is known at a particular PH. and the ratio of $[A^-] / [HA]$ can be determined spectrophotometrically from the graph plotted between absorbance and wavelength at different PH values.

Chemical Kinetics

Kinetics of reaction can also be studied using UV spectroscopy. The UV radiation is passed through the reaction cell and the absorbance changes can be observed.

Quantitative Analysis of Pharmaceutical Substances

Many drugs are either in the form of raw material or in the form of formulation. They can be assayed by making a suitable solution of the drug in a solvent and measuring the absorbance at specific wavelength. Diazepam tablet can be analyzed by 0.5% H_2SO_4 in methanol at the wavelength 284 nm.

Molecular Weight Determination

Molecular weights of compounds can be measured spectrophotometrically by preparing the suitable derivatives of these compounds. For example, if we want to determine the molecular weight of amine then it is converted in to amine picrate. Then known concentration of amine picrate is dissolved in a litre of solution and its optical density is measured at λmax 380 nm. After this the concentration of the solution in gm moles per litre can be calculated by using the following formula:

$$C = \frac{\log I_o / I_t}{\varepsilon_{max} \times I}$$

"c" can be calculated using above equation, the weight "w" of amine picrate is known. From "c" and "w", molecular weight of amine picrate can be calculated. And the molecular weight of picrate can be calculated using the molecular weight of amine picrate.

As HPLC Detector

A UV/Vis spectrophotometer may be used as a detector for HPLC. The presence of an analyte gives a response which can be assumed to be proportional to the concentration. For more accurate results, the instrument's response to the analyte in the unknown should be compared with the response to a standard; as in the case of calibration curve.

Difference Spectra

A difference spectrum is the difference of two absorption spectra. There are two methods of obtaining the difference spectra:

- By subtraction of one absolute spectrum from another (reduced -oxidized).

- By placing one test compound in the reference cell (containing the compound in reduced state) and the other in the test cuvette (containing the same compound in oxidized state).

Difference spectra may have negative values. The absorption maxima and minima are different from the individual spectra of reduced and oxidized compound. Isobestic points are also wavelengths where absorbances under two different conditions (e.g. reduced and oxidized sate of analyte) are same. Presence of interfering substance can be checked at isobestic points.

An example: Ubiquinone (oxidized) and ubiquinol (reduced). A difference spectrum is obtained by subtracting the absorbance values of ubiquinone from ubiquinol.

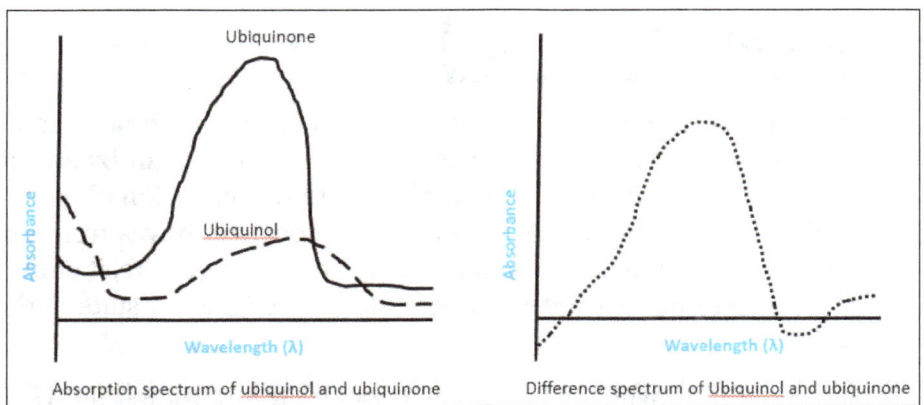

Absorption spectrum of ubiquinol and ubiquinone   Difference spectrum of Ubiquinol and ubiquinone

The main advantage of difference spectra is that it enables the detection of small changes in absorbance occurring in a biological system with a high background absorbance. Examples of such system are the mitochondria and chloroplast where it is possible to detect the changes in the oxidation states of its respiratory chains components.

Binding Spectra

Binding spectra is also known as substrate binding spectra. As the name suggests, this is used to study the level of interaction between an enzyme and its specific substrate. Binding of a substrate or drug to enzymes can be detected by the changes in the absorbance or changes in the λ_{max}. Increase in absorbance is known as hyperchromic shift and decrease in absorbance is known as hypochromic shift. An example of hyperchromic shift is noted when DNA is subjected to heat or placed under alkaline conditions, it causes disruption of the hydrogen bonds of each DNA duplex to yield single-stranded structures and thereby enhances UV light absorption. With the renaturation, the absorbance will decrease and causes hypochromic shift.

Binding of substrate also cause shifts in the λ_{max}. Increase in λ_{max} is known as bathochromic shift. Conjugation of double bonds lowers the energy required for the electronic transition and therefore chromophore absorbs at longer wavelength. Red color in the visible spectrum has a longer wavelength than most other colors and thus this effect is also commonly called a red shift. Decrease in λ_{max} is hypsochromic shift (blue shift). Decrease in conjugation increases the demand for higher energy wavelength to be absorbed by the chromophore and this causes decrease in λ_{max}.

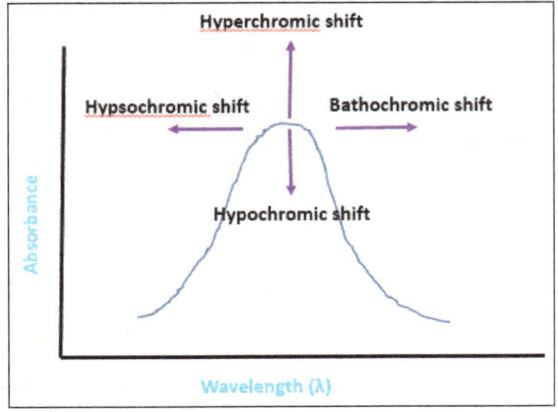

Fluorescence Spectroscopy

Fluorescence spectroscopy or fluorimetry or spectrofluorimetry is a technique to detect and analyze the fluorescence in the sample. Fluorescence is the emission of light by a substance (flour) that has absorbed light or other electromagnetic radiation. In this emission phenomenon, a beam of light (usually UV light) excites the electron in a molecule which moves from ground state to higher energy excited state. When the electron falls back to the ground state, it emits fluorescence. Fluorescence spectroscopy is mainly concerned with electronic (ground state and excited state) and vibrational states.

In molecular species, energy transition may occur in different vibrational levels of a particular excited state because the energy of the vibrational level of excited state matches with the energy of vibrational level of ground state and therefore in such energy transition, some energy is lost as heat (also known a non-radiative transition) until it reaches the lowest vibrational level of the excited state. After losing some energy as non-radiative transition and reaching the lowest vibrational level of the excited state, the electrons follow radiative transition. Radiative transition occurs when electrons falls back from higher energy excited state to lower energy ground state within the molecule, then energy emitted is measured as light. Therefore, in most cases, the emitted light has a longer wavelength (lower energy) than the absorbed radiation (higher energy). During radiative transition, the electrons or molecules may descend into any of several vibrational levels in the ground state; as a result the emitted photons will have different energies, and thus different frequencies. It is a form of luminescence when the emitted light is in the visible range. A fascinating example of fluorescence is when the absorbed radiation is in the ultraviolet region (invisible to the human eye) of the electromagnetic spectrum and the emitted light is in the visible region.

Similar phenomenon occurs in some atomic or molecular species. There are some chromophores which are inflexible and rigid molecules and therefore, may have limited range of vibrational energy levels. In such molecules, the vibrational energy level of the excited state often does not overlap with those of the ground state. When chromophores of this type absorb light, it is not possible for them to return to the ground by simply losing their excess energy as heat. Instead, they undergo a radiative transition in which the absorbed energy is reemitted as light with the same frequency. This process of re-emitting the absorbed photon is "resonance fluorescence" and this is seen in molecular fluorescence.

1. Stokes Shift: Stokes shift is named after Irish physicist George G. It is the difference between the wavelength of absorption maxima and the emission maxima.

Wavelength of absorbed radiation (having low wavelength units and higher energy) is denoted by a. Wavelength of emitted (fluorescence) radiation (having higher wavelength units values and lower energy is denoted by b.

$$\text{Stokes shift} = b - a$$

Good results are achieved with the compounds having the greater Stokes shift. Greater the Stokes shift, lesser will be the interference as the excitation and the emission spectra do not overlap.

Chromophores which exhibit the phenomenon of fluorescence are called fluors or fluorophores.

Fluorophores are organic molecules of 20-100 Daltons. Fluorescent molecules absorb the electro-magnetic radiation in visible region and emit the radiation at a higher wavelength in the visible. Example: ethidium bromide (493 nm/620 nm). Most commonly fluorescent molecules absorb the electromagnetic radiation in the UV range and emits in visible range. Example: green fluorescent protein (360 nm/508 nm).

2. Intrinsic fluors: The native compound exhibits the property due to the presence of aromatic groups in amino acid side-chains in the case of proteins for example tyrosine, tryptophan and phe-nylalanine. Cofactors such as FMN, FAD and NAD also exhibit fluorescence.

3. Extrinsic fluors: Non-fluorescent compounds can be detected by coupling a fluorescent probe (or fluor). Examples are 1- Anilino-8-naphthalene suphonate, fluorescein (for protein), ethidium bromide and acridine orange (for DNA).

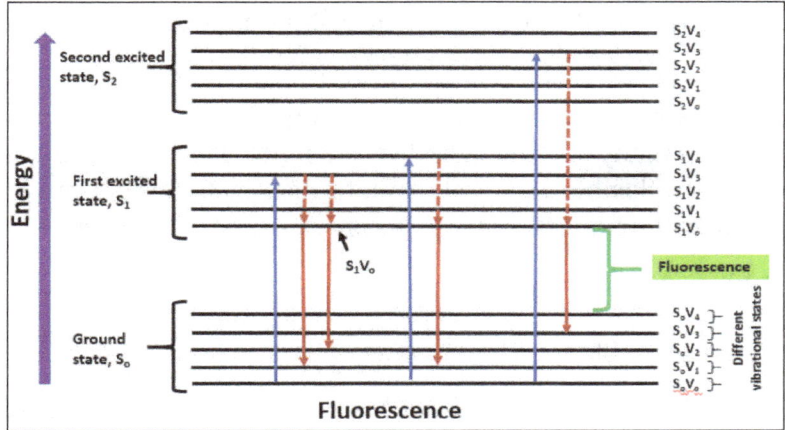

Fluors have characteristics emission spectrum (fluorescence) or as well as characteristic absor-bance spectrum which depends upon its structure and chemical environment.

- Most electrons will occupy the ground state and lowest vibrational level (S_0V_0) at room temperature.

- Electrons are elevated to the high energy excitation state S_1, S_2, etc by the absorption of pho-tons provided by the electromagnetic radiation. The excitation occurs in less than 10^{-15} s.

- The life time of excited state is very short, ranging from 0.5 to 8 ns (0.5 to 8 x 10^{-9} s) or in some cases it may range up to 2s (this situation can arise as a consequence of a phenome-non associated with electrons called magnetic spin).

- Non radiative transition of electrons leads to a rapid loss of energy in the form of heat. This occurs by the collision degradation resulting in the lowest vibrational energy in the lowest excited state (S_1V_0).

- Electrons after reaching the lowest vibrational level of the excited state return to the ground state in less than 10^{-8} s and the emitted energy is stated as fluorescence.

- In fluorescence emission measurement, the excitation wavelength is fixed but the detection wavelength varies. In fluorescence excitation measurement, the detection is fixed but the emission wavelength varies.

Quantum efficiency (Q) = quanta fluoresce/quanta absorbed or number of photons emitted/number of photons absorbed. It is independent of the excitation wavelength.

Spectrofluorimetry exhibits accuracy with fluorescent samples present at low concentration. At low concentrations, the intensity of fluorescence (I_f) is related to the intensity of the incident radiation (I_o) by:

$$I_f = 2.3 \, I_o \, \varepsilon_\gamma c / Q \text{ , i.e. } I_f \, \alpha \, c$$

Where,

- c=Concentration of the fluorescing solution (molar).

- l = Light path in fluorescing solution (cm).

- ε_γ = Molar extinction coefficient for the absorbing material at wavelength γ (dm^3 mol^{-1} cm^{-1}).

Depending upon the final vibrational and rotational energy levels of electrons in ground state, the emitted radiation will be of many closely related wavelengths and will exhibit as band spectra. Aliphatic molecules are usually flexible and tend to photo-dissociate rather than fluorescence, whereas aromatic compounds with delocalized π-electrons sometimes fluoresce.

Disadvantages

- Fluorescence is susceptible to pH, temperature and solvent polarity.

- Whether a particular compound will fluoresce or not is the main problem being encountered.

- Fluorescence quenching. This occurs when emitted fluorescence is lost to other molecules by collision interaction. Quenching is more in concentrated samples and therefore assay is used for concentrated solutions. To increase the sensitivity and accuracy of spectrofluorimeter, very low concentrated samples are used which decrease the collisions and hence the quenching.

- Many interfering materials such as detergents, filter paper and some tissues material may affect the fluorescence.

Instrumentation

- Light source: Mercury lamp emits light near peak wavelengths. Xenon arc exhibits continuous emission spectrum with constant intensity in 300-800 nm range, but can also use for just above 200 nm.

- Monochromators: Most common type of monochromators utilizes a diffraction grating. Two monochromators are used. One monochromator (excitation monochromator) is used for the selection of the excitation wavelength from incident beam. Fluorescent sample will emit the fluorescence in all the directions. Angle of 90° is chosen for emitted florescence and second monochromator (emission monochromator) is used for determination of fluorescence spectrum. The excitation wavelengths which are frequently being selected are in the ultraviolet region and the emission wavelength is in the visible region.

- Detector: Detector can be a single-channeled (detects the one wavelength at a time) or multichanneled (detects all emitted wavelength) both having advantages and disadvantages. Detector is a sensitive photocell, (e.g.: red sensitive photomultiplier for wavelengths greater than 500nm).

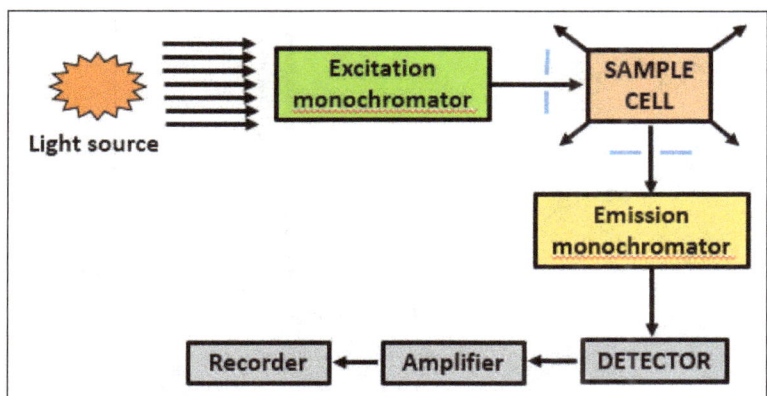

There are two setup for the illumination of the sample:

- 90° illumination: Pre-filter effects arise due to the absorption of radiation before reaching to the fluorescent molecule and post-filter effects arises due to the decrease in fluorescence emitted by the fluorescent molecule before escaping the cuvette. These effects increase with the increase in sample concentration. The use of micro cuvettes alleviates this effect to some extent.

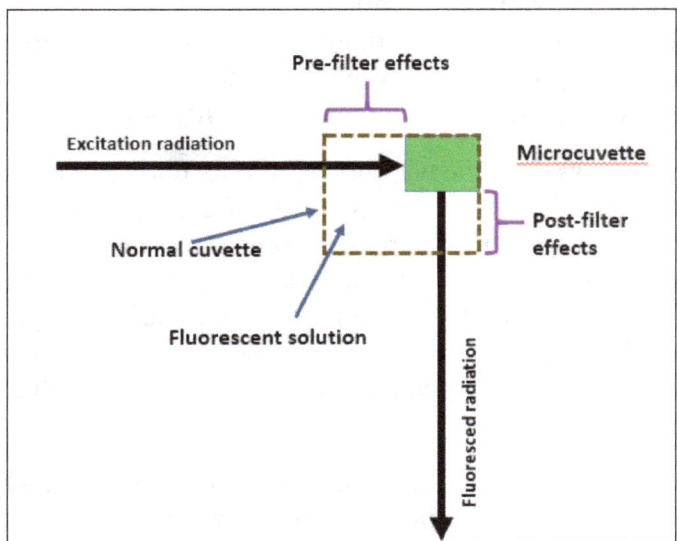

Diagram showing pre and post filter effects.

- Front face illumination: This type of illumination setup removes pre and post-filter effects. In front face illumination, cuvette with one optical face is used and the excitation and the emission occur at the same face. This set up is less sensitive than 90° illumination.

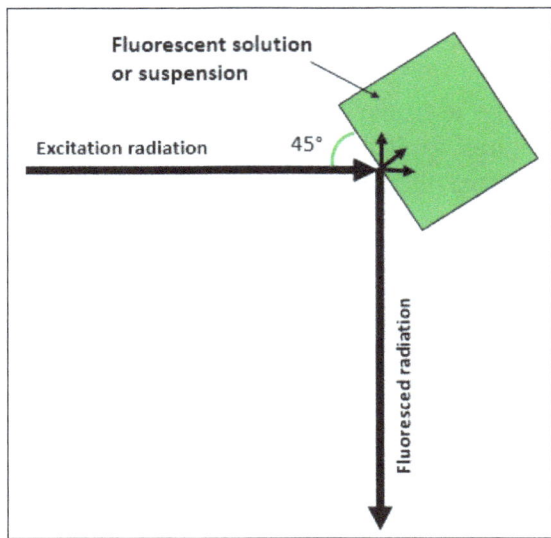

Front face illumination.

Applications

1. Fluorescent probes: Probes are useful in both qualitative and quantitative detection. It helps in the detection of biological compound which is present in very low concentration in a mixture. They are applied to characterize folding intermediates and surface hydrophobicity.

2. Protein and peptide structure: The intrinsic fluors such tryptophan, tyrosine and phenylalanine present in the protein are responsible for the fluorescence exhibited by the proteins. Proteins are generally excited at 280 nm and fluorescence is measured at 295 nm. The fluorescence of folded protein is contributed by all individual aromatic amino acids present in it. Among these, tryptophan exhibits strong fluorescence whereas tyrosine and phenylalanine exhibits less fluorescence. The emission fluorescence of tyrosine is solvent dependent. As the polarity of the solvent surrounding the tryptophan decreases, the fluorescence intensity of the tryptophan increase.

Tyrosine emits fluorescence less than the tryptophan and its fluorescence is quenched by the tryptophan present in its vicinity. Phenylalanine gives weak fluorescence and its fluorescence is only observed when both tyrosine and tryptophan are absent. Any conformational change in the protein therefore changes the absorbance. Cofactors such as FMN, FAD, NAD exhibits the fluorescence and are also applied in the protein structural studies. The binding and release of cofactors, inhibitors, substrates at sites close to the fluor, cause changes in the conformational change and thus changes the fluorescence spectra. It can also be used to study the denaturation and aggregation of protein and peptides.

3. Membrane Structure: The intensity fluorescence of a fluorescently labelled molecule is dependent upon the solvent/environment in which it is present. Changes in the pH or solvent polarity affects the conformation and therefore structure changes can be monitored by the changes in the fluorescence. Extrinsic fluor, ANS (1-Anilino-8- naphthalene suphonate) probe can be used to monitor the changes in the mitochondrial membranes during energy transduction. Hydrophillic and hydrophobic probes can be used for the membrane structure studies as they can orient themselves in hydrophilic and hydrophobic regions of the membrane and gives the information

regarding the properties of the membrane and its surface. Phospholipids containing 12-(9- anthro-anoyl)-stearic acid and 2-(9-anthroanoyl)-palmitic acid into membranes yields the information about the thickness of the membrane. 12-(9-anthroanoyl)-stearic acid and 2-(9-anthroanoyl)-palmitic acid when present in the membranes indicates the regions 0.5nm and 1.5 nm, respectively, from the phosphate head groups of the lipid bilayer.

4. Fluorescence recovery after photobleaching (FRAP): FRAP technique is used for measuring the lateral diffusion in layers or thin membrane by fluorescent probes. The sample under the study is fluorescently labelled and fluorescence in measured in sample and image is observed and captured with the help of optical microscope equipped with the time line camera. Light source is focused on the small patch of the sample and exposed to high intensity illumination (radiation) which causes photobleaching of fluorescent probes. Photobleaching permanently lose the ability of fluor to fluoresce. This turns the patch in to dark color, fluorescence intensity in this area decreases and the image of the sample is continuously observed in the microscope. With time, the adjacent and near-by fluorescing probes will slowly diffuse in to the dark patch as Brownian motion proceeds. Depending upon the speed of diffusion and time, the dark patch will fluoresce again as the fluorescent probes moved in to the bleached area of non-fluorescent probes (beached probes). This technique is very useful for studying the diffusion; fluorescently labelled phospholipids or proteins may be incorporated into a biological membrane and subjected to the similar treatment. The motion of these phospholipids or proteins in the membrane can be studied by monitoring with low intensity radiation. FRAP can also be used to study the protein binding in cell membrane, cell surface characterization, studying free energy in phospholipid layer.

5. Fluorescence resonance energy transfer (FRET): Energy may be transferred from donor to acceptor flour through FRET or electronic energy transfer or dipole-dipole coupling. For this to happen, the distance between the donor and acceptor is critical and both the fluors must be situated closely, there must be overlap between the donor fluorescence spectrum and acceptor fluorescence spectrum. When the donor fluor is present alone, it will fluoresce. Placing the acceptor fluor in the vicinity of donor fluor, quenches the fluorescence emitted by the donor flour. This emitted radiation is sufficient for the electronic transitions in the acceptor flour, and thus emits the fluorescence of different intensity.

This technique detects very small changes in distance, detects molecular interactions in different systems, localization of metals in metalloproteins, detects the interaction the between the proteins, measurement of conformational changes during binding of enzymes with substrate and receptors with ligand, used to measure the distance between the two domains in the same protein, gives information about lipid rafts in the cell membranes.

6. Fluorescence immunoassay (FIA): FIA is a sophisticated technique and is used to detect the antigen and antibody interactions by using the fluorescent probes to label either antigen or antibody. Antigen is detected by the binding of primary antibody. Excess of the primary antibody can be removed by washing. The antigen-antibody complex is then detected by the secondary antibody labelled with the fluor. Excess of the secondary antibody can be removed by washing. The fluor is excited at a particular wavelength and the fluorescence is detected by the spectrofluorimetry. High background fluorescence is the major disadvantage of this technique. Two approaches are followed to reduce the background fluorescence and increase the sensitivity. First, fluors having large stokes shifts should be preferred, example: europium chelates. And secondly

well designed fluorimeters, which delays the detection of emitted light and mean while the background fluorescence declines.

7. Fluorescence activated cell sorter (FACS): FACS is a type of flow cytometry. It is the method of physical separation or sorting of cells from a mixture of cells (cell suspension) into different compartments which is based on the fluorescence and light scattering emitted by the cells tagged with different fluor. A cell suspension is allowed to pass through a narrow nozzle in a stream of liquid. This fast flowing stream of liquid is broken in to droplets, each droplet containing a single cell and is achieved by vibration. One cell in one droplet then passes through the fluorescence detection apparatus, which senses the fluorescence characteristic of the particular cell. Additionally, each droplet also passes through an electric charging ring which give the charges to the droplet, the charges depends upon the fluorescence emitted by the cell. These charged droplets then pass through an electrostatic deflection system, which senses the charges on the droplet and physically separate/deflects the droplets in to sequentially arranged containers. This techniques is used in the counting the cells, separation of cells, purification of cells from a mixture.

8. Microspectrofluorimetry: A spectrofluorimter is equipped with microscope which enables to observe the binding of antibody to a single bacterial cell or subcellular organelle and also helps in the identification of cancerous cells from normal cell as they express different set of protein for which fluoresce labelled specific antibody can be applied. Since some malignant cells have more nucleic acid content than do normal cell have, therefore these malignant cells take up more acridine orange dye which binds to the DNA and can be differentiated from the normal cells.

Circular Dichroism Spectroscopy

Circular dichroism (CD) spectroscopy is a spectroscopic technique where the CD of molecules is measured over a range of wavelengths. CD spectroscopy is used extensively to study chiral molecules of all types and sizes, but it is in the study of large biological molecules where it finds its most important applications. A primary use is in analysing the secondary structure or conformation of macromolecules, particularly proteins as secondary structure is sensitive to its environment, temperature or pH, circular dichroism can be used to observe how secondary structure changes with environmental conditions or on interaction with other molecules. Structural, kinetic and thermodynamic information about macromolecules can be derived from circular dichroism spectroscopy.

Measurements carried out in the visible and ultra-violet region of the electro-magnetic spectrum monitor electronic transitions, and, if the molecule under study contains chiral chromophores then one CPL state will be absorbed to a greater extent than the other and the CD signal over the corresponding wavelengths will be non-zero. A circular dichroism signal can be positive or negative, depending on whether L-CPL is absorbed to a greater extent than R- CPL (CD signal positive) or to a lesser extent (CD signal negative). An example circular dichroism spectrum of a sample with multiple CD peaks is shown below, demonstrating how CD varies as a function of wavelength, and that a CD spectrum may exhibit both positive and negative peaks.

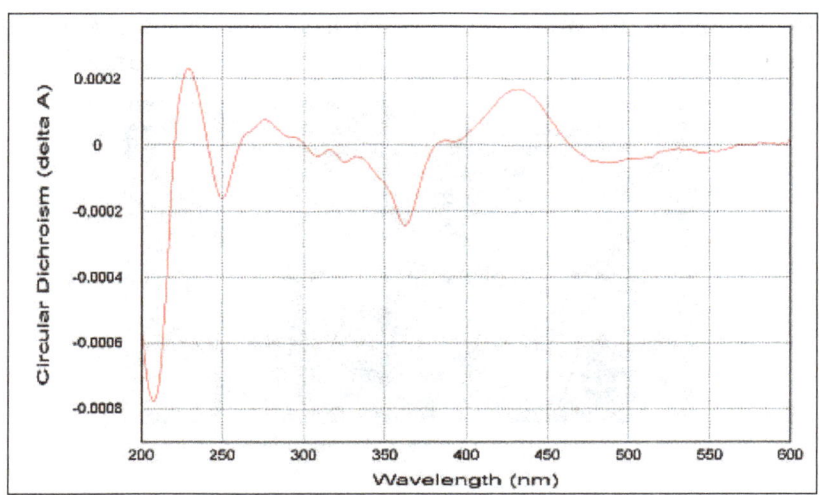

Circular dichroism spectra are measured using a circular dichroism spectrometer, such as the Chirascan, which is a highly specialised derivative of an ordinary absorption spectrometer. CD spectrometers measure alternately the absorption of L- and R-CPL, usually at a frequency of 50 kHz, and then calculate the circular dichroism signal.

The Basics of Polarisation

To really understand circular dichroism, one must first understand the basics of polarisation. Linearly polarised light is light whose oscillations are confined to a single plane. All polarised light states can be described as a sum of two linearly polarised states at right angles to each other, usually referenced to the viewer as vertically and horizontally polarised light.

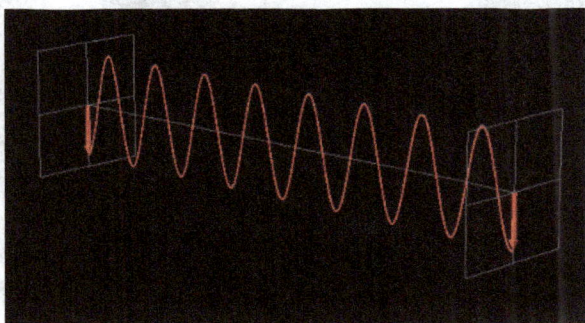

Figure: Vertically Polarised Light.

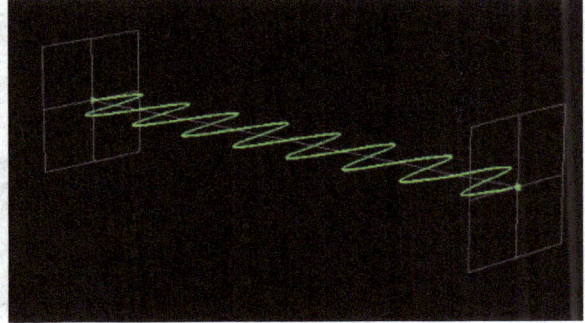

Figure: Horizontally Polarised Light.

If for instance we take horizontally and vertically polarised light waves of equal amplitude that are in phase with each other, the resultant light wave (blue) is linearly polarised at 45 degrees.

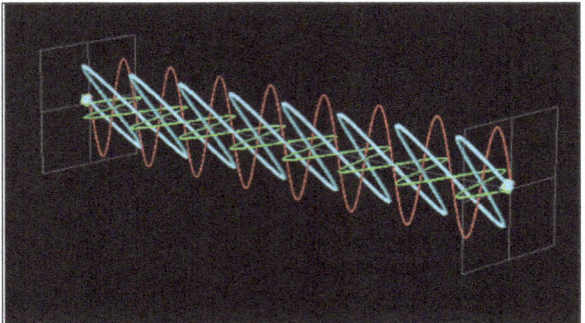

Figure: 45 Degree Polarised Light.

If the two polarisation states are out of phase, the resultant wave ceases to be linearly polarised. For example, if one of the polarised states is out of phase with the other by a quarter-wave, the resultant will be a helix and is known as circularly polarised light (CPL). The helices can be either right-handed (R-CPL) or left-handed (L-CPL) and are non-superimposable mirror images.

The optical element that converts between linearly polarised light and circularly polarised light is termed a quarter-wave plate. A quarter-wave plate is birefringent, i.e. the refractive indices seen by horizontally and vertically polarised light are different. A suitably oriented plate will convert linearly polarised light into circularly polarised light by slowing one of the linear components of the beam with respect to the other so that they are one quarter-wave out of phase. This will produce a beam of either left- or right-CPL.

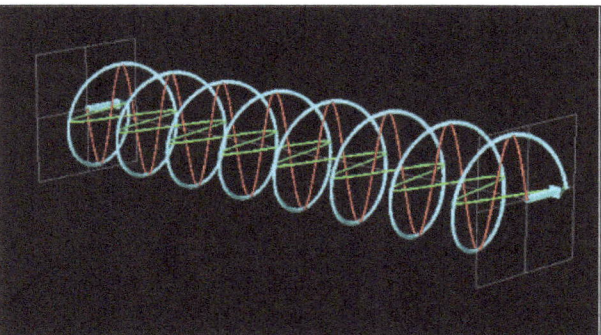

Figure: Left Circularly Polarised (LCP) Light.

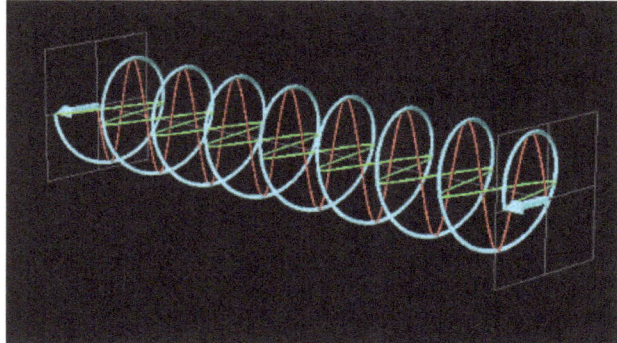

Figure: Right Circularly Polarised (RCP) Light.

The difference in absorbance of left-hand and right-hand circularly polarised light is the basis of circular dichroism. A molecule that absorbs LCP and RCP differently is optically active, or chiral.

Chiral Molecules

The Origin of Optical Activity

Chiral molecules exist as pairs of mirror-image isomers. These mirror image isomers are not super-imposable and are known as enantiomers. The physical and chemical properties of a pair of enantiomers are identical with two exceptions: the way that they interact with polarised light and the way that they interact with other chiral molecules.

Circular Birefringence and Optical Rotation

Chiral molecules exhibit circular birefringence, which means that a solution of a chiral substance presents an anisotropic medium through which left circularly polarised (L-CPL) and right circularly polarised (R-CPL) propagate at different speeds. A linearly polarised wave can be thought of as the resultant of the superposition of two circularly polarised waves, one left- circularly polarised, the other right-circularly polarised. On traversing the circularly birefringent medium, the phase relationship between the circularly polarised wave's changes and the resultant linearly polarised wave rotates. This is the origin of the phenomenon known as optical rotation, which is measured using a polarimeter. Measuring optical rotation as a function of wavelength is termed optical rotatory dispersion (ORD) spectroscopy.

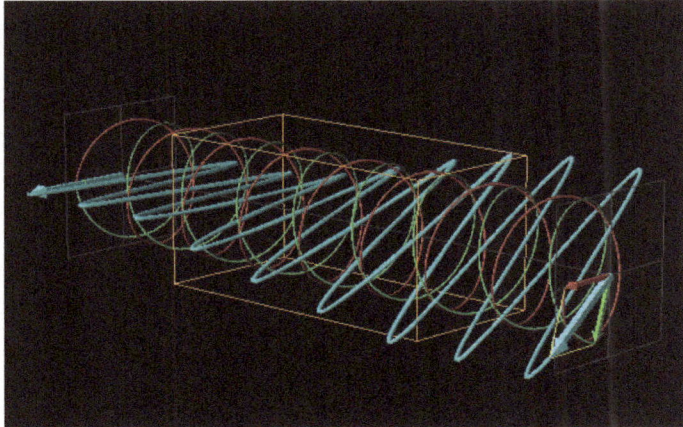

Figure: Circular birefringence - The orange cuboid represents the sample.

Circular Dichroism

Unlike optical rotation, circular dichroism only occurs at wavelengths of light that can be absorbed by a chiral molecule. At these wavelengths Left-and right-circularly polarised light will be absorbed to different extents. For instance, a chiral chromophore may absorb 90% of R- CPL and 88% of L-CPL. This effect is called circular dichroism and is the difference in absorption of L-CPL and R-CPL. Circular dichroism measured as a function of wavelength is termed circular dichroism (CD) spectroscopy and is the primary spectroscopic property measured by a circular dichroism spectrometer such as the Chirascan.

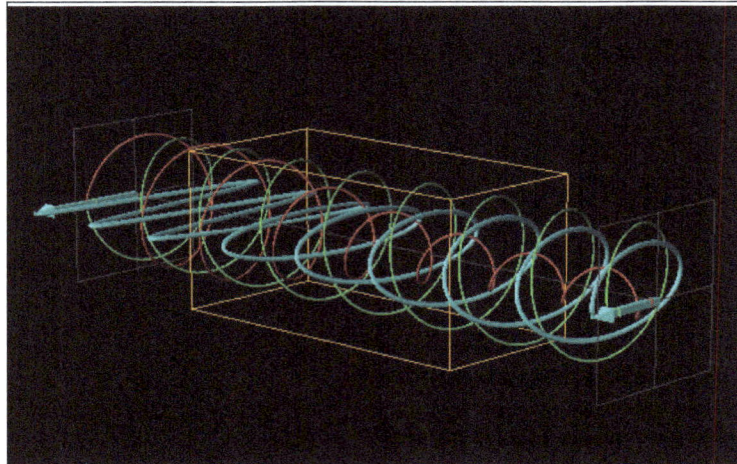

Figure: Circular dichroism - the orange cuboid represents the sample.

Optical rotation and circular dichroism stem from the same quantum mechanical phenomena and one can be derived mathematically from the other if all spectral information is provided. The relationship between optical rotatory dispersion, circular dichroism, absorption spectra and chirality are shown below, with a comparison of the two enantiomers of camphor sulphonic acid.

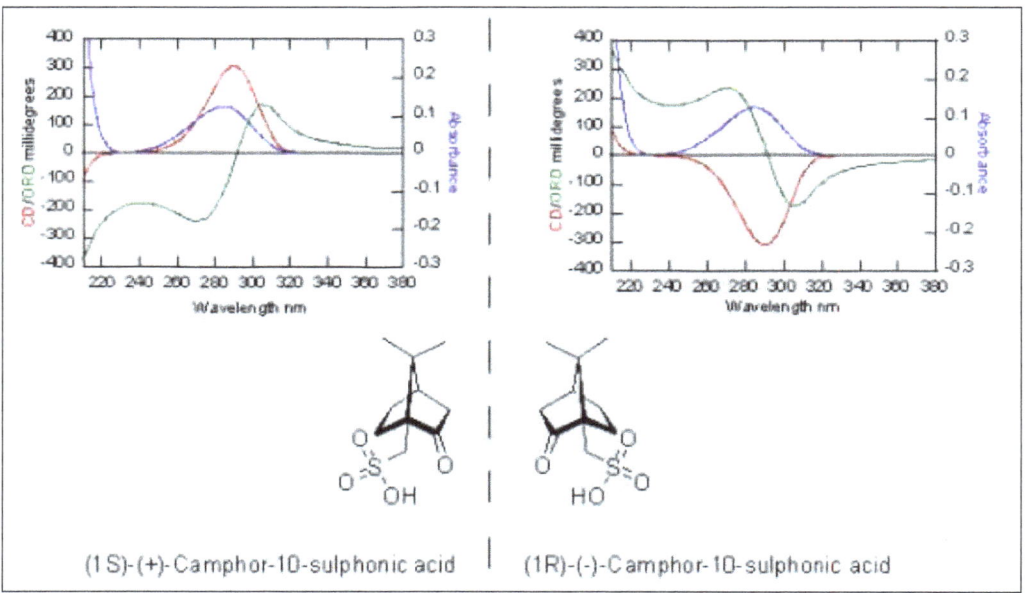

Figure: CD, ORD and Absorbance spectra of R and S forms of camphor sulphonic acid.

Although ORD spectra and CD spectra can theoretically provide equivalent information, each technique has been used for very distinct applications. Optical rotation at a single wavelength is used as a general measurement tool for chiral molecules, to determine concentration and as a determinant of chiral purity compared to a known standard. The simplicity and low-cost of the experiment and instrumentation makes it ideal for this application. Circular dichroism spectra on the other hand are better spectrally resolved than ORD spectra, and consequently more suitable for advanced spectral analysis.

Chirality and Biology

Circular Dichroism and the Study of Biological Molecules

Circular dichroism is a consequence of the interaction of polarised light with chiral molecules. The vast majority of biological molecules are chiral. For instance, 19 of the 20 common amino acids that form proteins are themselves chiral, as are a host of other biologically important molecules, together with the higher structures of proteins, DNA and RNA. The highly chiral chemistry of biological molecules lends itself well to analysis by circular dichroism and the study of biological molecules is the main application of the technique.

A large subset of the use of circular dichroism in biochemistry is in the understanding of the higher order structures of chiral macromolecules such as proteins and DNA. The reason for this is that the CD spectrum of a protein or DNA molecule is not a sum of the CD spectra of the individual residues or bases, but is greatly influenced by the 3-dimension structure of the macromolecule itself. Each structure has a specific circular dichroism signature, and this can be used to identify structural elements and to follow changes in the structure of chiral macromolecules.

The most widely studied circular dichroism signatures are the various secondary structural elements of proteins such as the α-helix and the β sheet. This is understood to the point that CD spectra in the far-UV (below 260nm) can be used to predict the percentages of each secondary structural element in the structure of a protein. Some of the common protein secondary structural elements and the CD spectra associated with them are shown below:

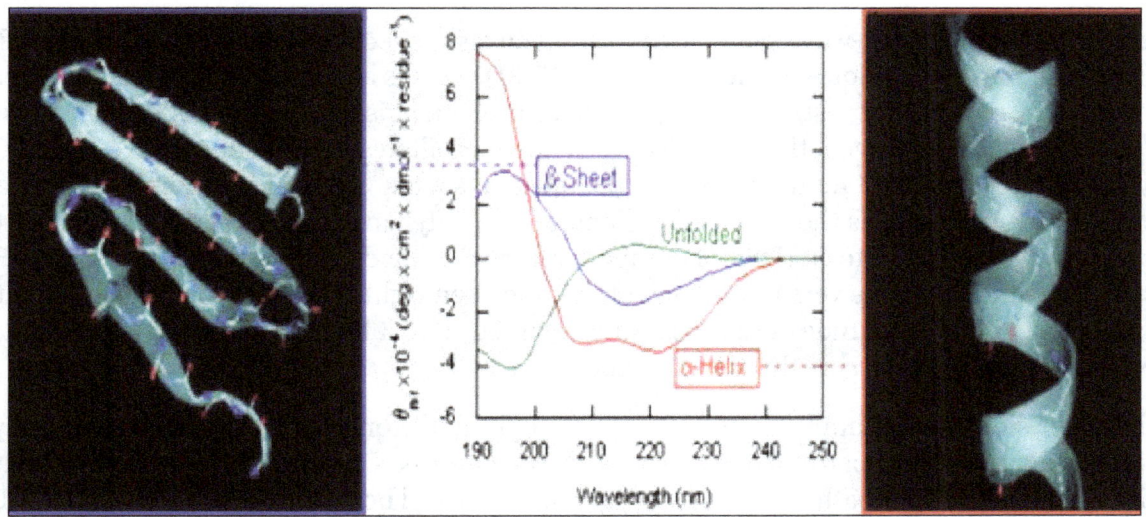

Figure: The secondary structure conformation and the CD spectra of protein structural elements.
Right is an example of the backbone conformation of a peptide in an α-helix and left is the conformation
of a peptide in a β-sheet. In the centre are the associated CD spectra for these different conformations.

There are many algorithms designed for fitting the circular dichroism spectra of proteins to provide estimates of secondary structure. The protein secondary structure CD analysis software distributed with the Chirascan is CDNN.

Secondary structure prediction is only part of the power of circular dichroism spectroscopy. Changes in circular dichroism spectra are very good proxies for changes in the structure of a molecule. Couple this with the facts that (i) spectra can be recorded in minutes and (ii) single wavelength kinetics can be recorded from milliseconds onwards, CD is a particularly powerful tool to follow dynamic changes in protein structure. For instance changes induced by changing temperature, pH, ligands, or denaturants are all commonly used.

A powerful application of circular dichroism is to compare two macromolecules, or the same molecule under different conditions, and determine if they have a similar structure. This can be used simply to ascertain if a newly purified protein is correctly folded, determine if a mutant protein has folded correctly in comparison to the wild-type, or for the analysis of biopharmaceutical products to confirm that they are still in a correctly folded active conformation.

CD Spectrometer Operating Principles

Circular dichroism (CD) is the difference in light absorbance between left- (L-CPL) and right- circularly polarised light (R-CPL) and circular dichroism spectrometers (spectrophotometers) are highly specialised variations of the absorbance spectrophotometer.

A circular dichroism spectrophotometer is also commonly termed a circular dichroism spectropolarimeter or a circular dichrograph. Most modern circular dichroism instruments operate on the same principles, which are demonstrated in the slide show at the bottom of the page. There is a source of monochromatic linearly polarised light which can be turned into either left- or right-circularly polarised light by passing it through a quarter-wave plate whose unique axis is at 45 degrees to the linear polarisation plane.

Instead of a static quarter-wave plate, a circular dichroism spectrophotometer has a specialised optical element called a photo-elastic modulator (PEM). This is a piezoelectric element cemented to a block of fused silica. At rest, when the piezoelectric element is not oscillating, the silica block is not birefringent; when driven, the piezoelectric element oscillates at its resonance frequency (typically around 50 kHz), and induces stress in the silica in such a way that it becomes birefringent. The alternating stress turns the fused silica element into a dynamic quarter-wave plate, retarding first vertical with respect to horizontal components of the incident linearly polarised light by a quarter-wave and then vice versa, producing left - and then right- circularly polarised light at the drive frequency. The amplitude of the oscillation is tuned so that the retardation is appropriate for the wavelength of light passing through the silica block.

On the other side of the sample position there is a light detector. When there is no circularly dichroic sample in the light path, the light hitting the detector is constant. If there is a circularly dichroic sample in the light path, the recorded light intensity will be different for right- and left-CPL. Using a lock-in amplifier tuned to the frequency of the PEM, it is possible to measure the difference in intensity between the two circular polarisations (vAC). The average total light intensity across many PEM oscillations (vDC) can be used to scale the size of the lock-in amplifier signal to take into account variations in total light level. Both signals can be recorded and from them the circular dichroism signal can be calculated easily by dividing the vAC component by the vDC signal.

$$CD = \left(\frac{vAC}{vDC}\right) \cdot G$$

G is a calibration-scaling factor to provide either ellipticity or differential absorbance.

CD Spectrometer Performance

The Design and its Effect on Operation

The limit of detection of a circular dichroism (CD) spectrophotometer (or any spectrophotometer) is determined by its signal-to-noise (S/N) characteristics: the better the S/N, the better its limit of detection.

The signal-to-noise ratio is limited by photon shot noise, which is the statistical variation about an average in the number of photons per unit time detected by the light detector. The quantised nature of photons and their random arrival at the detector means that although the average number of photons detected per second may be say 5, the number in any particular one-second interval may be 0, 2, 7 or some other number. Thus, a measurement must be made over a sufficiently long period of time to determine the true average and the time taken to determine the true average will be inversely proportional to S/N. It is therefore important to design a circular dichroism spectrophotometer to maximise its S/N characteristics.

A general relationship between the contributing factors to the signal-to-noise in an optical spectrometer can be written as:

$$\frac{\text{Signal}}{\text{Noise}} = \left(Q \cdot I \cdot t \right)^{1/2}$$

Where, Q = detector quantum efficiency, I = light intensity, t = time scale of the measurement.

From this it is apparent there are three ways to improve the signal-to-noise of a circular dichroism spectrophotometer: increase the intensity of the incident linearly-polarised monochromatic light, increase the quantum efficiency of the detector, or spend more time collecting and averaging data points.

The first two factors, light intensity and detector performance, are those that can be influenced by the design of a circular dichroism spectrophotometer and work together to lower the last factor, the time required to carry out a measurement. The higher the light throughput and better the detector efficiency, the less time it takes to collect quality data or, equally, the higher the quality of data that can be collected in time-limited experiments such as stopped-flow measurements.

Increasing the intensity of the incident light is the main avenue for increasing the performance of circular dichroism spectrophotometers and this finds its ultimate expression in the use of synchrotron light-sources for CD spectroscopy. Synchrotron facilities provide tremendous light intensity across a very wide spectral range of wavelengths but access to them is expensive and limited and their use is restricted to the more cutting-edge applications of circular dichroism. For the vast majority of CD experiments, a high-intensity bench-top source is the only practical option: Applied Photophysics Chirascan has been designed from the ground up to maximise the light throughput from its Xe arc-lamp source to the sample.

The Chirascan™ monochromator uses two synthetic, single-crystal quartz prisms instead of the diffraction gratings that most people are familiar with from normal absorbance spectrophotometers. Quartz prisms are more efficient than diffraction gratings for a very wide range of wavelengths, particularly in the UV. Quartz is also birefringent and the prisms not only disperse light into the component wavelengths but also, because of their birefringence, disperse the linearly polarised components, one of which is selected for conversion to circularly polarised light. A further advantage of prisms is they do not pass second-order multiples of the desired wavelength, which is a major source of stray-light in grating-based monochromators.

Unlike the dispersion of a grating, which is linear and highly customisable, the dispersion of a prism is non-linear and is set by the properties of the prism material. Consequently the optics and mechanics of a prism monochromator have to be more complex than a grating monochromator, with the need to constantly vary the slit-width as a function of wavelength to maintain a constant band-pass, and a complex relationship between prism movement and wavelength. However, the large wavelength dispersion in the UV means that wider slits can be used even at a small spectral band-pass, which means greater light collection efficiency throughout the UV region. The large majority of circular dichroism applications are carried out in the UV and it is at these wavelengths that the characteristics of a prism are a major advantage.

In addition to the high intensity to improve signal-to-noise, the light from the monochromator must have very low stray-light content and a very high purity of linear polarisation to provide accurate measurements. All of these three key elements have been optimised in the Chirascan™ circular dichroism spectrophotometer and have been achieved by key design features of the Chirascan monochromator.

The second influence on the S/N is the quantum efficiency of the detector used. i.e. its efficiency in turning an incident photon into an electronic signal. In circular dichroism spectrophotometers, the detector of choice in the last few decades has been the photomultiplier tube. The quantum efficiency of these detectors, which are traditionally used in spectroscopy for UV and visible light detection, has remained fairly static over that period. Recently, advances in photodiode technology have resulted in new, high-gain, large area solid state detectors, which provide significant improvements in quantum efficiency in the ultra-violet, visible and near-infra-red regions when compared with photomultiplier tubes. One of these new high-performance solid state detectors is featured in the new Chirascan-plus and it gives further S/N improvements over the photomultiplier-based Chirascan spectrometer. The quantum efficiency improvement outlined below results in significant signal-to-noise improvements over and above the already high performance of the standard Chirascan™.

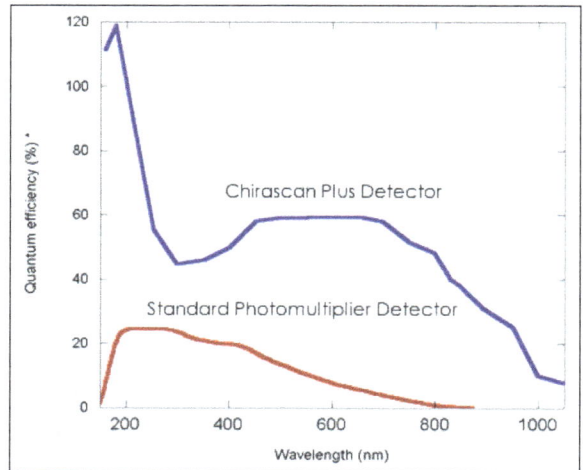

Figure: Quantum efficiency of Chirascan Plus detector vs a standard photomultiplier tube.

CD Signatures of Structural Elements

Circular Dichroism Signatures of Secondary Structural Elements

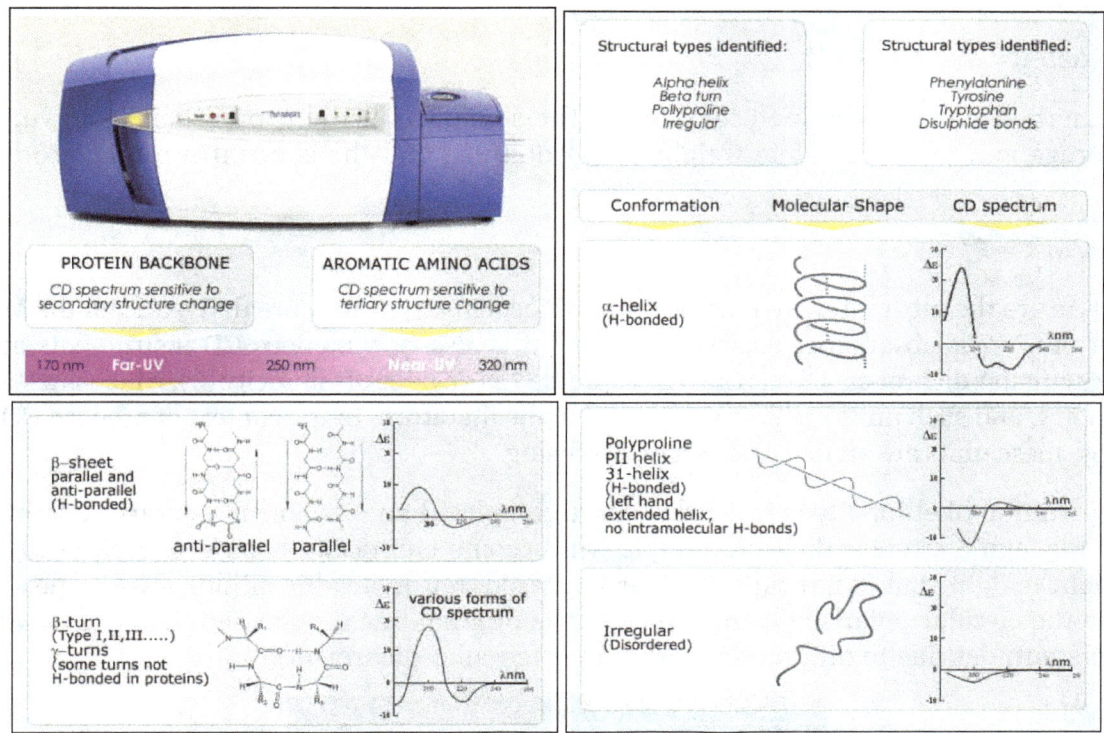

CD Units and Conversions

All Relationships Explained

Circular dichroism (CD) is usually understood and actually measured as the differential absorbance of left (ALCP) and right circularly polarised (ARCP) light, and so can be expressed as:

$$\Delta A = ALCP - ARCP$$

Taking into account cell pathlength and compound concentration, we can arrive at a molar circular dichroism $(\Delta\varepsilon)$.

$$\Delta\varepsilon = \varepsilon LCP - \varepsilon RCP = \Delta A / (C \times l)$$

Where, εLCP and εRCP are the molar extinction coefficients for LCP and RCP light respectively, C= Molar concentration, and l = Pathlength in centimeters.

Another important unit is mean residue molar circular dichroism $\Delta\varepsilon MR$. This is a unit specific for proteins, and reports the molar circular dichroism for individual protein residues instead of whole protein molecules. This allows easy comparison of different proteins with vastly different molecule weights. There are two ways to calculate this depending on how much information is known about the protein.

The concentration of protein (C) in molar is multiplied by the number of amino acids (N) in the protein to provide the mean residue concentration (CMR):

CMR = C x N

$\Delta\varepsilon MR = \Delta A/(CMR \times l)$

An estimate can be determined for CMR if the sequence of the protein isn't known, using the average amino acid residue weight of 113 daltons, and the concentration of protein (P) in gL-1.

CMR = P/113

ΔA and $\Delta\varepsilon$ are the most intuitive units for many biochemists, as they are derived from the familiar concept of UV/Vis absorbance spectroscopy, and it is also how modern CD instruments actually measure circular dichroism. CD can also be expressed as degrees of ellipticity (θ), which is a legacy of polarimetry, and such units are frequently used in the literature. In the context of modern CD spectroscopy, these units are archaic and can be confusing.

The description of ellipticity is somewhat more complex than ΔA. Linearly polarised light when passed through a circular dichroic sample will become elliptically polarised. Elliptically polarised light is light that is not fully circular polarised, but instead is elliptical in shape. This is because the circular polarised components of the original linear polarised light are now not of equal magnitudes due to differential absorbance (circular dichroism).

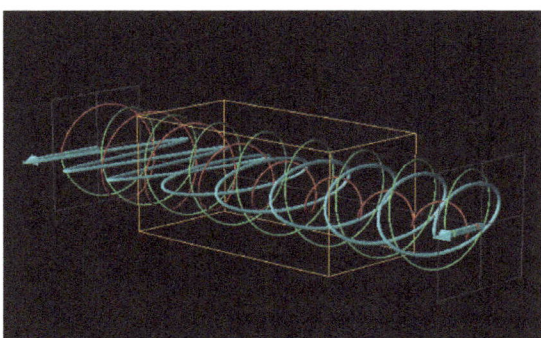

Figure: Circular dichroism as ellipticity - The orange cuboid represents the sample.

The degree of ellipticity (θ) is defined as the tangent of the ratio of the minor to major elliptical axis, and is illustrated below:

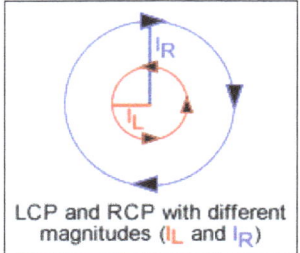
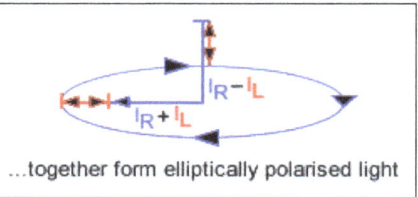
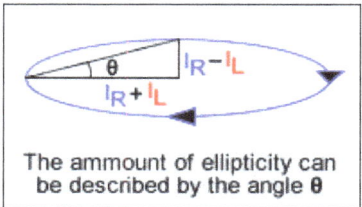

Figure: Linear polarised light has 0 degrees of ellipticity (θ), while fully LCP or RCP will have + or - 45 degrees respectively.

The advantage of circular dichroism ellipticity as a measurement unit is that it is more easily related to optical rotation measurements and polarimetry. Both ellipticity and optical rotation are measurements of changes in polarisation state of a linear polarised analyzer beam, and both have the same units and similar amplitudes for a given sample. This similarity aids in comparison of optical rotation and circular dichroism measurements, a useful ability when circular dichroism spectroscopy first started to be widely used, back in the 1960's.

Fortunately it is very easy to inter-convert between θ and ΔA:

$\Delta A = \theta / 32.982$

Due to the small size of many measurements, θ is often quoted as millidegrees (m°) or 1/1000 of a degree.

Molar ellipticity can be manipulated in the same way as ΔA. For instance taking into account concentration and cell pathlength according to Beer Lamberts law, we can derive a measurement of molar ellipticity [θ]. Following polarimetric conventions, molar ellipticity is reported in degrees cm2 dmol−1, or degrees M-1 m-1 which are equivalent units as shown below.

$$M^{-1}m^{-1} = \frac{1000 \ cm^3}{mol \ 100 \ cm} = \frac{10 \ cm^2}{mol} = cm^2 dmol^{-1}$$

Molar ellipticity can be calculated using the following equation:

$[\theta] = 100 \times \theta / (C \times l)$

C is the concentration in molar, and l the cell pathlength in cm. The factor of 100 converts to pathlength in meters.

Molar Circular dichroism and molar ellipticity can be converted directly by:

$\Delta \varepsilon = [\theta] / 3298.2$

This factor is a hundred fold larger than between raw absorbance and ellipticity due to the conversion between molar extinction defining pathlengths in centimeters and ellipticity having pathlength defined in meters.

Another important unit is mean residue ellipticity [θ]MR. This is a unit specific for proteins, and reports the molar ellipticity for individual protein residues instead of whole protein molecules. This allows easy comparison of different proteins with vastly different molecule weights. There are two ways to calculate this depending on how much information is known about the protein.

The concentration or protein (C) in molar is multiplied by the number of amino acids (N) in the protein to provide the mean residue concentration (CMR):

$CMR = C \times N$

$[\theta]MR = 100 \times \theta / (CMR \times l)$

An estimate can be determined for CMR if the sequence of the protein isn't known, using the average amino acid residue weight of 113 daltons, and the concentration of protein (P) in gL-1.

$$CMR = P/113$$

Atomic Spectroscopy

Atomic spectroscopy includes the techniques of atomic absorption spectroscopy (AAS), atomic emission spectroscopy (AES), atomic fluorescence spectroscopy (AFS), X-ray fluorescence (XRF), and inorganic mass spectroscopy (MS). AAS, AES, and AFS exploit interactions between UV-visible light and the valence electrons of free gaseous atoms. In XRF, high-energy charged particles collide with inner-shell electrons of atom, initiating transitions with eventual emission of X-ray photons. For inorganic MS, ionized analyte atoms are separated in a magnetic field according to their mass to charge (m/z) ratio.

General Principles

Every element has a characteristic atomic structure, with a small, positively charged nucleus surrounded by a sufficient number of electrons necessary to maintain neutrality. Electrons settle into orbitals within an atom and one of the electrons can also jump from one energy level to the higher level by acquiring the necessitated energy. This energy is provided by colliding with other atoms, such as heating-AES, photons derived from light-AAS and AFS, or high-energy electrons-XRF. Possible transitions happen, when the required energy reaches to the difference between two energy states (ΔE). A neutral atom may exist at a low energy shell or ground state (E_0), or at any of a group of excited states depending on how many electrons have been jumped to higher energy levels (E') although it is normal to think for the first transition. Each element has a unique energy level and the ΔEs associated with transitions between those levels.

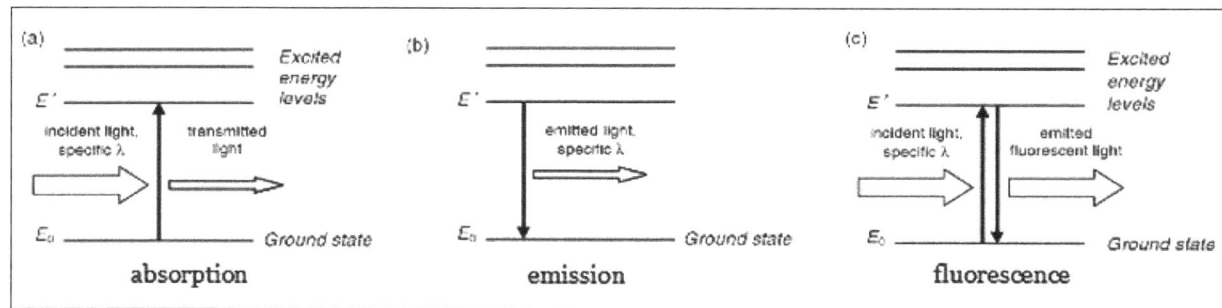

Figure: Energy level diagrams to show transitions associated with (a) AAS, (b) AES, and (c) AFS.
The vertical arrows indicate absorption or emission of light.

The ΔE for movements of valence electrons in most elements meets the energy equal to UV/visible radiation. The energy of a photon (E) is computed with the following equation:

$$E = h\upsilon$$

Where, h is Planck's constant (6.63×10^{-34} Js) and υ the frequency of the waveform corresponding to that photon. The relationship between wavelength and frequency is showed by the equation below:

$$\upsilon = \frac{c}{\lambda}$$

Where, c is the speed of light and λ the wavelength. Thus,

$$E = \frac{hc}{\lambda}$$

And, a specific transition, ΔE, is associated with a unique wavelength.

When light of a specific wavelength enters an analytical system, outer shell electrons of the corresponding atoms will be excited as energy is absorbed. As a result, the amount of light transmitted from the system to detector will be reduced, this is understood as AAS.

Under appropriate circumstances, outer shell electrons of vaporized atoms may be excited by heating. As these electrons return to the more stable ground state, energy is lost. As Figure shows, some of this energy is emitted as light, which can be measured with a detector, this is AES.

Some of the radiant energy absorbed by ground state atoms can be emitted as light as the atom returns to the ground state i.e. AFS. When high-energy photons strike to a massive particle, it can excite an inner shell electron of the atom. The forming inner orbital vacancy can be filled with an outer shell electron. The transition is created by an emission of an X-Ray photon. This process is called X-ray fluorescence (XRF).

The energy of the emission i.e. the wavelength is characteristic of the atom (element) from which it originated while the intensity of the emission is related to concentration of the atoms in the sample. The high temperature inductively coupled plasma has been successfully used as an effective ion source for a mass spectroscopy, the type of method of inductively coupled plasma-mass spectroscopy (ICP-MS) is routinely used for measurements of trace elements in clinical and biological samples. It follows from previous three equations that the wavelengths of the absorbed or emitted light are unique to a given element.

Instrumentation

Formation of the atomic vapor i.e. atomization is the major principle of emission, absorption, and fluorescence techniques. The most critical component of instruments used in atomic spectroscopy is the atomization sources and sample introduction devices with an associated spectrometer for wavelength selection and detection of light. Atomization involves the several key (the basic) steps: solvent removal, separation from anion and other elements of the matrix, and reduction of ions to the ground state atom. The design of an AFS instrument is similar to those for AAS and AES except that the light source and the detector are located at a right angle.

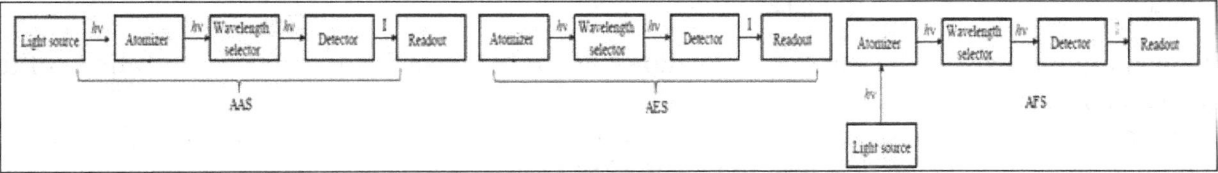

Figure: Schematic diagram of an AAS, AES, and AFS instrument.

A light source which emits the sharp atomic lines of the element to be determined is selected. There are two types of light sources used in these instruments: continuous sources and line sources. A continuous source, also called to as a broad-band source, emits radiation over a broad range of wavelengths. A line source, on the other hand, emits radiation at specific wavelengths, but this source of radiation is not as pure as radiation from a laser. Table provides a list of most common kinds of lamps considered to be light sources.

Table: The most common kinds of light sources.

Source	Wavelength region	Useful for
H_2 and D_2 lamp	Continuum source from 160-380 µm	Molecular absorption
tungsten lamp	Continuum source from 320-2400 µm	Molecular absorption
Xe arc lamp	Continuum source from 200-1000 µm	Molecular fluorescence
nernst glower	Continuum source from 0.4-20 µm	Molecular absorption
globar	Continuum source from 1-40 µm	Molecular absorption
nichrome wire	Continuum source from 0.75-20 µm	Molecular absorption
hollow cathode lamp	Line source in UV/Visible	Atomic absorption
Hg vapor lamp	Line source in UV/Visible	Molecular fluorescence
laser	Line source in UV/Visible/IR	Atomic and molecular absorption, fluorescence, and scattering

The atomizer is any device which will produce ground state atoms as a vapor into the light path. Many atomizers utilized for AFS are similar to those used for AAS and AES. The atomizers most commonly used in these techniques are flames and electrothermal atomizers. The flame provides for easy and fast measurements with little interference and is preferred at any appropriate concentration for the analyte. Flame atomizers contain a pneumatic nebulizer, an expansion chamber, and an air-acetylene laminar flame with a 10 cm path length. The typical pneumatic nebulizer for sample introduction is insufficient, and although elements such as Na and K can be determined in biological samples by flame AES, flame atomization is more suitable for AAS and AFS. AAS measurements can detect concentrations at approximately 1 µg/ml (ppm) or more. Devices are being developed to overcome these limitations of the typical nebulizer.

Atomization can be reached to 100% and the devices can also generate the sample as a pulse flow rather than the continuous flow. Most systems use a graphite tube which is heated electrical energy, a technique called graphite furnace atomization, although other materials are sometimes employed. A programmed sequence of the furnace temperature is used in electrically heated graphite tube. With this atomizer, 10–50 µl of test solution is dried, organic material is destroyed, and the analyte ions dissociated from anions for reduction to ground state atoms. This atomizer also produces temperatures up to 3000 K which allows forming an atomic vapor of refractory elements such as aluminum and chromium. Since the analyte is atomized and retained within a small volume furnace, this procures a dense atom population.

The technique is extremely sensitive as it allows one to detect a few µg/ml concentrations of the analyte. Although the technique is widely used for AAS, electrothermal atomization will provide a better performance for both AES and sample introduction into inductively coupled plasma. Traditional sources usually include arcs and sparks but modern instruments use argon or some other inert gas to create plasma. The plasma may be produced when gas atoms are ionized, $Ar + e^- - Ar^1 + 2e^-$—a process generated by seeding ions from a high-voltage spark—and is sustained from a radio frequency generator in the area of the induction coil. This is known as inductively coupled plasma (ICP). Plasma exists at temperatures of up to 10,000 K and the instrument prevents the torch from melting. XRF requires that sample should be irradiated by high energy photons. In most instruments, the source is the polychromatic primary beam from X-Ray tubes. Of interest to biological applications, however, it is the use of radioactive isotopes such as ^{244}Cm, ^{241}Am, ^{55}Fe, and ^{109}Cd.

An ideal wavelength selector has a high throughput of radiation and a narrow effective bandwidth. There are two major types of wavelength selectors —filters and monochromators. A simple example of an absorption filter is a piece of colored glass. Absorption filters provide effective bandwidths of 30–250 nm, although the throughput can be only 10% of the source's emission intensity at the low end of this range. Interference filters constructed of a several optical layers deposited on a glass or transparent material. Typically, effective bandwidth is 10–20 nm, with maximum throughputs of at least 40%. A monochromator is used to convert a polychromatic source of radiation at the entrance slit to a monochromatic source of restricted effective bandwidth at the exit slit. These devices are classified as either fixed-wavelength or scanning. The wavelength selects by manually rotating the grating in a fixed-wavelength monochromator. A scanning monochromator includes a drive mechanism that continuously rotates the grating, allowing sequential wavelengths to exit from the monochromator.

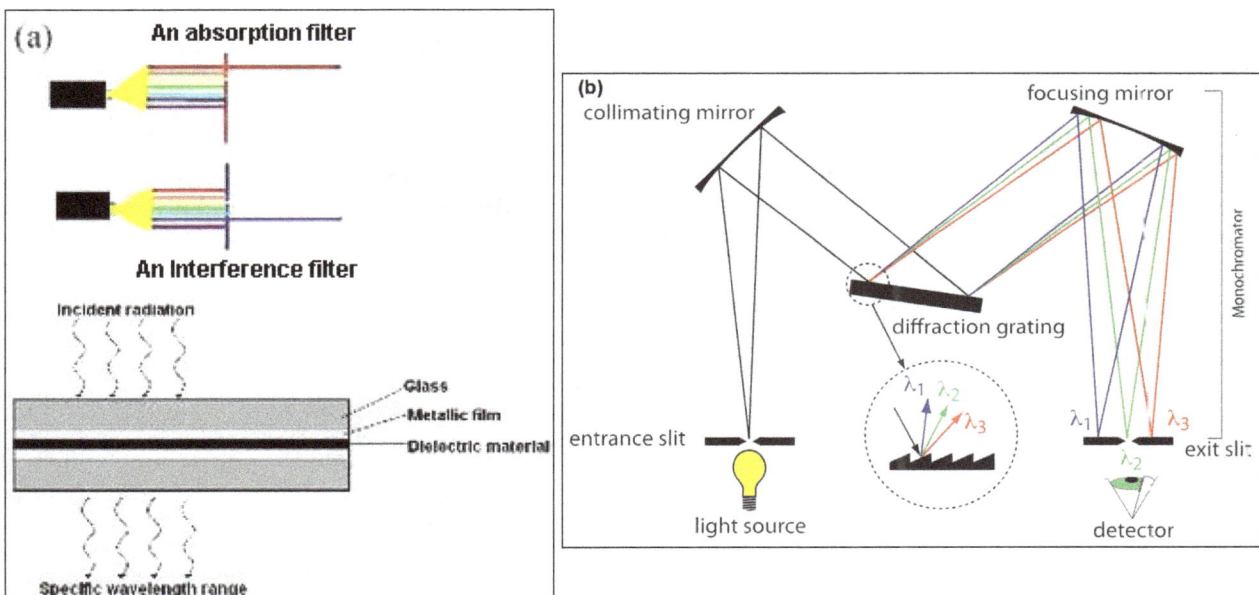

Figure: Schematic diagram of wavelength selectors: (a) filters and (b) a diffraction grating monochromator.

Detectors use a sensitive transducer that converts a signal comes from light energy into electrons An ideal detector produces signal, S, is a linear function of the electromagnetic radiation's power, P,

$$S = kP + D$$

Where, k is the detector's sensitivity and D is the detector's dark current, or the background current when no radiation of source reached to the detector.

Phototubes and photomultipliers include a photosensitive surface that absorbs radiation in the UV-visible, or near-IR, generating an electrical current proportional to the number of photons reaching the transducer. Other photon detectors use a semiconductor compound as the photosensitive surface. One advantage of the Si photodiode manufactured utilizing semiconductor process is that it is easy to miniaturize. Infrared photons do not have enough heat to generate a measurable current with a photon transducer.

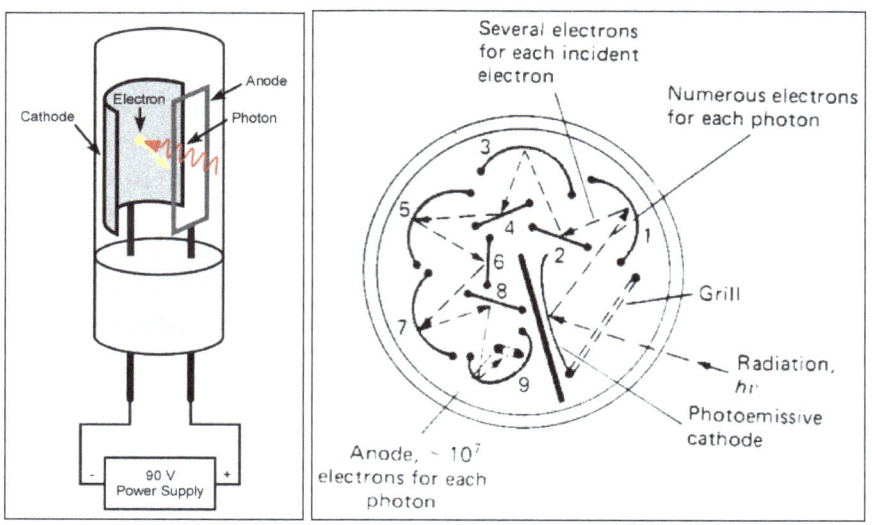

Figure: Diagram of a phototube and a photomultiplier tube.

A transducer's electrical signal is sent to a signal processor where it is displayed in a form that is more convenient to explain. The analog meters, digital meters, recorders, and computers equipped with data acquisition boards are good examples of signal processors. A signal processor is used in calibrating the detector's response, amplificating the transducer's signal, removing noise by filtering, or mathematically transforming the signal.

Table: Examples of detectors for spectroscopy.

Detector	Class	Wavelength Range	Output signal
phototude	Photon	200-1000 μm	Current
photomultiplier	Photon	110-1000 μm	Current
Si photodiode	Photon	250-1100 μm	Current
photoconductor	Photon	750-6000 μm	Change in resistance
photovoltaic cell	Photon	400-5000 μm	Current or voltage
thermocouple	thermal	0.8-40 μm	Voltage
thermistor	thermal	0.8-40 μm	Change in resistance
pneumatic	thermal	0.8-1000 μm	Membrane displacement
pyroelectric	thermal	0.3-1000 μm	Current

Sample Preparation

An ideal sample preparation should remove interfering components from the matrix and to adjust of analyte to facilitate the actual measurement. Methods for destruction of the organic matrix by simple heating or by acid digestion have been developed and are thoroughly approved. Microwave heating is used for this purpose, with the specifically designed a compatible equipment to avoid dangerous of excessive pressure within reaction flask. Although the number of samples that can be processed is not large, microwave heating affords rapid digestion and low reagent blanks. More recent developments include continuous flow systems for automated digestion which has a direct link with the instrument.

Liquid-liquid portioning has been widely applied for pre-concentration procedure. Analyte atoms in a large volume of aqueous solution are complexed with a suitable agent and collected into a small volume of solvent. Vapor generation procedures permit the rapid introduction of 100% of the sample into the atomizer and are used for AAS, AES, AFS, and ICP-MS. Certain elements such as arsenic, selenium, and bismuth readily evolve gaseous hydrides and transferred by a flow of inert gas to an AES, and ICP-MS or to a heated silica tube positioned in the light path for AAS, AFS. The tube can be heated using the air-acetylene flame or an electric current. The obtained heat is enough to cause decomposition of the hydride and atomization of the analyte. Thus, there is no loss off analyte, which in all the atoms flow the light path with in few seconds and they are trapped within the silica tube that was retarded their dispersion. Any sample volume added to the reaction container, hydride generation AAS has detection limits a few nanograms of analyte. Mercury can quickly form a vapor in the ambient temperature, and this property is the basis for cold vapor generation. When a reducing agent is added to sample solution, Hg^{2+} converts to the elemental mercury. Agitation or bubbling of gas through the solution is used to enhance rapid vaporization of the atomic mercury and to improve the transfer of mercury to a flow through cell located in the light path. As with hydride generation, the detection limit is a few nanogram and some manufacturers have been developed common instrumentation to accomplish both procedures. Chromatographic or electrophoretic techniques have been also developed that are coupled directly to the atomic spectroscopic instrument to develop integrated analytical arrangements.

Detection Limits

The detection limits are important parameters of analytical techniques. Typical detection limit ranges for the major atomic spectroscopy techniques are shown in Figure. AAS detection limits are generally better in all cases where the element can be atomized. Detection limits for refractory elements such as bor, titanium, and vanadium are better by ICP than by AAS. Nonmetals and the halogens can only be determined by ICP. Optimum detection of nonmetals such as sulfur, nitrogen, and halogens by ICP-ES can only be achieved when a vacuum monochromator is used. For mercury and those elements that form hydrides, the cold vapor mercury or hydride generation techniques offer exceptional detection limits.

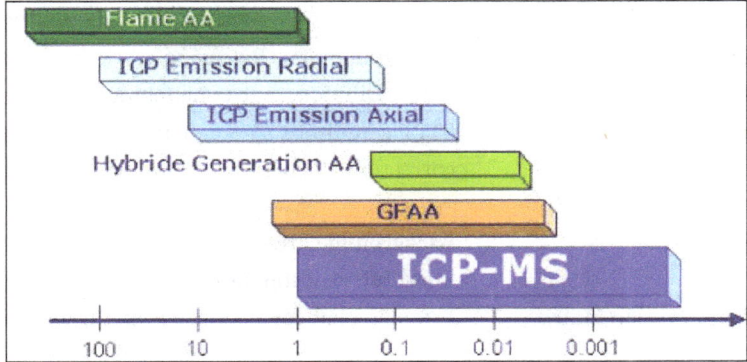

Figure: Typical detection limit ranges for the major atomic spectroscopy techniques.

Analytical Working Range

The analytical working range can be considered as the concentration range over which quantitative results can be obtained without recalibration for system. Selecting a technique with an analytical working range based on the expected analyte concentrations, minimizes the analysis times by allowing the samples with different analyte concentrations to be analyzed together. For example; ICP-MS, once considered only an ultratrace element technique, can now run concentration ranges from low parts-per-trillion (ppt) level up to high parts per million (ppm). A wide analytical working range also can reduce, for example handling requirements, minimizing potential errors. Figure shows typical analytical working ranges with a single set of instrumental conditions.

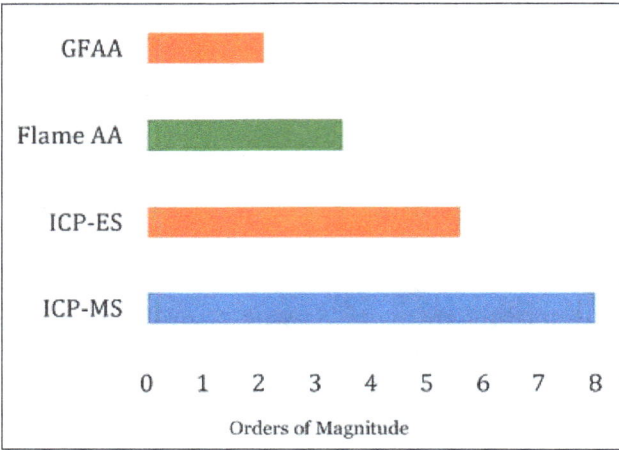

Figure: Analytical working ranges for the major atomic spectroscopy techniques.

Interferences

Spectroscopic interferences have been determined and documented, and methods have been used to correct or compensate for those interferences which may occur. For example; ICP-AES provides a wide dynamic range and minimal chemical interferences. A summary of the types of interferences seen with atomic spectroscopy techniques, and the corresponding methods of compensation are shown in table.

Table: Atomic spectroscopy interferences.

Technique	Type of interference	Method of compensation
Flame AA	Ionization Chemical Physical	Ionization buffer. Releasing agent or nitrous oxide-acetylene flame. Dilution, matrix matching, or method of additions.
GFAA	Physical and chemical Molecular absorption Spectral	STPF conditions. Zeeman or continuum source background correction. Zeeman background correction.
ICP emission	Spectral Matrix	Background correction or the use of alternate analytical lines. Internal standardization.
ICP-MS	Mass overlap Matrix	Inter element correction, use of alternate mass values, or higher mass resolution. Internal standardization.

Other Performance Criteria

Performance criteria for analytical techniques include the ease of use, required operator skills, and availability of documented methodology. Table summarizes comparative advantages and limitations of the most common atomic spectroscopy techniques.

Table: Comparison of spectroscopic techniques performance.

Criteria	Flame AA	Flame AE	AFS	ICP	X-RF
Costs	Low (~$10–15 K)	Moderate	Moderate	Highest (~$200 K)	Highest
Instrumental ion	Low	Low	High	High	High
Maintenance	Low	Low	Low	High	High
Sample preparation	Moderate	Low		Moderate	High
Speed	Slow	Medium		Fast	Fast
Operator skill	Lower	Moderate	Higher	Higher	Higher

Recent Developments and Applications

Analytical methods of atomic spectroscopy have been used for elemental analysis identification, and quantitation in varieties of samples. Recently, most all of the spectroscopic techniques available are used in the analysis of metals and trace elements in samples of industrial and environmental origin.

Progress continues to develop in analytical spectroscopy as improvements are made to sensitivity, limits of detection, and availability. Recent development depends on instrumental adjustments and slight modifications to allow new types of measurements. Advancements in materials science have revealed demand for new methods of measurement using instruments already accessible, pushing the boundaries of what was previously available. For example, some new and interesting miniaturized plasma sources and a new distance of flight (DOF) mass spectrometer have been to the fore in developments. In addition, several novel methods have been developed, such as laser ablation molecular isotopic spectrometry (LAMIS) for isotope ratio analysis, and stand-off LIBS techniques such as "underwater LIBS".

Fourier Transform Infrared Spectroscopy

Infrared spectroscopy is supposed to be one of the most robust analytical techniques. The greatest benefits of infrared spectroscopy is that it can almost analyse any sample in any state may be powder, film, liquid, gas, solution, paste, fibre.

Principle of Fourier Transfom Infrared Spectroscopy

Infrared spectroscopy works on the principle that all molecules absorb frequencies and it is a characteristic of structure of the molecule. These absorptions occur at resonant frequencies, i.e.

the frequency of the absorbed radiation matches the vibrational frequency. Infrared radiation does not have enough energy to induce electronic transitions as in UV. Absorption of IR is restricted to compounds with small energy differences in the possible vibrational and rotational states.

For a molecule to absorb IR, the vibrations or rotations within a molecule must cause a net change in the dipole moment of the molecule. The alternating electrical field of the radiation interacts with fluctuations in the dipole moment of the molecule. If the frequency of the radiation matches the vibrational frequency of the molecule then radiation will be absorbed, causing a change in the amplitude of molecular vibration.

- Molecular Rotations: Rotational transitions are of little use to the spectroscopy. Rotational levels are quantized, and absorption of IR by gases yields line spectra. However, in liquids or solids, these lines broaden into a continuum due to molecular collisions and other interactions.

- Molecular Vibrations: The positions of atoms in molecules are not fixed; they are subject to a number of different vibrations. Vibrations fall into the two main categories of stretching and bending. Stretching refers to change in inter-atomic distance along bond axis while bending refers to change in angle between two bonds. There are four types of bend:

FTIR

Fourier Transform Infrared Spectroscopy (FTIR) is technique which provides spectrum of absorption or emission of a solid or liquid or gas. In Fourier transform, mathematical process is required to convert raw data in to spectrum. In FTIR, data are collected simultaneously over a wide range of wave length. This feature is different from UV-Visible spectrophotometer where data are collected over narrow range of wave length at a time. The technique is useful for characterization of organic molecules which includes alcohol, alkane, alkene, alkyl halide, alkyne, amine, carbonyl, ether, nitrile, nitro, acid, aldehyde, amide, anhydride, ester, and ketone. The technique is also useful for analysing inorganics which includes paints, coatings, and resins. The technique is most commonly used for detection of organic contamination and identification of functional groups in organic compounds. In FT-IR, mid infrared waves are most commonly used.

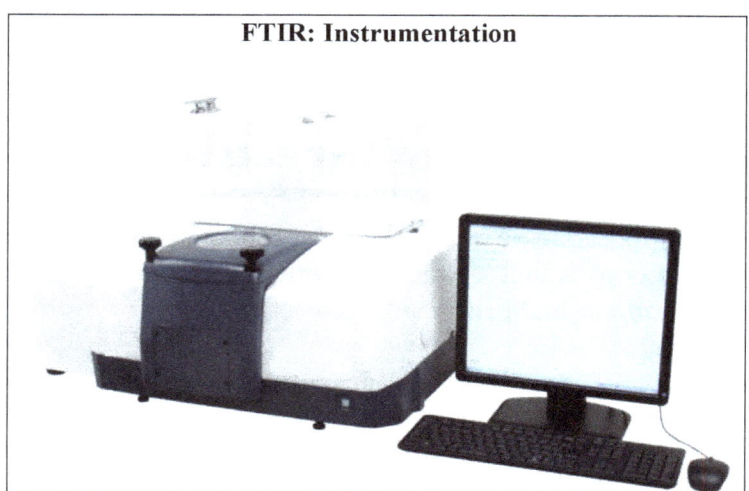

FTIR: Instrumentation

IR Sources

Most FTIR spectrometer are used in mid or near IR region. Wave length in mid IR region is 2-25 μm which corresponds to frequency of 5000-400 cm^{-1}. For this region, most common source is silicon carbide which is heated to 1200 K. Tungston-halogen lamp at high temperature is used for near IR which corresponds to 1-2.5 μm or 10000-4000 cm^{-1}. For far-IR in wave length beyond 50 μm (200 cm^{-1}), mercury discharge lamp is used.

Detectors

For mid IR, pyroelectric detectors are used. These detectors respond to changes in temperature which results from fall of radiation of different intensities. These detectors contain dueterated tri-glycine sulphate (DTGS) or lithium tantalate (LiTaO$_3$). The detector works at room temperature. Time of scan is few seconds. For higher sensitivity or faster response, cooled photoelectric detector is used. In mid-IR, liquid nitrogen cooled mercury cadmium telluride (MCT), is used. With this, interferogram can be measured in 10 milliseconds.

DTGS can be also used in near IR. For far IR, sensitive detector is required since both source and beam splitters are inefficient. The detector used in this region is liquid helium cooled silicon or germanium bolometer.

Beam Splitter

Beam splitter transmits 50% radiation and reflects remaining 50%. But individual beam splitter has limited range and therefore several beam splitters may be required for coving wide range. Germanium coated KBr is used for mid-IR. CaF$_2$ is in near IR region but it is less sensitive to moisture than KBr. Far IR beam splitters uses polymer films and covers limited range.

Attenuated Total Reflectance (ATR)

This accessory is required for measuring surface properties of solid or thin samples. It has penetration depth of 1-2 μm.

FTIR: Interferometer

FTIR uses different optical systems. One of the most common optical systems used is Michelson interferometer. It consists of light source, collimator mirror, beam splitter, fixed mirror, moving mirror and detector. Light source emits infrared radiation and these falls on collimator. Transmitted radiations from collimator move parallel to each other. The radiation falls on beam splitter which splits beam into two equal parts. One half of incident radiation is transmitted and falls on moving mirror while other half is reflected and falls on fixed mirror. As the name states that fixed mirror is fixed at defined distance from beam splitter and thus path length between beam splitter and fixed mirror is constant. On other hand moving mirror can be made to move away or towards beam splitter and therefore path length between moving mirror and splitter beam can be varied. Reflected light fallen on fixed mirror is reflected back on beam splitter which again split into two equal half and one half of it or 25% of original radiation is transmitted to sample or detector. 50% of initial radiation which is transmitted by beam splitter and falls on moving mirror is also reflected back beam splitter. At this point, radiation will again split into two; one half or 25% of original radiation is transmitted to sample or detector. In nut shell, it can be said that 50% of original radiation will fall on sample and of which 50% radiation (or 25% of original radiation) has moved between beam splitter and fixed mirror and remaining 50% (or 25% of original radiation) has moved between beam splitter and moving mirror.

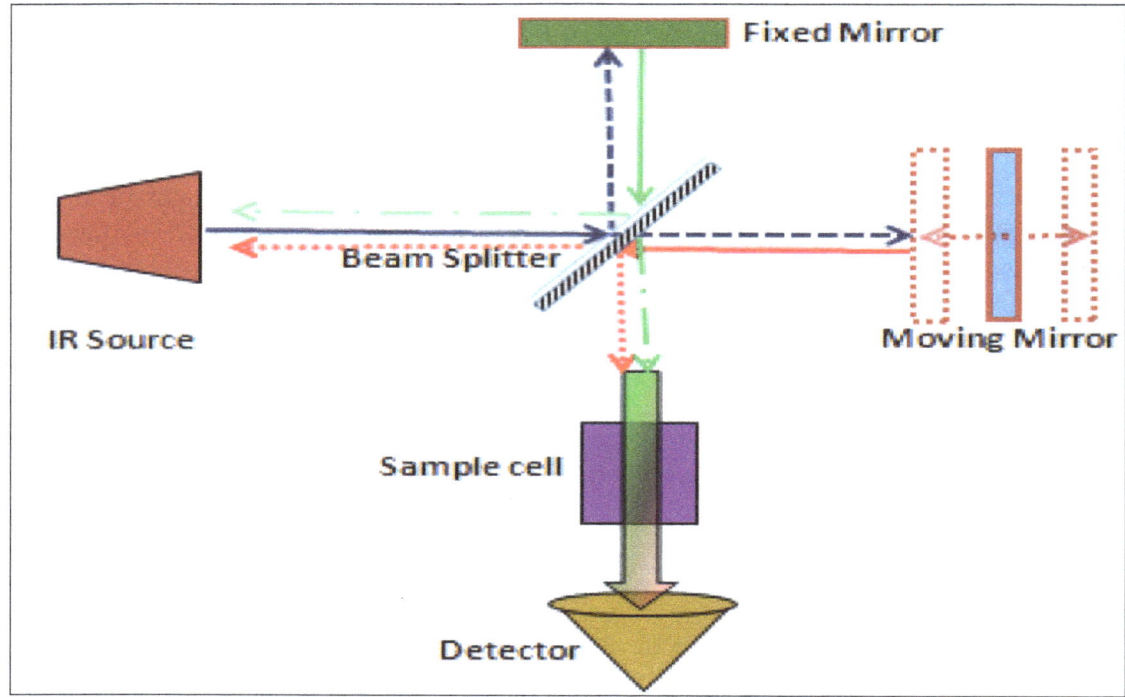

Let us assume, the distance between beam splitter and fixed mirror is L_1 and the distance between beam splitter and moving mirror is L_2:

- Case I: If L_1 is equal to L_2; then path length covered by radiation from beam splitter to fixed mirror and back to beam splitter is same as path length covered by radiation from beam splitter to moving mirror and back to beam splitter. Thus, radiation fallen on sample from both routes has covered equal distance prior to fallen on beam splitter and therefore both will be in same phase while leaving beam splitter. This is called constructive interference. This will enhance intensity of radiation fallen on sample or detector.

- Case II: If wave length of radiation is λ and moving mirror is displaced by $\lambda/4$, then path length difference will be $\lambda/2$ and two beams will be out of phase and interference is called destructive interference.

Inference

1. If x is difference in optical path length of two beams and is equal to zero or 1λ or 2λ or 3λ or 4λ or 5λ and so on; two beams will interfere constructively. We can say for constructive interference $x = n\lambda$ where n is an integer.

2. If x is equal to zero or $\frac{1}{2}\lambda$ or $\frac{3}{1}\lambda$ or $\frac{5}{2}\lambda$ or $\frac{7}{2}\lambda$ and so on; two beams will interfere destructively. We can say for destructive interference, $x = (n + \frac{1}{2})\lambda$ where n is an integer.

3. For monochromatic radiation, constructive interference will repeat when moving mirror is additionally displaced by distance equivalent to ½ λ or $\frac{3}{2}\lambda$ or $\frac{5}{2}\lambda$ or $\frac{7}{2}\lambda$.

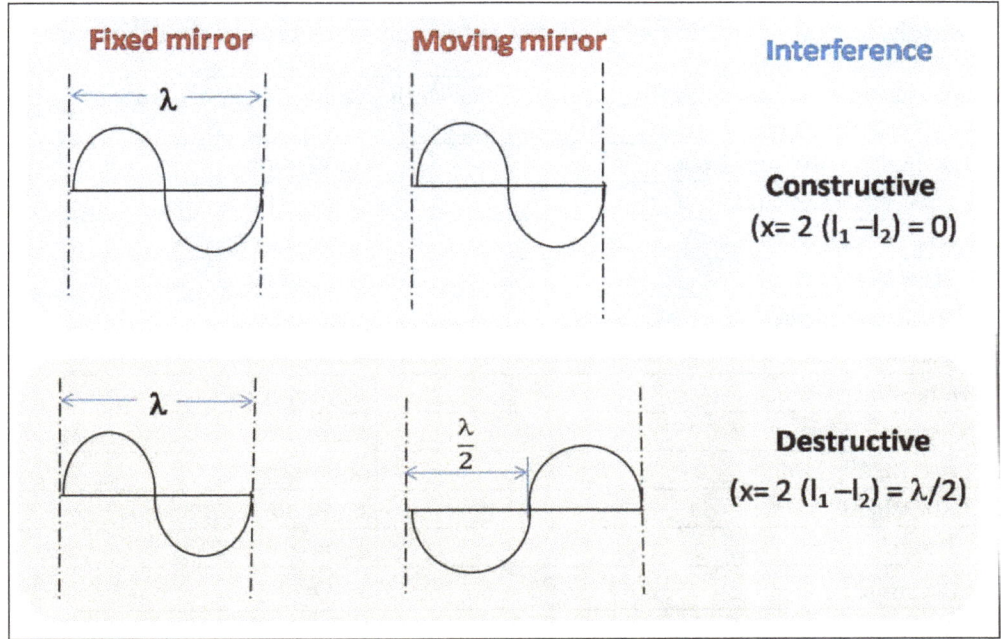

4. Intensity of light observed by the detector for monochromatic light is given by:

$$I(x) = 4RTS(\lambda).\frac{1}{2}Cos2\pi\sigma x = B(\sigma)Cos\,2\pi\sigma x$$

Where,

- 4RT is beam splitter efficiency, R is energy reflected by the beam splitter, T is energy transmitted by the beam splitter, $S(\lambda)$ is radiation energy from the light source, σ (cm⁻¹) is wave number. Wave number is number of complete wave in unit distance. It is reciprocal of wave length.

$$B(\sigma) = 4RTS(\lambda).\frac{1}{2}$$

5. If polychromatic light is emitted instead of monochromatic light, intensity of light observed by detector can be given by integration of equation 1 with respect to wave number ranging from 0 to ∞.

$$I(x) = \int_{0}^{\infty} B(\sigma)Cos\,2\pi\sigma\delta\,\sigma.$$

Here, (x) is Fourier cosine transform of spectrum B (σ). Inverse of Fourier cosine transform of I(x) will recover original spectrum B (σ).

$$B(\sigma) = \int_{-\infty}^{+\infty} I(x)Cos\,2\pi\sigma x\delta\,x.$$

FTIR detector measures I(x) i.e. interferogram which is Fourier transformed to obtain the spectrum.

FTIR: Applications

FTIR analysis is very fast and sensitive and is particularly beneficial in cutting-edge research applications. From FTIR spectra, the composition of any solid, liquid or gas can be predicted. The technique is widely used for identification of groups which gives bands at particular frequency.

Functional Group, type of vibration and molecule in which such group present	Characteristic Absorptions (cm^{-1})
O-H, H-bonded, stretch, Alcohol	3200-3600
O-H , free, stretch, Alcohol	3500-3700
C-O, stretch , Alcohol	1050-1150
C-H, stretch, Alkane	2850-3000
-C-H ,bending, Alkane	1350-1480
=C-H, stretch, Alkene	3010-3100
=C-H, bending, Alkene	675-1000
C = C, stretch, Alkene	1620-1680
C-F, stretch, Alkyl Halide	1000-1400
C-Cl, stretch, Alkyl Halide	600-800
C-H , stretch, Alkyne	3300
-C ≡ C- , stretch, Alkyn	2100-2260
N-H, stretch, Amine	3300-3500
N-H, bending, Amine	1600
C-N, stretch, Amine	1080-1360
C-H , stretch, Aromatic	3000-3100
C=C , stretch, Aromatic	1400-1600
C=O , stretch, Carbonyl	1670-1820
CN , stretch, Nitrile	2210-2260
N-O, stretch, Nitro	1345-1385 and 1515-1560
C=O, stretch, Acid	1700-1725
O-H , stretch, Acid	2500-3300
C-O , stretch, Acid	1210-1320
C=O, stretch, Aldehyde	1740-1720
=C-H, stretch, Aldehyde	2720-2750 and 2820-2850 &
C=O, stretch, Amide	1640-1690
N-H , stretch, Amide	3100-3500
N-H , bending, Amide	1550-1640
C=O, stretch, Anhydride	1740-1775 and 1800-1830 &
C=O, stretch, Ester	1735-1750
C-O, stretch, Ester	1000-1300
Acyclic, stretch, Ketone	1705-1725
α, β-unsaturated, stretch, Ketone	1665-1685

The total scope of FTIR applications is extensive. Some of the more common applications are:

- Quality verification of incoming/outgoing materials.

- Deformulation of polymers, rubbers, and other materials through thermogravimetric infra-red (TGA-IR) or gas chromatography infra-red (GC-IR) analysis.

- Microanalysis of small sections of materials to identify contaminants.

- Analysis of thin films and coatings.

- Monitoring of automotive or smokestack emissions.

- Failure analysis.

As an example, numerous components on automobiles are ideal for FTIR analysis: epoxies, oil coatings on parts, fuel, rubber seals and o-rings, tires, paints, fabrics (flame retardants) and exhaust emissions, to name a few.

Environment

Infrared spectroscopy is a valuable technique for monitoring air quality, testing water quality, and analyzing soil to address environmental and health concerns caused by increasing pollution levels. The technique offers a "green" method of testing and fast, accurate results with the added benefit of saving money on the cost of consumables.

Food

Food manufacturers can use the infrared attenuated total reflectance (ATR) technique for rapid determination of the trans-fat content of manufactured food products. This analysis is instrumental for compliance with food labeling requirements and to help promote healthy eating habits.

Companies in the food and feed industries are under increasing pressure to produce products that meet customer specifications while increasing plant production and profitability. Near-Infrared spectroscopy is a solution that helps companies optimize their production process and guarantee products are meeting specifications. FT-Near-Infrared (FT-NIR) is a convenient and easy-to-use technology that makes precise results accessible to even the most inexperienced user.

Forensics

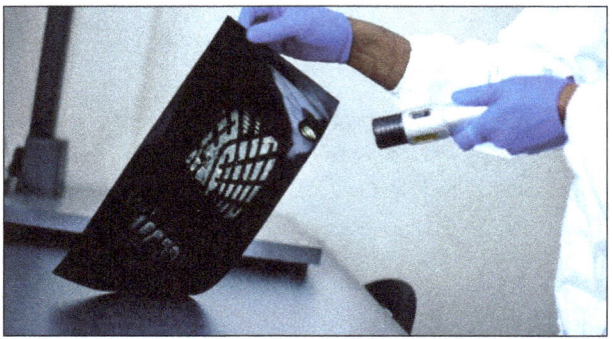

International drug enforcement agencies, police departments, and customs laboratories rely on spectroscopy to quickly identify illegal drugs, crime scene evidence, banned materials, and counterfeit goods. FTIR, FT-Raman, GC-IR, and IR microscopy techniques build a complete understanding of evidence samples and allow forensic scientists to confidently give expert testimony in court. These techniques can provide fast, easy and consistent analysis for:

- Seized drugs: controlled substances and cutting agents.

- Clandestine labs: chemical evaluation.

- Hit and run: paint and materials.

- Textile identification: fibers, coatings, and residues.

Pharmaceuticals

Pharmaceutical laboratories face strong regulatory requirements and market pressures at every step along the product development pipeline. FTIR is an excellent technique for pharmaceutical analysis because it is easy to use, sensitive, fast, and helps ensure regulatory compliance through validation protocols. Applications include:

- Basic drug research and structural elucidation.

- Formulation development and validation.

- Quality control processes for incoming and outgoing materials.

- Packaging testing.

Polymers and Plastics

FTIR spectroscopy is used to quickly and definitively identify compounds such as compounded plastics, blends, fillers, paints, rubbers, coatings, resins, and adhesives. It can be applied across all phases of the product lifecycle including design, manufacture, and failure analysis. This makes it a useful tool for scientists and engineers involved in product development, quality control, and problem solving. Key areas where infrared analysis adds value include:

- Material identification and verification.

- Copolymer and blend assessment.

- Additive identification and quantification.

- Contaminant identification—bulk and surface.

- Molecular degradation assessment.

Quality Control

Infrared spectroscopy is an ideal analytical tool for both routine quality control (QC) analysis to verify if materials meet specification, and analytical investigations to identify the causes of failures when they occur. The utility of infrared for these purposes arises from the simplicity of sample analysis and data acquisition, coupled with the information-rich spectra that it provides. Thanks to its compact design and ruggedness, FTIR instrumentation can be located in the analytical laboratory or near the production line. With its low cost, speed, and ease of analysis, FTIR is a method of choice for many industrial applications.

General

While FTIR is frequently used for polymer testing and pharmaceutical and forensic analysis, the application of the technique is virtually limitless, offering both qualitative and quantitative analysis of a wide range of organic and inorganic samples. Whether you are a new user or an experienced spectroscopist, you can obtain high-quality spectral data to accelerate your research, routine QA/QC testing, or investigative needs.

Raman Spectroscopy

There are numerous forms of light-matter interaction: fluorescence and phosphorescence are examples of absorption and subsequent emission of light by matter. Elastic scattering of light, such as Rayleigh scattering by atoms, molecules or phonons, and Mie/Tyndall scattering by dust particles are examples where the wavelength of the light is unchanged. Inelastic scattering such as Brillouin scattering by acoustic waves in crystals, Compton scattering by charged particles and Raman scattering by molecules or phonons are examples where the wavelength of the light does change. Raman scattering of light by molecules was first predicted using classical quantum theory by Smekal in 1923 and experimentally observed by Raman and Krishnan in 1928.

There are now more than 25 different types of known Raman spectroscopy techniques, such as spontaneous Raman, hyper-Raman scattering, Fourier transform Raman scattering, Raman-induced Kerr effect spectroscopy and stimulated/coherent Raman scattering.

Fifty years after its first observation, Raman spectroscopy started to become a prominent analysis technique among other optical metrology techniques, such as those involving absorption of infrared light; particularly when water and other useful polar solvents were present, because these media typically strongly absorb light in the infrared region. For example, in 1974, Fleischmann et al. used Raman spectroscopy to distinguish two types of adsorbed pyridine (a basic cyclic heterodyne compound molecule) on the surface of a silver electrode to mitigate absorption effects. This experiment was incidentally the first serendipitous observation of SERS.

Raman spectroscopy is now an eminent technique for the characterisation of 2D materials (e.g. graphene and transition metal dichalcogenides) and phonon modes in crystals. Properties such as number of monolayers, inter-layer breathing and shear modes, in-plane anisotropy, doping, disorder, thermal conductivity, strain and phonon modes can be extracted using Raman spectroscopy.

The biological and medical fields of research are greatly impacted by the development of Raman spectroscopy as it is a label-free (does not require fluorescent marker molecules) chemically selective hyperspectral imaging technique. For instance, studying the transdermal delivery of drugs into skin often ordains ex vivo and invasive analysis techniques. Ex vivo transdermal delivery studies are unfavourable because skin regeneration stops, the immune response ceases, and metabolic activity is usually lost. Hence, the performance of transdermal drug delivery ex vivo is not an accurate reflection of the in vivo situation. However, non-invasive in vivo measurements can be performed using Raman spectroscopy to gain detailed information about the molecular composition and concentration gradients in the skin. In many biological processes, living microorganisms such as bacteria act as biocatalysts. Raman spectroscopy can probe inhomogeneity in the properties and physiological status of individual cells in biocatalytic processes. Raman spectroscopy has also been used to identify and differentiate benign and malignant breast cancer lesions by probing their unique chemical compositions.

For biological samples, approximately 90% of the peaks are found in the 'fingerprint' spectral region, covering ($\Delta\tilde{\nu}$ ~ 500 cm^{-1} to ~ 1800 cm^{-1}; $\Delta\tilde{\nu}$ is the wavenumber shift), with the remaining found in the higher energy CH/OH stretching vibrational modes covering ($\Delta\tilde{\nu}$ ~ 2700 cm^{-1} to ~ 3300 cm^{-1}).

Coherent Raman Spectroscopy

Coherent light-scattering events involving multiple incident photons simultaneously interacting with the scattering material was not observed until laser sources became available in the 1960s, despite predictions being made as early as the 1930s. The first laser-based Raman scattering experiment was demonstrated in 1961.

SRS is a coherent process providing much stronger signals relative to spontaneous Raman spectroscopy as well as the ability to time-resolve the vibrational motions. SRS is relevant to numerous areas of research such as plasma physics, atomic interferometry, supercontinuum generation, imaging of biomolecules in food products, imaging chemistry inside living cells, bulk media and nanoscale specimens. The exchange of photon orbital angular momentum by SRS in plasma is gaining interest, particularly in the context of inertial fusion research. Supercontinuum generation is a complex nonlinear phenomenon that is characterized by the dramatic spectral broadening of intense light pulses passing through a nonlinear material. Knight et al. demonstrated flat ultrabroadband octave-spanning white-light supercontinuum generation by SRS and parametric four-wave mixing with 60-ps pump pulses of sub-kilowatt peak power in a photonic crystal fibre. Kasevich and Chu demonstrated a matter-wave interferometer with laser-cooled sodium atoms using the mechanical effects of stimulated Raman transitions. SRS has even been used to observe time-resolved vibrational spectra of the primary isomerisation of retinal in the visual pigment rhodopsin.

Since its resurgence in 1999, CARS has become a prominent vibrational mode imaging tool in biological medicine. As anti-Stokes photons are blue shifted from the pump and Stokes frequencies, they are more easily detected in the presence of single-photon fluorescence. CARS microscopy has been successfully applied to live-cell imaging, skeletal stem cells, tracing toxic nanomaterials in biological tissues, volumetric imaging of human somatic cell division, flow cytometry, detection of brain tumours and tracking organelle transport in living cells. Zirak et al. has developed a CARS endoscope for in vivo imaging and demonstrated the instrument with murine adipose tissue and human nervus suralis samples. Evans et al. have combined CARS with video rate microscopy to chemically image tissue in vivo. Potma and Xie have directly visualised lipid phase segregation in single lipid bilayers with CARS. CARS can even be used as a high temporal and spatial resolution thermography technique and has found applications in electronic and opto-electronic device characterisation and even turbomachinery.

Orientational order is a salient feature of many soft matter systems. Detail in structural molecular organisation is a prevailing goal in the field of biology, biomedicine, material sciences and molecular physics. Polarisation-resolved optical microscopy is becoming a powerful tool to address molecular orientational distributions into the focal volume of a microscope. In coherent nonlinear optics, polarised second harmonic generation, polarised third harmonic generation and polarised four-wave mixing have already been used to recover orientational information on endogeneous proteins and lipids in biological tissues. In addition to the orientational information, coherent Raman scattering (CRS) processes are sensitive to molecular bond vibrations, allowing chemical specificity without the need for fluorescence labelling/dyes. CARS microscopy can be used to image chemical and orientational order of liquid crystalline (commonly used in display technology) samples. Polarisation-resolved hyperspectral SRS microscopy has also been demonstrated as a label-free biomolecular imaging technique with teeth. In addition, polarised-CARS have been used to study the molecular order of lipids in myelin at sub-diffraction scales in mice.

Enhanced Raman Spectroscopy

The sensitivity of Raman spectroscopy can be enhanced through various techniques such as resonance Raman spectroscopy, TERS or SERS. SERS is particularly interesting since it allows an enhancement of several orders of magnitude of the Raman signal by modifying the surface upon which an analyte material is to be placed. The enhanced light-matter interaction in TERS and SERS is tuneable (to some extent) by modifying the surface nanostructure of metallic films on dielectric surfaces. The wavelength of charge density oscillations, known as plasmons, is dependent on these surface nanostructures and can enhance the light-matter interaction locally.

Fleischmann et al. first observed SERS in 1974 when investigating pyridine on the rough surface of a silver electrode. The sensitivity of SERS makes it well-suited to study electron transfer reactions, which lie at the heart of numerous fundamental processes: electro-catalysis, solar energy conversion, energy storage in batteries, and biological events such as photosynthesis. SERS has also been identified as a valuable technique for the detection of explosives/chemical weapons, unmodified DNA, aerosol pollutants and pathogens.

TERS is a technique that provides spectral information with a spatial resolution on the nanometre scale. Since the first reports of TERS emerged in 2000, TERS has become a powerful technique for studying thin crystalline materials, carbon nanotubes, single strands of RNA/DNA, redox reactions, mapping of individual molecules, semi-conductor nanostructures and microcavities.

Fundamental Principles

When light interacts with matter, the oscillatory electro-magnetic (EM) field of the light perturbs the charge distribution in the matter which can lead to the exchange of energy and momentum leaving the matter in a modified state. Examples include electronic excitations and molecular vibrations or rotational-vibrations (ro-vibrations) in liquids and gases, electronic excitations and optical phonons in solids, and electron-plasma oscillations in plasmas.

Spontaneous Raman

When an incident photon interacts with a crystal lattice or molecule, it can be scattered either elastically or inelastically. Predominantly, light is elastically scattered (i.e. the energy of the scattered photon is equal to that of the incident photon). This type of scattering is often referred to as Rayleigh scattering. The inelastic scattering of light by matter (i.e. the energy of the scattered photon is not equal to that of the incident photon) is known as the Raman Effect. This inelastic process leaves the molecule in a modified (ro-)vibrational state. In the case of a crystal lattice, the energy transfer creates a quantum of vibration in the lattice known as a phonon (a quasi-particle). Raman scattering in crystals can also lead to paramagnetic ions, surface plasmons and spin waves. The shift in angular frequency of the scattered light can be described by the following equation:

$$\omega_{scat} = \omega_p \pm \omega_{osc},$$

Where, subscripts osc denotes the lattice or molecule vibration, p denotes the incident photon (often referred to as the pump photon) and scat denotes the scattered light. The binary operator ($\pm$) is determined by energy conservation. When the energy of the scattered photon is lower than that of the incident photon (i.e. red shifted), the process is referred to as Stokes Raman scattering. Conversely, when the energy of the scattered photon is higher than that of the incident photon (i.e. blue shifted), the process is referred to as anti-Stokes Raman scattering. The Raman process must also conserve momentum, which is expressed in wave vector form as:

$$\vec{k}_{scat} = \vec{k}_p \pm \vec{q},$$

Where, $\vec{k}_{scat,}$ $\vec{k}_p$ and $\vec{q}$ are the wave vectors of the scattered light, the incident light and the phonon or molecular (ro-)vibration, respectively.

In molecules and crystals, the charge distribution has an equilibrium state to which it tends. An externally applied field can modify or perturb the charge distribution but only in accordance with the molecule or crystal's ability to form dipoles which may be anisotropic. This anisotropic property of molecules and crystals is called the polarisability and dielectric susceptibility, respectively. The classical approach theorises that the existence of the Raman Effect is associated with the modulation of the polarisability (for molecular (ro-)vibrations) or dielectric susceptibility (for crystal lattice vibrations) due to the oscillatory nature of their interatomic displacement. For crystal lattice vibrations, consider the polarisation vector of the material $\bar{P}$. If the suffixes j and k represent the vector components in the x, y and z directions, the jth component of $\bar{P}$ (to first-order) is related to the oscillatory electric field vector $\bar{E}$ associated with the light by:

$$P_j^{(1)} = \varepsilon_o \chi_{jk}^{(1)} E_k,$$

Where, ε_o is the permittivity of free space, χ_{jk} is the dielectric susceptibility of the material (a rank two tensor) and the convention of summation over repeated indices is implied; the superscript (1) signifies that this is the first-order contribution to the polarisation. The polarisability tensor is a function of the nuclear coordinates which, by extension, means that it will also depend on the (ro-) vibrational frequency. Assuming the modulation is small, the dependence can be expressed in a Taylor series with respect to the coordinates of vibration as follows:

$$\chi_{jk}^{(1)}\left(\bar{k}_p,\omega_p\right) \approx \chi_{jk}^{(1)}\left(\bar{k}_p,\omega_p\right)_{\bar{u}=0} + u_1\left(\frac{\partial\chi_{jk}^{(1)}\left(\bar{k}_p,\omega_p\right)}{\partial u_1}\right)_{\bar{u}=0} + u_1 u_m\left(\frac{\partial^2\chi_{jk}^{(1)}\left(\bar{k}_p,\omega_p\right)}{\partial u_1 \partial u_m}\right)_{\bar{u}=0} + ...,$$

Where, $\bar{u}$ is the nuclear displacement vector, the indices j, k, l and m indicate different spatial coordinates with repeated indices in any of the terms implying the summation of the constituents of that index. If we write the electric field associated with the light as follows:

$$\bar{E}\left(\bar{r},t\right) = \bar{E}\left(\bar{k}_p,\omega_p\right)\cos\left(\bar{k}_p \bullet \bar{r} - \omega_p t\right),$$

And the nuclear displacement as follows:

$$\bar{u}\left(\bar{r}, t\right) = \bar{u}\left(\bar{q},\omega_{osc}\right)\cos\left(\bar{q} \bullet \bar{r} - \omega_{osc} t\right),$$

An explicit expression for time dependence of $P_j^{(1)}$ can be found by substitution of these two mathematical equations of the monochromatic light and displacement. The numerous resulting terms pertain to optical processes such as Rayleigh scattering, optical absorption and Raman scattering. The term which pertains to the first-order Raman scattering is derived from the second term on the right-hand side of Eq.

$$\chi_{jk}^{(1)}\left(\bar{k}_p,\omega_p\right) \approx \chi_{jk}^{(1)}\left(\bar{k}_p,\omega_p\right)_{\bar{u}=0} + u_1\left(\frac{\partial\chi_{jk}^{(1)}\left(\bar{k}_p,\omega_p\right)}{\partial u_1}\right)_{\bar{u}=0} + u_1 u_m\left(\frac{\partial^2\chi_{jk}^{(1)}\left(\bar{k}_p,\omega_p\right)}{\partial u_1 \partial u_m}\right)_{\bar{u}=0} + ...,$$

and yields:

$$P_j\left(\bar{r},t,\bar{u}\right) = \frac{1}{2}\varepsilon_o\left(\frac{\partial\chi_{jk}^{(1)}\left(\bar{k}_p,\omega_p\right)}{\partial u_l}\right)_{\bar{u}=0} u_l\left(\bar{q},\omega_{osc}\right)E_k\left(\bar{k}_p,\omega_p\right)\times$$

$$\left\{\cos\left[\left(\bar{k}_p + \bar{q}\right)\bullet\bar{r} - \left(\omega_p + \omega_{osc}\right)t\right]\bullet + \cos\left[\left(\bar{k}_p - \bar{q}\right)\bullet\bar{r} - \left(\omega_p - \omega_{osc}\right)t\right]\right\}$$

This term contains sum (anti-Stokes) and difference (Stokes) frequencies and demonstrates conservation of momentum as per Eqs. $\omega_{scat} = \omega_p \pm \omega_{osc}$ and $\bar{k}_{scat} = \bar{k}_p \pm \bar{q}$.

The quantum mechanical description of the Raman process states that the (ro-)vibrational energy of the molecules/phonons are discrete quanta. Figure shows an energy level diagram illustrating the Raman processes with Stokes emission at ω_S and anti-Stokes emission at ω_{AS}.

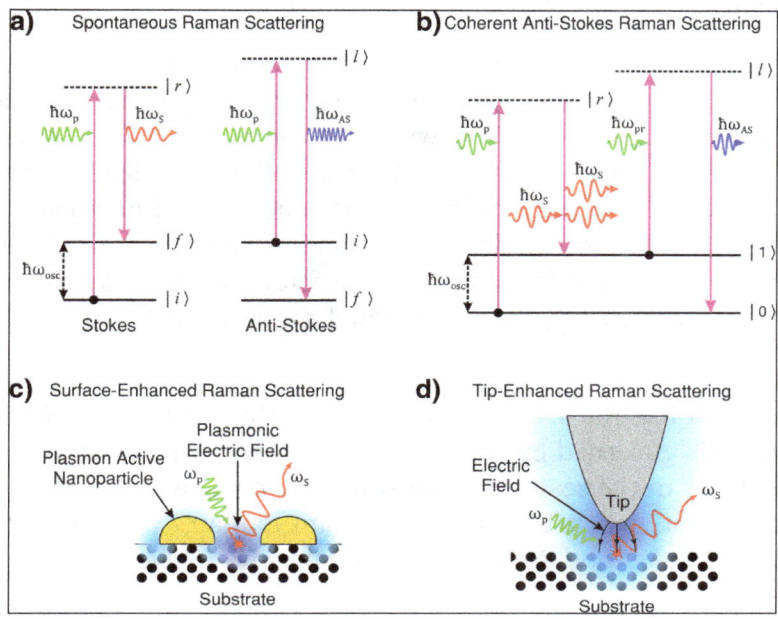

Figure shows a) Energy transfer process in Stokes (left) and anti-Stokes (right) Raman scattering, in both scattering processes, the lifetime of the excited state is probabilistic and spontaneous. In Stokes Raman scattering, the initial (ro-)vibrational energy $|i\rangle$ of the scattering material is less than that of the final state $|f\rangle$, the scattered light has less energy than the pump light. In anti-Stokes scattering, the initial (ro-)vibrational energy $|i\rangle$ of the scattering medium is greater than that of the final state $|f\rangle$, the scattered light has more energy than the pump light. b) Coherent anti-Stokes Raman scattering (CARS). CARS is a four-wave mixing process of pump, Stokes, probe and anti-Stokes light in which the emission of anti-Stokes light is coherently induced through an intermediate (ro-)vibrational energy state population inversion. c) Surface-enhanced Raman scattering (SERS). The incident pump light induces a surface plasmon resonance. The resultant enhancement of the oscillatory electro-magnetic (EM) field strength (shown in blue) on the surface intensifies the light-matter interaction and consequently increases the intensity of the Raman scattered light. d) Tip-enhanced Raman scattering (TERS). The incident pump light induces a tip-surface plasmon resonance associated with the plasmonically active tip. The resultant enhancement of the oscillatory EM field strength (shown in blue) is localised to the vicinity of the tip apex. The lighting rod effect (illustrated by curved black arrows) intensifies the light-matter interaction in the tip region and provides high-resolution (beyond the diffraction limit of light).

In Raman scattering, the intermediate states of the perturbation imposed by the incident pump photon ($|r\rangle$ and $|l\rangle$ in Figure) generally do not correspond to electronic states of the system and are said to be virtual energy states. These virtual intermediate states do not represent a well-defined energy state of the system. As the frequency of the pump photon approaches the energy of the electronic states, the strength of the Raman Effect increases due to resonance effects and is termed pre-resonance Raman. If the intermediate state corresponds to a discrete electronic energy state, the interaction is described as resonance Raman scattering and the signal strength is expected to exceed that of virtual-intermediate-state Raman scattering by orders of magnitude. If the energy of the incident light is in the range of dissociative energy levels, the process is described as continuum resonance Raman scattering.

Raman scattering transitions between certain quantum states are forbidden. In materials with inversion symmetry (i.e. centrosymmetric crystal structure), the initial and final states must have the same parity and are mutually exclusive with absorptive transitions (optically active transitions). In other words, transitions can be either Raman active or optically active. For linear molecules, the symmetric stretching modes of vibration or bending are Raman active and are optically inactive; those with anti-symmetric modes are Raman inactive and optically active (i.e. mutually exclusive). This rule is general and for nonlinear molecules, mutual exclusion is relaxed. In materials without inversion symmetry, (ro-)vibrational mode transition can be both Raman and optically active.

The Stokes Raman signal for molecules is more intense than the anti-Stokes signal as the population of energy states is governed by thermal statistics. For bosonic systems, such as phonons in crystals, the probability of the scattering target occupying a given vibrational quantum energy state obeys Bose-Einstein statistics. Under non-resonant Raman scattering and thermal equilibrium, the ratio of the anti-Stokes and Stokes scattered intensity is given by:

$$\frac{I_{AS}}{I_S} = \left(\frac{\omega_p + \omega_{osc}}{\omega_p - \omega_{osc}} \right)^4 e^{\left(-\frac{\hbar \omega_{osc}}{kT} \right)}$$

Where, I_S and I_{AS} are the intensity of the Stokes and anti-Stokes light, respectively, $\hbar$ is Planck's constant divided by 2π, k is the Boltzmann constant and T is the temperature associated with the scattering species. This equation is sometimes used to measure the temperature via Raman spectroscopy. This relation becomes inaccurate for resonance Raman scattering because the Stokes and anti-Stokes processes occur at different pump photon frequencies.

In the case of spontaneous Raman scattering, the Raman Effect is very weak; typically, 1 in 10^8 of the incident radiation undergoes spontaneous Raman scattering. The transition from the virtual excited state to the final state can occur at any point in time and to any possible final state based on probability. Hence, spontaneous Raman scattering is an incoherent process. The output signal power is proportional to the input power, scattered in random directions and is dependent on the orientation of the polarisation. For example, in a system of gaseous molecules, the molecular orientation relative to the incident light is random and hence their polarisation wave vector will also be random. Furthermore, as the excited state has a finite lifetime, there is an associated uncertainty in the transition energy which leads to natural line broadening of the wavelength as per the Heisenberg uncertainty principle ($\Delta E \Delta t \geq \hbar/2$). The scattered light, in general, has polarisation properties that differ from that of the incident radiation. Furthermore, the intensity and polarisation are dependent on the direction from which the light is measured. The scattered spectrum exhibits peaks at all Raman active modes; the relative strength of the spectral peaks are determined by the scattering cross-section of each Raman mode. Photons can undergo successive Rayleigh scattering events before Raman scattering occurs as Raman scattering is far less probable than Rayleigh scattering.

Nonlinear Susceptibility

The polarisation described by $P_j^{(1)} = \varepsilon_0 \chi_{jk}^{(1)} E_k$, is in agreement with first-order (i.e. linear) optics and describes the single-photon scattering process (two-wave mixing process). In wave mixing processes with more than two waves, nonlinear optical polarisation must be considered due to the products of the mixed electric field components. Nonlinear optical polarisation can be described by the following:

$$P_j = \varepsilon_0 \left[\chi_{jk}^{(1)} E_k + \chi_{jkl}^{(2)} E_k E_l + \chi_{jklm}^{(3)} E_k E_l E_m + \ldots \right],$$

Where, $e\chi^{(2)}$ is the second-order susceptibility (rank three tensor), $\chi^{(3)}$ is the third-order susceptibility (rank four tensor) and the sum over repeated subscript indices is again implied. Each of the terms in Equation above can be written in shorthand by $\bar{P}^{(1)}$, $\bar{P}^{(2)}$, $\bar{P}^{(3)}$ etc. The physical processes that occur because of the second-order polarisation, $\bar{P}^{(2)}$, tend to be distinct from those arising from the third-order polarisation, $\bar{P}^{(3)}$. This polarisation can have electric dipole, quadrupolar, octupolar, (etc.) contributions. Under the electric dipole approximation, the second-order polarisation can only occur in crystals that are noncentrosymmetric (lack inversion symmetry). Hence, $\chi^{(2)}$ vanishes for media such as fluids (e.g. liquid/gas) and amorphous solids (e.g. glass). Third-order nonlinear optical interactions (i.e. those described by a $\chi^{(3)}$ susceptibility) can occur for both centrosymmetric and noncentrosymmetric systems. Electric quadrupolar, octupolar, (etc.) $\chi^{(2)}$ contributions do not disappear under inversion symmetry.

Stimulated Raman Scattering

While spontaneous Raman scattering is an incoherent process, SRS is a coherent four-wave non-linear optical mixing process. The modes of oscillation are in phase forming a coherent modulation of polarisation in the sample with susceptibility $\chi^{(3)}(\omega_S; \omega_p + \omega_S - \omega_p)$. The scattered light is also coherent. The SRS process is dependent on the spontaneous Raman cross-section, the spectral line width, the path length of the light-field-matter interaction, the input intensity and optical feedback (light generation) of Stokes frequency light.

When photons of frequency ω_p and ω_S simultaneously interact with a molecule or crystal lattice in the ground state, the system vibrates with an induced frequency: $\omega_{osc} = \omega_p - \omega_S$. Unlike spontaneous Raman scattering, the de-excitation (relaxation) time to and energy of the final state are determined by the stimulation effect. The interaction results in the transfer of energy from the pump photon to the molecule/lattice, and the molecule/crystal scatters a new photon with frequency and phase matching that of the incident light of frequency ω_S. Figure shows the process schematically.

It is common to employ an external radiation source tuned to the Stokes frequency in tandem with the pump laser beam to provoke this effect. This technique can lead to exponential gain in the Stokes signal, by transferring energy from the pump radiation, and rapid population of the final (ro-)vibrational state $|f\rangle$. However, if the intensity of the incident light of frequency ω_p is sufficient, the generation of Stokes frequency photons within the material can self-promote SRS without the need for an external ω_S source. The intensity threshold of incident light in organic liquids, such as ethanol, for this kind of self-generated SRS typically requires an incident peak intensity of pump light > 10^9 W/cm² for an optical path length of a few centimetres. However, the SRS threshold can be significantly reduced by extending the length of the pump and Stokes field interaction with an optical resonator, such as internal reflection in a droplet of liquid. The example shown in Figure is the SRS spectrum taken with droplets of ethanol directly compared to the spontaneous Raman spectrum of bulk ethanol. The droplets act to confine the light by internal reflection which feeds back the Stokes light as a self-SRS inducing optical resonator.

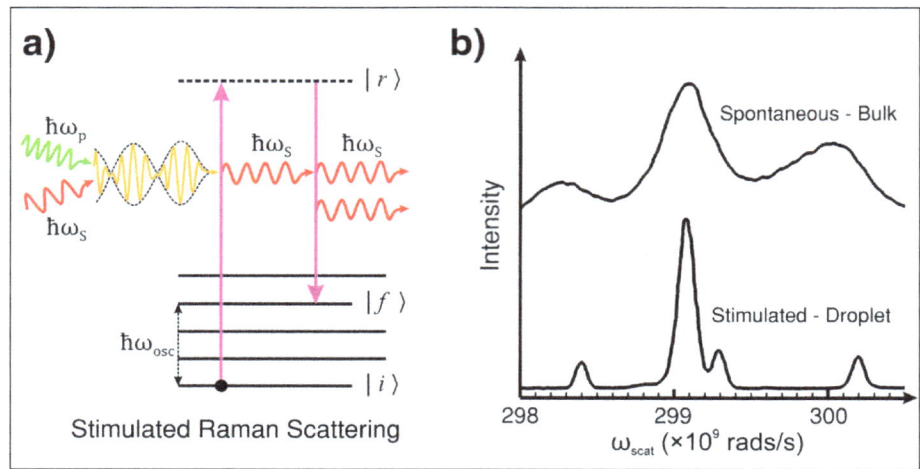

Figure shows a) Energy level diagram of stimulated Raman scattering (SRS). SRS is the induced emission of Stokes light by the coherent interaction of the pump and Stokes light with the material. Unlike spontaneous Raman scattering where the lifetime of the state $|r\rangle$ and the energy of the final state $|f\rangle$ are probabilistic, in SRS, the (ro-)vibration of the molecule or lattice is coherently driven by the difference frequency of the pump and Stokes light. b) Comparison of spontaneous Raman scattering and SRS of bulk and droplet ethanol. The spontaneous measurements were performed in a cuvette (bulk ethanol). The SRS measurements were performed in a droplet of ethanol which acted as an optical resonator for the Stokes light.

Coherent Anti-stokes Raman Scattering

CARS is a third-order nonlinear four-wave optical mixing process. Figure b shows the energy level diagram for the process. A pump beam and probe beam of frequency ω_p and ω_{pr} are mixed with a third beam of frequency ω_S (Stokes frequency) and incident on the sample. The frequency difference $(\omega_p - \omega_S)$ needs to match the frequency associated with the Raman active (ro-)vibrational mode $\omega_{osc} = \omega_p - \omega_S$. The frequency of the Stokes beam is usually adjusted/tuned to satisfy this criterion. Next, a probe photon of frequency ω_{pr} provides a perturbation for the anti-Stokes scattering process to occur at frequency $\omega_{AS} = \omega_p - \omega_S + \omega_{pr}$. A macroscopic third-order polarisation, $P^{(3)}$, is induced due to the coherent superposition of the microscopic dipole oscillations. Hence, CARS is governed by the third-order susceptibility of the form: $\chi^{(3)}(\omega_{AS}; \omega_p - \omega_S + \omega_{pr})$.

There are numerous treatments and approaches to formulating expressions for $\chi^{(3)}$. If one assumes that the excitation field is much weaker than the intramolecular forces, then a perturbative approach can be adopted. If this is not the case, non-perturbative treatments can be considered. By considering the density matrix equation of the system and expressing the external field interaction as a perturbation in the Hamiltonian, the semi-classical nonlinear optics theory generates an expression for $\chi^{(3)}$ with 48 terms, each of which contribute to the third-order susceptibility. A generalised expression for dominant terms in resonant CARS is given by the following:

$$\chi^{(3)} = \frac{A_R}{\omega_{osc} - (\omega_p - \omega_S) - i\Gamma_R} + \chi_{NR}^{(3)},$$

Where, Γ_R is the half width at half maximum for the Raman line; A_R is a constant representing the Raman scattering cross-section. The first term is the contribution due to CARS vibrational resonance as in Figure $(\omega_{osc} = \omega_p - \omega_S)$. The second term is the nonresonant background signal and is independent of the Raman shift $(\omega_{osc} \neq \omega_p - \omega_S)$. The nonresonant background occurs because not all quantum pathways of the scattering process involve a resonance with a (ro-)vibrational state. This nonresonant contribution interferes with the resonant part of the signal. The nonresonant background causes distinctive distortions of CARS spectra in comparison with spontaneous Raman spectra and has prevented CARS from becoming a widespread technique.

The incident light beams of differing frequency move in and out of phase with each other in both time and space. Hence, the CARS signal reaches its first maximum when the field-sample interaction length scale is less than the coherence length scale to yield constructive interference. For plane-wave pump and Stokes beams, the intensity of the anti-Stokes signal is as follows:

$$I_{AS} \alpha |\chi^{(3)}|^2 I_p I_{pr} I_S \left(\frac{\sin\left(\Delta\vec{k} \bullet \frac{\bar{z}}{2} \right)}{\frac{|\Delta\vec{k}|}{2}} \right)^2$$

Where, $\bar{z}$ is the sample thickness (vector normal to the lattice cell surface), $\vec{k}$ is the wave vector of light, $\Delta\vec{k} = \vec{k}_p - \vec{k}_S + \vec{k}_{pr} - \vec{k}_{AS}$ is the wave vector mismatch (the velocity difference between the four waves) and I_i is the intensity of the wave denoted by the subscript. Phase matching is achieved when $\Delta\vec{k} = 0$ and the intensity of the anti-Stokes signal is maximised because the energy and momentum transfer processes correspond to allowed transitions. As the magnitude of $\chi^{(3)}$ is linearly proportional to the number oscillators involved in the process, the intensity of the anti-Stokes signal is quadratically proportional to the number/concentration of oscillators.

Researchers typically employ the pump beam to provide the second virtual excitation (i.e. the probe light shown in Figure; i.e. $\omega_{pr} = \omega_p$ and $\omega_{AS} = 2\omega_p - \omega_S$). The intensity of the CARS signal is therefore quadratically proportional to the intensity of the pump beam. The CARS signal is monodirectional due to the phase-matching condition. However, high numerical aperture (NA) lenses or microscope objectives (confocal light) are commonly employed to satisfy the phase-matching condition without the need for complex mechanical systems to achieve phase-matched beam alignment.

Raman resonances typically have coherence times of ~1ps. Hence, the pump and Stokes beams are typically pulsed in picosecond timescales to obtain coherent excitation and to inhibit multiphoton effects. The CARS process takes place in the immediate vicinity of the beam's focal spot. The signal produced is typically 10^6 times that of spontaneous Raman scattering. CARS microscopy offers non-invasive characterisation and imaging of (ro-)vibrational spectra with high sensitivity and spectral resolution as well as three dimensional sectioning capabilities.

Surface Plasmons and Polaritons

Surface plasmons can occur at the interface between a dielectric and conducting material, such as a metal or degenerate semi-conductor. They are the light-induced coherent oscillations of surface conduction electrons about their equilibrium position. The nanoscale volume of opposing charge that remains acts as a restoring force on the electrons. The result can be described with a damped simple harmonic oscillator model, in which the oscillations of the free-charge carriers have an associated resonance. Surface plasmons can be excited by EM radiation and plasmonics is the study of these light-matter interactions.

Plasmonic nanoparticles that are much smaller than the wavelength of the incident light can support non-propagating surface plasmons that oscillate with a frequency known as the local surface plasmon resonance (LSPR). The wavelength of surface plasmons is much shorter than that of the associated propagating light for a given frequency. The LSPR wavelength is dependent on the nanoparticle's size, shape, material, external dielectric properties and inter-nanoparticle separation.

Surface plasmons that propagate are referred to as surface plasmon polaritons (SPPs). They are essentially light waves that are trapped at the interface due to their interaction with the free electrons of the conducting material. For a planar dielectric-conducting interface, polaritons propagate in 2-dimensional space along the surface interface for length scales of tens to hundreds of micrometres. They decay evanescently in the direction perpendicular to the surface interface with $1/e$ decay lengths of up to 200 nm. The field intensity in this evanescent decay region is amplified by orders of 10 to 100 relative to the incident radiation. Hence, light-matter interactions with adsorbed molecules on the surface are also enhanced.

In the case of LSPR, the surface roughness or surface nanoparticles cause local concentrations of charge carriers which further amplify the evanescent EM field due to the lightning rod effect. Even larger field-enhancements (up to 10^6) can be observed in gap plasmons (in the gap between two neighbouring plasmonic nanoparticles). This enhanced near-field effect gives rise to the technique known as SERS.

Surface-enhanced Raman Scattering

Raman is generally a very weak process; it is estimated that approximately one in every 10^8 photons undergo Raman scattering spontaneously. This inherent weakness poses a limitation on the intensity of the obtainable Raman signal. Various methods can be used to increase the Raman throughput of an experiment, such as increasing the incident laser power and using microscope objectives to tightly focus the laser beam into small areas. However, this can have negative consequences such as sample photobleaching. Placing the analyte on a rough metal surface can provide orders of magnitude enhancement of the measured Raman signal, i.e. SERS.

Two mechanisms have been proposed to explain the increase in Raman signal provided by SERS. The first is via EM enhancements where local surface plasmons concentrate the local electric field near the surface of the metal in 'hot spots' located on the sharp edges of nanostructures or in regions of concentrated free-charge carriers due to the lighting rod effect. Figure, illustrates the SERS process. This process can increase Raman generation by a factor of 10^8 to 10^{11}. The second method is chemical enhancement via charge transfer between the metal surface and the analyte, which enhances Raman scattering by a factor of approximately 10^2 to 10^3. However, the charge transfer mechanism only applies to specific molecules, whereas the EM mechanism is applicable for all analytes.

The ubiquity of EM enhancements has led to the development of numerous SERS substrates, which can be divided into two groups: metallic nanostructures fabricated on a solid substrate and colloidal suspensions of plasmonic nanoparticles. The most common materials used to fabricate SERS substrates are gold and silver because of their good plasmonic response. Gold also benefits from chemical stability as it is a noble metal. Other metals are also being investigated, such as aluminium for UV Raman spectroscopy.

Tip-enhanced Raman Scattering

The diffraction limit of light restricts the focus spot size in standard optical techniques (such as Raman spectroscopy) to be at least half of the wavelength of the light according to Abbe's criterion. Light from the sample is composed of both propagating and non-propagating radiation. The non-propagating evanescent waves remain in the vicinity of their sources and do not participate in image formation in the far field. Instead, they extend laterally on the sample among the plasmon-active sites. Hence the spatial resolution is restricted by the size of the focal spot of the light. Even with a focal spot size of a half-wavelength (~ 250 nm for visible light), any objects that are much smaller than the half-wavelength would appear as a defuse shape.

TERS is a relatively new optical nanoimaging technique that combined Raman spectroscopy with scattering (or apertureless) scanning near-field optical microscopy. TERS offers spatial resolution far beyond the diffraction limit of the probing light. In the context of the a priori description, this is achieved by forcing the near-field evanescent light into the far-field image formation. At the present date, the spatial resolution of TERS is typically reported to be 10–30 nm and is largely assumed to scale with the size of the tip's apex. Incremental improvements to this resolution have been reported. Enhancement factors for TERS are significantly weaker than SERS due to the relative size of the probed signal volume. The enhancement factor (relative to spontaneous Raman scattering) is typically reported to be 10^3 to 10^6. As with SERS, two field enhancement mechanisms are thought to contribute to the Raman signal: EM and chemical enhancement.

TERS is implemented by positioning a plasmon-active (plasmonic) nanotip approximately 50 nm above the sample's region of interest. The Raman probe light is focused onto the tip-surface cavity to induce LSPR within the tip' apex and (in some circumstances) the sample surface. The surface plasmons may then enhance evanescent or near-field light with the incident probe light and/or the Raman scattered light. Hence, the LSPRs both confine and enhance the light field in the vicinity of the tip's apex. The enhanced local EM field is most concentrated at the tip apex due to the lightning rod effect. This evanescent light at the tip apex can then excite or stimulate Raman, two-photon or second harmonic scattering from a nanoscale volume of the sample under the tip. A Raman image of the sample surface can be obtained by raster scanning the sample under the nanometric tip.

Experimental Considerations

Instrumentation

The nonresonant Raman Effect is a very weak process. Hence, monochromatic, narrow-beam and high-intensity lasers are preferable to produce quality Raman spectra. The exploitation of micro-electronics, such as stepper motor drives, photon counters, digital data acquisition and computational processing systems can further enhance the quality of spectra. As spontaneous Raman spectroscopy is naturally an incoherent process, continuous-wave laser sources are commonly used because pulsed lasers require higher peak powers for sufficient signal-to-noise ratio, which can photobleach/damage samples.

The choice of wavelength of the laser source depends on the required application. Lower visible wavelengths and UV cause strong photoluminescence in organic materials, which can mask the Raman peaks. Therefore, a longer visible or near-IR wavelength (500–830 nm) laser source is often suited for studying organic materials, because of the reduced photoluminescence. However, the Raman signal intensity is inversely proportional to the wavelength of the pump light. Hence, longer wavelengths of light require longer acquisition times.

Raman spectroscopy is most often performed using laser sources at $\lambda = 785$ nm. This wavelength source is often selected as it balances the competing factors between Raman signal intensity, fluorescence, detector sensitivity and cost, and cost-effective/compact high-quality laser sources. However, visible lasers in the blue and green (e.g. $\lambda = 532$ nm) are becoming more common in Raman spectroscopy.

Raman scattering is measured in terms of the wavelength shift from the source wavelength. Ideally the illumination source for Raman measurements should be purely monochromatic, in other words, a single wavelength. However, all laser sources possess a spectrum of wavelengths known as a linewidth. The linewidth of a laser is usually measured in Hertz and is typically > 1 MHz for solid-state lasers used in Raman applications. A narrow linewidth is preferable for Raman spectroscopy because the measured shift in the Raman scattering process is limited by the laser's linewidth.

Laser sources for Raman spectroscopy need to be stable in wavelength and power over extended periods of time and from use to use. Raman spectra are usually collected over long integration times and for many acquisitions. If the wavelength of the source drifts during a measurement, then the Raman peaks will drift as well, because Raman is measured as a shift relative to the pump light. Wavelength drift is also problematic from measurement to measurement as it causes peaks to shift, in turn making comparisons between measurements difficult. The output power stability of the source is important for similar reasons. If the laser power drifts from measurement to measurement, then quantitative comparisons cannot be made easily.

Spectral purity is another key criterion for Raman laser sources. The spectral purity of laser sources often requires side-mode suppression better than 60 dB. In many cases, side-mode suppression is sufficient if > 60 dB spectral purity is reached at ~1–2 nm from the laser wavelength peak. However longer wavelength (near-IR) Raman spectroscopy requires side-mode suppression ratios within a few hundreds of pm from the main peak.

Most modern Raman systems use solid-state laser sources rather than gas lasers because of their spectral quality and stability. There are three main categories of continuous-wave solid-state laser sources used in Raman spectroscopy: Diode-pumped single-longitudinal mode (SLM) lasers; single-mode diode lasers (distributed feedback (DFB) or distributed Bragg reflection (DBR)); and volume Bragg-grating (VBG) frequency-stabilised diode lasers. These laser sources have varying optical characteristics.

Diode-pumped SLM lasers are readily available in compact form from the UV to the near-IR. Power levels of several Watts are achievable at 1064 nm in the near-IR. In the visible range, numerous lines in the blue-green-red region (457 to 660 nm) are available with output powers of ~ 100 mW. In the UV spectral range, power outputs of 10–50 mW at 355 nm are available. Hermite-Gaussian laser beam modes are described by their transverse electro-magnetic mode (TEM): $TEM_{m,n}$, where m and n represent the Hermite-Gaussian mode index. Diode-pumped SLM lasers provide excellent TEM_{oo} mode beams, precise wavelengths with low drift, and a single-frequency linewidth > 1 MHz. The spectral purity of diode-pumped SLM lasers is typically > 60 dB in terms of their side-mode suppression ratio. Weak emissions that neighbour the laser's main peak several nanometres in spectral shift can occur in diode-pumped SLM lasers. However, these neighbouring lines can be mitigated with dielectric band-pass filters. The wavelength of diode-pumped SLM lasers is typically stable to within 4 pm over a temperature change of 30 °C.

Single-mode diode lasers are compact and cost-effective pump illumination sources with single-frequency linewidth (> 1 MHz), single-TEM beam quality and output powers of up to ~ 100 mW. Wavelengths of λ = 785, 830, 980 and 1064 nm are most common in Raman spectroscopy. The side-mode suppression ratio is typically limited by sideband emission to ~ 50 dB at ~ 100 pm from the main peak.

VBG frequency-stabilised diode lasers use a narrow-linewidth VBG element with a diode-laser emitter to achieve narrow-line emission. These lasers are often used for applications requiring narrow-line emission at wavelengths that are not available for DFB or DBR laser sources. Frequency-locking multi-TEM diode lasers can be used to increase the output power of the narrow-linewidth emission. The stability of the output wavelength and linewidth requires careful thermo-mechanical control and high-precision alignment inside VBG frequency-stabilised diode lasers. Linewidths can range from single-frequency emission to ~ 10s of pm, depending on the wavelength and the output power. The side-mode suppression ratio is limited to ~ 50 dB, ~ 250 pm from the main peak emission. However, this can be improved using filters.

In confocal Raman imaging applications, it is necessary to use diffraction-limited TEM_{oo} beams for optimum spatial resolution. However, this is relaxed for probe-based quantitative Raman analysis. In addition, confocal Raman setups require laser beam isolation as samples may generate optical feedback that is well aligned to the incident pump light. This counter-propagating feedback can induce power and noise instability and can even damage the laser source. Optical isolators are often integrated into the laser system itself because careful alignment must be achieved in the output after the isolator.

The spectrometer is a core component of any set-up used for measuring Raman spectra. The spectrometer should match the wavelength(s) of the laser source(s) used. The spectral range and resolution required will depend on the application. For example, the spectral range is determined by the position of the Raman peaks of interest (i.e. at large $\Delta \tilde{v}$ ~ 3000 cm–1 or low $\Delta \tilde{v}$ ~ 1 cm–1). If the application requires closely spaced Raman peaks to be resolved, then spectral resolution is key.

The spectral resolution of a spectrometer is largely determined by the slit width at the spectrometer entrance, the focal length of the spectrometer, the dispersion, the size of the grating (or prism) and the size and sensitivity/quality of the detector. There is a trade-off between the overall spectral range and resolution when considering the design of the experiment for a given application. In the case of weak Raman signals, optimising the signal-to-noise ratio is a priority.

Spectral filtering plays a vital role in the acquisition of Raman spectra. Firstly, the incident laser light must be spectrally pure, which is accomplished with a narrow-linewidth laser source. However, if the laser light is delivered to the sample by an optical fibre, then it is inevitable that Raman generation will occur in the fibre. Therefore, it is important to use a narrow band-pass filter to reject any Raman signal generated in delivering the laser to the sample. Narrow band-pass filters can provide transmission > 90 % at the laser wavelength while suppressing light to an optical density of OD > 5 at wavelengths differing by just 1% from the laser wavelength.

Importantly, light collected for detection requires filtering to block the laser wavelength. If the laser light is not filtered out, it can go on to generate Raman in the detection arm of the set-up and drown out the desired Raman signal when it reaches the spectrometer. The type of filter required depends on whether Stokes, anti-Stokes or both are to be measured. To only detect anti-Stokes Raman, a short-pass filter should be used as anti-Stokes Raman light has a higher energy and hence shorter wavelength than the laser source. To only detect Stokes Raman, a long-pass filter should be used as the Stokes Raman light has a lower energy and hence longer wavelength than the laser source. Long pass edge filters with edge-transition widths of < 3 nm and edge steepness < 40 cm^{-1} are available. To detect both Stokes and anti-Stokes Raman light, a notch filter centred on the laser wavelength should be used as it allows both shorter and longer wavelengths to be detected. Notch filters with OD > 6 at the laser line wavelength are available. Multi-notch filters are also available and can block multiple laser lines simultaneously. Holographic notch filters significantly outperform dielectric notch filters, providing excellent attenuation of the Rayleigh line while passing light as near as 50 cm^{-1} from the Rayleigh line. Acousto-optic modulators can also be used in conjunction with an excitation laser to select emissions with a desired wavelength (as a filter) or as a time-gated illumination system in tapping mode atomic force microscopy (AFM)-based TERS.

The quantum efficiency of standard room-temperature silicon-based CCD devices for Raman signal detection degenerates above $\lambda = 800$ nm. For longer wavelengths, indium gallium arsenide array devices can be used, but these are less sensitive with higher noise levels and cost.

The visible to near-infrared wavelength range ($\lambda = 500–830$ m) is particularly suitable for inorganic materials (e.g. graphene, carbon nanotubes (CNTs) and fullerenes) and SERS. UV lasers are attractive for organic materials (e.g. pathogens, proteins, DNA, and RNA). For materials with strong fluorescence that require near-IR illumination, it is common to use a 1064-nm wavelength.

Spontaneous and Coherent Raman Scattering Setups

Spontaneous Raman spectroscopy is most commonly used for modes with forbidden single-photon absorption or emission experiments. SRS is sometimes used for wavelength shifting of coherent light, light amplification, pulse compression, phase conjugation and beam combining. Unlike spontaneous Raman scattering, SRS is highly directional and offers enhanced signal strength and the ability to time-resolve the evolution and dephasing of coherent (ro-)vibrational motion.

Figure a) shows a typical Raman setup based on a confocal geometry used by Wiedemeier et al. Confocal setups of this type are commonly used and employ an infinity-corrected objective lens (large numerical aperture (NA) lens) to focus the pump light. Wiedemeier et al. used a diode-pumped solid-state laser as a monochromatic light source centred at 532 nm. Confocal mode is achieved by the use of a pinhole module in front of the spectrometer to spatially filter the light. The pinhole only passes light that originates from the focal plane to the detector. For detection of the Raman signal, a holographic-imaging spectrometer with an attached CCD camera is used. A holographic transmission grating with high light throughput served as a dispersive element, which enables large spectral ranges in a comparatively short time period to be acquired. Raster scanning of the sample in a confocal setup needs to be precise. Hence, a piezo actuated nano-positioner is used for positioning of the specimen.

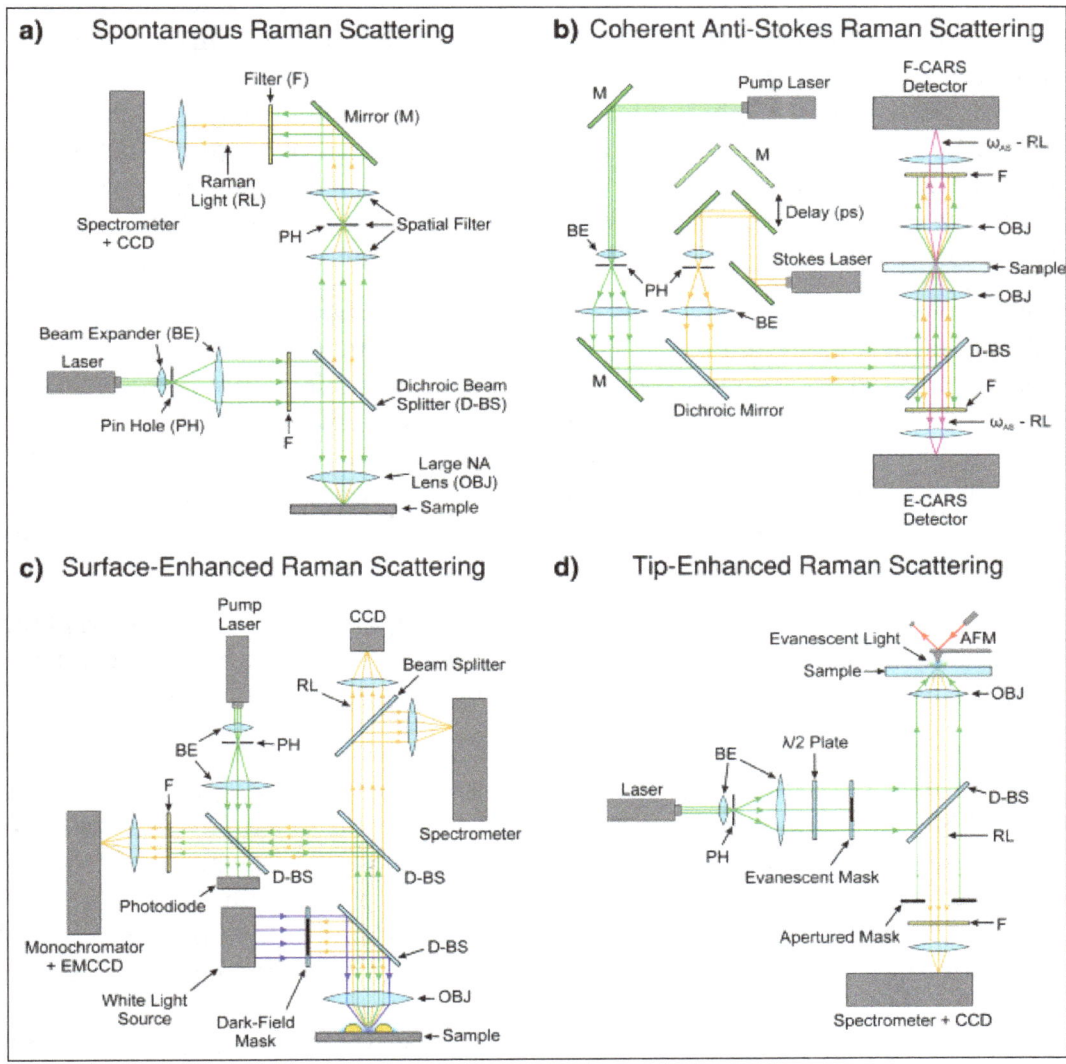

Figure shows a) Typical confocal Raman spectroscopy setup. The pump laser is spatially filtered through a pinhole. The back-scattered Raman light is spatially filtered and spectrally filtered through a notch filter. The Raman light is analysed by a spectrometer and a charge-coupled device (CCD). Hyperspectral images are obtained by raster scanning the sample. b) Typical CARS setup.

Two laser sources provide the pump and Stokes light and are synchronised through a picosecond path difference mirror setup. In this setup, the incident light is focused through an optically transmissive sample substrate. Both the forward scattered light (F-CARS) and epi-scattered light (E-CARS) are spectrally filtered by band-pass filters and are subsequently detected by two avalanche photodiodes. CARS images are obtained by raster scanning the sample. c) Typical SERS setup. The pump laser is coupled into a dark-field microscope in which the Raman light is edge-filtered and detected through a monochromator and EMCCD. The white-light source and dark-field mask provides the means for dark-field spectroscopy. The dark-field spectra of each plasmonically active nanoparticle are recorded through a secondary spectrometer (top right in c). An imaging CCD camera is used to automatically find and centre each nanoparticle. d) Typical TERS setup. The pump laser light is spatially filtered and passed through a half-wave plate. The evanescent mask ensures that only high numerical aperture (NA) pump light is incident on the sample such that total internal reflection occurs at the substrate-sample interface. This ensures that the tip apex is only illuminated by the evanescent light to achieve nanoconcentrated light in the vicinity of the tip. The reflected Raman light is filtered by an apertured mask (to remove any residual large NA pump light) and a notch filter. The Raman light is analysed by a spectrometer and a CCD. Hyperspectral images are obtained by raster scanning the sample. F, filter; M, mirror; RL, Raman light; CCD, charge-coupled device; PH, pinhole; BE, beam expander; D-BS, dichroic beam splitter; OBJ, Large numerical aperture (NA) lens; EMCCD, electron-multiplying charge-coupled device.

Spontaneous anti-Stokes scattering is weaker than Stokes Raman scattering due to the relatively low probability of thermal excitation. Hence, anti-Stokes Raman spectroscopy is typically used with stimulated or coherent spectroscopy. CARS spectroscopy offers a 10^5 increase in conversion efficiency, spectral and spatial discrimination against fluorescence and, most importantly, does not require a monochromator. Due to the required coherence of the process, high-peak power pulsed tuneable laser sources are employed. These peaks are readily available using picosecond or femtosecond light lasers, the choice of which is determined by the spectral resolution required and the timescale of interest.

Avoiding direct electronic excitations in the sample is an important consideration as photochemical damage (due to photobleaching) can occur in samples. Djaker et al., for example, use near-infrared laser sources to mitigate photobleaching in their samples of polystyrene beads.

Figure b) shows a typical CARS setup that measures both forward scattered light (F-CARS) and back- or epi-scattered light (E-CARS). The system has two synchronised picosecond pulse trains. The pump and Stokes beams are generated by two picosecond Ti:Sapphire lasers operating at 80 MHz and are tuneable from 700 to 1000 nm to cover the entire spectrum of molecular (ro-) vibrations in biological systems (up to $\Delta\tilde{v} \sim 3000\,cm^{-1}$). The ps pulse duration is adjustable by a Gires-Tournois interferometer. The Ti:Sapphire lasers are pumped by a frequency-doubled CW Nd:Vanadate laser that provides monochromatic light at 532 nm. The two pulse trains were polarised with pulse duration of 3 ps, corresponding to a spectral width of $1.76\,cm^{-1}$. The pump and Stokes beams are synchronously pulse picked through two Bragg cells to reduce the repetition rate of the pulse trains to several hundred kilohertz, thus avoiding photodamage of the sample while still maintaining high-peak power for CARS generation. The pump and Stokes beams are temporally synchronised by a SynchroLock system, which electronically adjusts the time delay between the two pulse trains. A small part of the output of the lasers are launched in optical fibres coupled to photodiodes and connected to a SynchroLock controller, which measures the lasers frequency or phase difference between the master and the slave; the timing jitter was reported to be ~250 fs. The spectral resolution was estimated to be 2.5 cm^{-1}, which is high enough to resolve Raman spectral features of biological samples. The use of a broadband Stokes wave enables the acquisition of a full CARS spectrum in only one measurement, with this configuration being known as multiplex or broadband CARS.

The two pulse trains are spatially filtered, collinearly combined and expanded through beam expanders. They are then sent into an inverted microscope and focused onto the sample by a water-immersion objective lens with a large NA. The E-CARS signal is collected by the same objective lens while the F-CARS signal is collected by a condenser lens with a lower NA. The E-CARS and F-CARS signals are filtered through a set of band-pass filters and detected by two avalanche photodiodes with a 200 μm × 200 μm active area. The CARS images are collected by raster scanning the sample, using an XYZ piezo flexure stage.

Several methods have been developed to suppress the nonresonant background associated with CARS. E-CARS is relatively insensitive to the nonresonant background of sample solvents. Polarisation-sensitive CARS can differentiate the resonant and nonresonant signals by their polarisation. However, these two techniques reduce the anti-Stokes signal strength. Time-resolved CARS, temporal or spectral interferometry CARS and frequency-modulated CARS can also suppress the nonresonant background. However, the setup in terms of both optics and electronics is challenging.

SERS Specific Considerations

A variety of nanostructures, such as bowtie antennas, nano-rings, nanovoids, nanoparticle aggregates, nanoflower, nanorod arrays and nanowells can be used for SERS. Each nanostructure can have a number of plasmonic resonances, and matching the excitation laser to these wavelengths can greatly enhance the SERS intensity. Matching the plasmonic resonance to the pump laser can be done either by tuning the laser wavelength or by tuning the LSPR of the nanostructures.

The difficulty faced in producing SERS substrates is consistency in fabrication and repeatability in measurements due to the inhomogeneity and randomness of SERS active hot spots. For SERS substrates produced by top-down methods, such as electron beam lithography, the main challenge is scaling the fabrication. Conventional top-down methods limit the active area of the SERS substrate and are not conducive to large-area manufacturing. Bottom-up fabrication methods have their own set of problems. Even though bottom-up approaches allow wafer scale fabrication, consistency across the wafer is usually lacking. This inconsistency hinders the repeatability of measurements, which is problematic for quantitative analysis. Colloidal SERS schemes suffer from complications introduced by stabilising agents at the surface of the nanoparticles, which help to keep nanoparticles in suspension. These stabilising agents can either impede or augment the measured Raman signal. The chemical synthesis for nanoparticle colloids also requires precise optimisation. The poor reproducibility of nanoparticle colloidal synthesis hampers batch-to-batch consistency.

Often, only very few sites exhibit the highest SERS enhancement and the variability in size and shape can alter the plasmonic properties from the desired LSPR. Figure c, shows a setup which combines SERS with dark-field spectroscopy. The dark-field spectrometer analyses the light scattered from the nanostructures (illuminated by the white-light source) to select nanostructures with the desired plasmonic properties.

TERS Specific Considerations

Scanning probe microscopy (SPM) techniques, such as atomic force microscopy (AFM), scanning tunnelling microscopy (STM) or shear force microscopy (SFM), are usually the tools of choice for TERS. TERS has the ability to simultaneously measure topography by the conventional SPM mode of the system and obtain corresponding spectral information from a sample with nanometric spatial resolution and high sensitivity. Certain SPM techniques ordain probe modifications for the plasmonically induced nanoscale evanescent light to activate/enhance the Raman signal. The tips can either be made of a metal or coated with a thin layer of metal to modify them for TERS. When the apex of a metallic or a metal-coated nanotip is illuminated with focused light at the LSPR wavelength, local surface plasmons around the tip apex are excited, and evanescent light is produced at the tip apex. This evanescent light can generate Raman scattering from a sample placed right under the tip apex. The process of Raman scattering takes place in the near-field and the spectral signal is scattered and converted back to the far-field by the tip apex, which is then collected by the usual optics and spectrometer in the far-field. Figure d) shows such a TERS setup with a modified AFM. The setup consists of largely similar equipment shown in Figure a. An inverted microscope illuminates the sample from underneath and the tip is placed at the top surface of the sample. The Raman back-scattered signal is then directed to the spectrometer. An evanescent mask blocks the central part of the laser beam inhibiting the low NA component of the incident light, so that only the high-NA component of the incident light reaches the sample so that total internal reflection occurs. This limits the transmitted light that falls onto the tip and, hence, only the evanescent light participates in the Raman scattering signal. Suppressing the participation of transmitted far-field light reduces the unfavourable background signal.

Polarisation-dependent TERS can be performed with light polarisation parallel to the tip apex in addition to the in-plane linear and radial polarisations. Polarisation dependent TERS is enabled by the large incidence angle from the high-NA objective lens and the use of devices that modify the polarisation state of the light such as a $\lambda/2$ waveplate. The Raman scattered light is then collected in the low NA region through an apertured mask, which inhibits any residual laser light. As the tip apex approaches the sample within the focal spot, evanescent light is created at the tip's apex. Since the intensity distribution within laser focus is not uniform, it is very important to lock the relative position of laser focus to the tip.

The strength and resolution of TERS depends on the ability of the tip to enhance and confine the light field at the tip's apex, respectively. In STM systems, the tips are made of solid metal and the substrates need to be conductive in order to control the tunnelling current. The STM tip resembles a long and smooth nanocone, with an apex diameter of ~ 20 nm. The length of the tip (~ tens of micrometres) makes them plasmonically unfavourable for visible light enhancement. However, the tunnelling gap between the tip and the sample can be tuned to the desired LSPR wavelength, creating a strong hotspot within the gap. Some of the more advanced STM systems allow high-vacuum and low-temperature measurements. As the substrate in STM needs to be conductive (often opaque in the visible wavelength range), the setup shown in Figure d would not be suitable. Hence, a side illumination and side collection configuration is more common with STM-based TERS. To prevent the objective from mechanically interfering with the STM tip, a lens with a long working distance is required. It is therefore not trivial to tightly focus the incident light on the tip apex. A parabolic mirror can be used to mitigate mechanical interference and tightly focus the incident light to the tip apex as well as to collect the Raman signal.

The spatial resolution in TERS is comparable to the size of the metallic nanostructure at the tip apex. The gain in spatial resolution comes at a cost to overall signal enhancement (relative to SERS) due to the reduction of the Raman active volume.

In AFM systems, the tips are usually semiconductor cantilevers, with an apex diameter of ~5 nm. Figure shows five examples of AFM-based TERS tips that have been demonstrated in the literature. The semiconductor tips are usually coated with metal either by thermal evaporation under high-vacuum or electroless metal plating (mirror reaction) techniques. Figure a) shows an example of a smooth AFM TERS tip. As the substrate does not need to be conductive, AFM-based TERS can be performed in either bottom-up transmissive illumination or in side/top reflective illumination configurations; the transmissive configuration in Figure d) is more common.

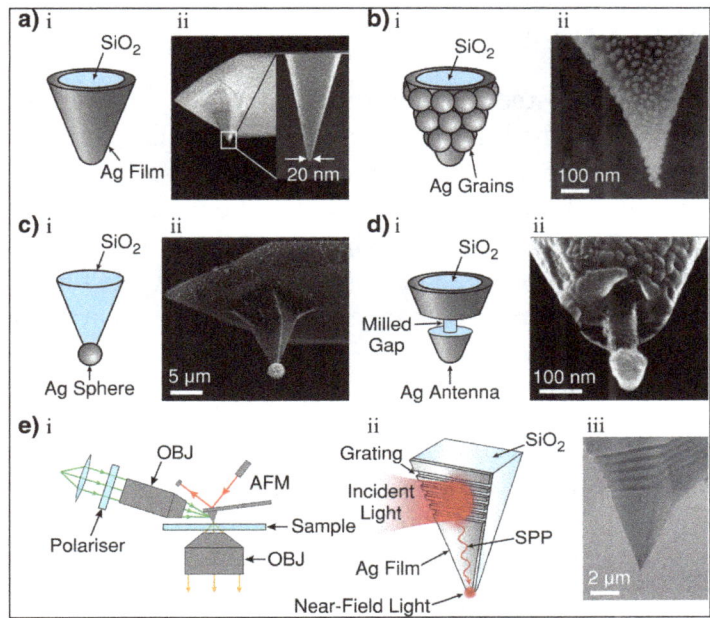

Figure a) i) Smooth metallic (silver; Ag) film-coated dielectric (silicon-dioxide; SiO_2) atomic force microscope (AFM) tip. a) ii) SEM image of a Ag-coated AFM tip. After Ag coating by thermal evaporation, a thin granular Ag layer is deposited onto the tip b) i), Rough Ag-nanoparticle-coated SiO_2 AFM tip. b) ii) SEM image of rough Ag-grain-coated SiO_2 AFM tip formed during the thermal evaporation process. c) i) Single Ag nanoparticle attached to the apex of a SiO_2 AFM tip. c) ii) SEM image of an AFM tip after photoreduction to selectively fabricate an Ag nanoparticle at the tip apex. d) i) Ag-coated SiO_2 AFM tip with a focused ion beam (FIB) milled gap. d) ii) SEM image of antenna fabricated by FIB milling of annular ring and subsequent Ag thermal evaporation from under the tip. The mushroom shape shadows the annular ring from Ag coating. e) i) Illustration of side illumination TERS for surface plasmon polariton (SPP) nanofocusing. OBJ, objective. e) ii) Schematic of the tip structure for SPP nanofocusing which is composed of a SiO_2 pyramidal structure (AFM tip) and a Ag film on the surface. The incident light is coupled to the surface by the FIB-fabricated grating nanostructure. e) iii) SEM image of a Ag-coated SiO_2 tip with a FIB-fabricated grating structure.

The surface of AFM tips becomes nanostructured during the coating process resembling aggregated nanoparticles on the semiconducting tip. These nanostructures are suitable for the resonant excitation of LSPR and SPPs. The smooth tip shown in Figure a) has been fabricated by subsequently depositing a thin granular layer of additional metal. Other researchers have tested AFM tips with a metallic nanoparticle attached to the tip apex, or segregation in the tips coating to form an antenna by focused ion beam lithography. Tips can also be created by electrochemical deposition.

For transparent dielectric substrates, a thin metal film (thin enough to be transparent) can be coated onto the substrate to further enhance the field in the tip-sample gap. It is also possible to perform TERS in liquids with AFM-based systems, which is favourable for biological specimens which require liquid environments to function. Performing TERS in liquid with STM systems is much more difficult. SFM-based TERS is also an attractive technique and maintains many of the properties of AFM-based TERS with the exception of the tip material which resembles similar TERS properties of STM-based TERS.

Some TERS setups have demonstrated vastly improved signal-to-noise ratio in TERS by SPP nanofocusing. This technique focuses the laser onto a plasmon-coupling nanostructure (in the form of a grating) on the upper area of the tip, usually at a distance of ~ 10 µm from the tip apex. Figure e shows a typical nanofocused SPP-based TERS setup (i), the process of SPP nanofocusing by coupling the incident light to a focused ion beam-fabricated grating (ii), and an example SEM image of a SPP-nanofocusing tip (iii). The excited plasmons then propagate toward the tip apex through the process of adiabatic compression and create a confined EM field at the tip apex.

Tuning the Plasmon Resonance

The size, shape, composition of the nanostructures and inter-nanostructure spacing all affect the wavelength of the surface plasmon resonance. Metals are most often used as the conducting medium for surface plasmons; however, semiconductors also possess plasmonic characteristics. Gold shows strong enhancement factors in the red spectral region, silver in the blue-green spectral region and aluminium in the UV and deep UV spectral regions. The blue-green spectral region is the most commonly used Raman spectroscopy range. However, silver is prone to oxidation which degrades the plasmonic characteristics within a few hours of exposure to atmosphere. For this reason, silver is often mixed with other metals, such as titanium.

The range of plasmon resonance can be tuned by the thickness and choice of coating metal, e.g. tungsten, gold, silver or aluminium. In TERS, the grain size of the metal coating corrugations is roughly comparable to the wavelength of the LSPR/SSP. Unlike STM tips, it is possible to control the LSPR/SPP wavelength by adjusting the size of the nanoparticles. The surface plasmon resonance wavelength is also dependent on the refractive index of the dielectric material. In AFM-based TERS, for example, the silicon cantilever tip can be heated to ~ 1000 °C in the presence of water vapour to oxidise the silicon into silicon dioxide. As SiO_2 has a lower refractive index than Si, the surface plasmon resonance is blue shifted.

The size and shape of the metal-coated AFM tip apex can also be modified to tune the LSPR. Fabricating a single metallic nanoparticle attached to the tip's apex or segregated antenna-shaped tip has been demonstrated as a means to finely tune the surface plasmon resonance in AFM-based TERS. However, the most commonly used tips for AFM-based TERS are the tips that have disconnected metal nanoparticles evaporated on a semiconductor cantilever in the standard coating process described a priori.

Analysis Methods

Units

By convention, Raman spectra are considered in terms of the wavenumber $\tilde{v}$ in units of cm^{-1}. The conversion from angular frequency is as follows:

$$\tilde{v} = \frac{\omega}{2\pi c_0},$$

Where, c_0 is the speed of light in vacuum and ω is the angular frequency. Raman spectra are usually plotted in terms of the wavenumber shift from the incident excitation radiation. This shift is defined as follows:

$$\Delta\tilde{v} = \tilde{v}_p - \tilde{v}_{scat},$$

Where, $\tilde{v}_p$ is the wavenumber of the pump beam with angular frequency ω_p and $\tilde{v}_{scat}$ is the wavenumber of the scattered light accordingly. For Stokes Raman scattering, $\tilde{v}_{scat} = \tilde{v}_p - \tilde{v}_{osc}$ (where $\tilde{v}_{osc}$ is the molecule or lattice vibration wavenumber) and $\Delta\tilde{v}$ is positive. By contrast, for anti-Stokes Raman scattering, $\tilde{v}_{scat} = \tilde{v}_p + \tilde{v}_{osc}$ and $\Delta\tilde{v}$ is negative.

Raman spectra are (by standard) presented with the wavenumber shift linearly increasing from right to left on the horizontal axis. The vertical axis ordinate is linear and proportional to intensity. However, researchers also present Raman spectra with wavenumber shift denoted simply as wavenumber and/or increasing from left to right instead of right to left.

Spontaneous Raman Spectra

Figure below shows the Rayleigh and the Raman spectrum of carbon tetrachloride (liquid) excited by an argon ion laser, $\tilde{v}_1 \sim 20,487\,cm^{-1}$ (487.99 nm). This spectrum is presented according to recommendations of the International Union of Pure and Applied Chemistry. It contains a strong band at $\tilde{v}_1 \sim 20,487\,cm^{-1}$ due to the Rayleigh scattering of the incident laser radiation and a number of weaker bands with wavenumbers, $\tilde{v}_1 \pm \tilde{v}_{osc} : \tilde{v}_{osc} = 218, 314, 459, 762$ and $790\,cm^{-1}$. The Stokes Raman lines are shown on the left-hand side of the plot; the anti-Stokes Raman lines are shown on the right. The $\tilde{v}_{osc}$ values relate to the fundamental vibrations of the carbon tetrachloride molecule. In the original work by Raman and Krishnan, the same spectrum was measured using mercury arc radiation ($\tilde{v}_1 = 22,938\,cm^{-1}$, 435.83 nm). In this seminal work, the anti-Stokes bands at $\tilde{v}_1 + 762$ and $\tilde{v}_1 + 790\,cm^{-1}$ were not observed. Hence, after the invention of the laser, Rayleigh and Raman scattering experiments are preferably performed using monochromatically intense lasers.

Figure a) Spontaneous Stokes and anti-Stokes Raman spectrum of carbon tetrachloride (liquid) excited by an argon ion laser, $\tilde{v}_p = 20487$ cm^{-1}. The spectrum is presented according to recommendations of the International Union of Pure and Applied Chemistry. b) i) Raman spectra of thin multi-layer (nL) and bulk MoS2 films. The solid line for the 2 L spectrum is a double Voigt fit through data (circles for 2L, solid lines for the remainder). b) ii) Frequencies of E_{2g}^1 and A_{1g} Raman modes (left vertical axis) and their difference (right vertical axis) as a function of the number of layers. b) iii), iv) spatial maps (23 μm × 10 μm) of Raman frequency of E_{2g}^1 (iii) and A_{1g} (iv) from a sample of thin MoS2 films deposited on a SiO_2/Si substrate. b) v) Atomic displacements of the four Raman-active modes and one infrared-active mode (E_{1u}) in the unit cell of the bulk MoS2 crystal as viewed along the [1000] direction. c) Microscopic image of nebulised ammonium sulphate aerosol particles on: i), Klarite; iii), silicon wafer. ii), iv) Raman mapping image of sample (i) and (iii), respectively. d) i) Pseudo colour broadband CARS image of tumour and normal brain tissue, with nuclei highlighted in blue, lipid content in red and red blood cells in green. d) ii) Broadband CARS image and axial scan with nuclei highlighted in blue and lipid content in red. d) iii) Broadband CARS image with nuclei highlighted in blue, lipid content in red and CH3 stretch–CH2 stretch in green. NB, normal brain; T, tumour cells; RBC, red blood cells; L, lipid bodies; WM, white matter. d) iv) Single-pixel spectra. e) Raman thermography measurements across the active region of a high electron mobility transistor on SiC substrate with both E_2 and A_1 (LO) phonons considered to compensate for thermal stress. Device temperature rise determined using either E_2 or A_1 (LO) phonon mode alone (neglecting thermal stress) is shown in the top left insert. f) (left) illustration of the manipulation of a straight isolated carbon nanotube (CNT) lying on a glass substrate by the sharp apex of an AFM tip. f) (right) two-dimensional image of a CNT constructed by colour-coding the frequency position of the G+ vibrational mode in TERS spectra. The colour variation shows the strain distribution along the CNT at high-spatial resolution.

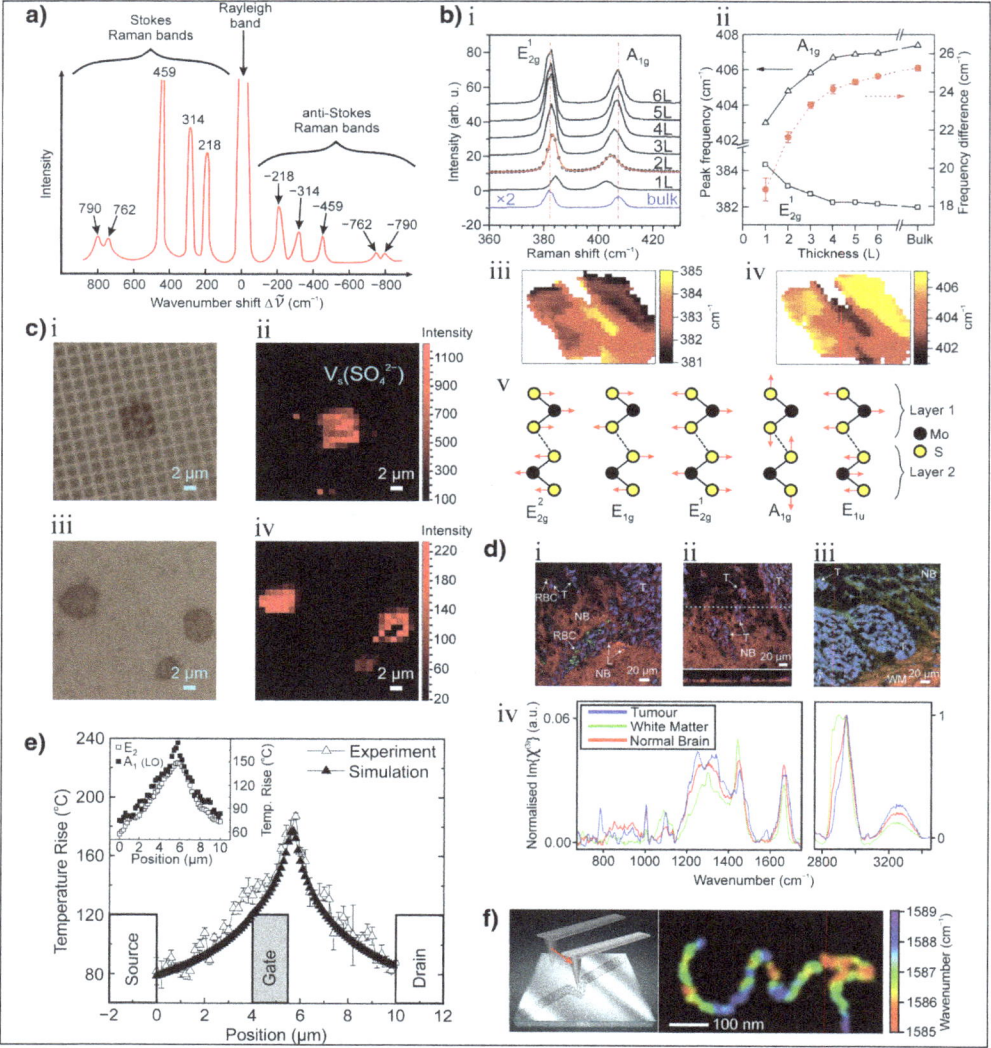

Layered Two-dimensional Systems

Raman spectroscopy can be used to determine the layer thickness in two-dimensional materials with atomic level precision, using either the inter-layer or intra-layer vibrational modes. Lee et al. demonstrated the technique with two intra-layer Raman modes of molybdenum disulphide (MoS_2). Figure b shows representative Raman spectra for single- and few-layer MoS_2 samples. Among the four Raman-active modes of bulk 2H phase MoS_2 crystal, Lee et al. only observed the E_{2g}^1 and A_{1g} modes near $\Delta \tilde{v} = 400\,cm^{-1}$. The other modes were not observed either because of the selection rules for the scattering geometry (E_{1g}) or because of the limited rejection of the Rayleigh scattering radiation (E_{2g}^2). Single-layer MoS_2 exhibits a strong in-plane vibrational mode at $\Delta \tilde{v} \sim$ $384\,cm^{-1}$, corresponding to the E_{2g}^1 mode of the bulk $2H$-MoS_2 crystal. For all film thickness, the Raman spectra in Figure b i show strong in-plane E_{2g}^1 and out-of-plane A_{1g} vibration signals. As the sample thickness increases, the E_{2g}^1 mode red shifts and the A_{1g} mode blue shifts. For films of four of more layers, the E_{2g}^1 and A_{1g} modes converge on the bulk values. Spatial maps of a MoS_2 film sample for the E_{2g}^1 mode is shown in figure b) iii); that of the A_{1g} mode is shown in Figure b) iv). These maps demonstrate that the frequency of the two modes only slightly vary in regions of the sample with a given layer thickness. Hence, Raman spectra can provide a convenient and reliable means of determining the layer thickness in two-dimensional crystalline materials with atomic level precision.

Enhanced Raman Scattering through SERS

Ault et al. were the first to use SERS to enhance the Raman scattering signal of previously undetectable secondary organic aerosol particles on Ag nanoparticle-coated quartz substrates. Fu et al. demonstrated enhancement factors of 6 for the Raman spectra of ammonium sulphate (AS) at the Raman active mode $\tilde{v}_s\,(SO_4^{2-})$ at $970\,cm^{-1}$ with Klarite. Figure c shows a microscope image of a large AS particle on the surface of Klarite, the corresponding Raman mapping image is in Figure c) ii). Figure c) iii) shows another sample of AS particle but on a silicon wafer. The corresponding Raman mapping image is shown in Figure c) iv). Aside from the three larger AS particles, small (sub-micron) AS particles are apparent in Figure c) iii). However, in the absence of SERS, these smaller particles are undetectable. On the other hand, the SERS Raman mapping image shows vastly enhanced signal intensity, as is evident from the scale bars, to the point where a number of small spots yield a signal at the $\tilde{v}_s\,(SO_4^{2-})$ Raman mode. Such spots most likely correspond to small AS particles that are observable in Figure c) ii) but are not apparent in Figure c) i).

Insights into Cellular Structure with CARS

CARS microscopy is relevant to the chemical, materials, biological and medical fields and can provide unparalleled insights into cellular structures. Spontaneous Raman and infrared micro/spectroscopy can provide adequate chemical specificity and sensitivity to delineate a variety of neoplasms but require long integration times and have a coarse spatial resolution, which may limit accurate tumour-boundary identification and early-stage tumour detection. However, coherent Raman imaging techniques have demonstrated high-speed, high-spatial-resolution imaging, but with contrast limited to single or few Raman peaks. Figure d presents images of orthotopic xenograft brain tumours from within a murine brain. Figure d) i) shows a broadband CARS image with nuclei in blue ($\Delta \tilde{v} = 730\,cm^{-1}$), lipid content in red ($\Delta \tilde{v} = 2850\,cm^{-1}$)

and red blood cells in green ($\Delta\tilde{v}$ = 1548 cm⁻¹ + 1565 cm⁻¹: C-C stretch from haemoglobin). The large tumour mass and a projection of neoplastic cells within healthy tissue are clearly shown. Figure ii) shows several small regions of main tumour mass migrating into the healthy brain matter. Figure iii) shows the boundary between normal brain tissue, white matter and tumour masses, which contrasts lipids in red ($\Delta\tilde{v}$ = 2850 cm⁻¹); CH_3 stretch-CH_2 stretch ($\Delta\tilde{v}$ = 2944 – 2850 cm⁻¹), a general contrast; and nuclei in blue ($\Delta\tilde{v}$ = 785 cm⁻¹). The image shows the fibrous texture of the white matter and strands of myelination around cancer cell clusters. Figure d) iv) presents a set of single-pixel spectra from an intra-tumoural nucleus, the white matter and normal brain, respectively. The spectra indicate that lipids are most concentrated in the white matter and least in the tumour regions.

Raman Thermography

Advances in electronic and opto-electronic semiconductor devices, such as high electron mobility transistors (HEMTs), have led to thermal management challenges. Conventional thermal characterisation approaches such as infrared thermography are often no longer applicable for the accurate characterisation of high-power density devices due to limited spatial resolution which can result in the underestimation of the device peak temperature. Batten et al. have demonstrated temperature profiling in AlGaN/GaN HEMTs using Raman thermography by exploiting the E_2 and A_1 (LO) phonon modes. Both the E_2 and A_1 (LO) modes shift to lower frequency when operating the device. Figure e) shows a comparison of the temperature rise in a AlGaN/GaN HEMT on a SiC substrate from Raman thermography and thermal simulations. The device was operated at a source-drain voltage of 40 V and a power density of 25 W/mm and had a thermal resistance of 8 °C/(W/mm).

Measuring Strain on the Nanoscale using TERS

TERS microscopy is an effective means of imaging nanostructures beyond the spatial resolution of the so-called light diffraction limit. Nanostructures such as DNA molecules, carbon nanotubes (CNTs), silicon devices, dye molecules and single molecules can be imaged using TERS. The technique can even be used to measure the local molecular strain in nanostructured materials. For example, AFM can be used to manipulate CNTs with nanoscale precision to develop a local strain. Figure f) (left) illustrates the process of CNT manipulation using contact-mode AFM. Although local strain in CNTs has previously been studied using AFM and transmission electron microscopy, TERS microscopy is the only optical technique that can provide images of such local structural distribution of nanomaterials. When a straight CNT is deformed by manipulation, a local breakdown in symmetry is induced. This causes the selections rules of Raman scattering to become relaxed, allowing forbidden Raman modes to become visible in the vicinity of the local curvature. The position of the characteristic G-mode Raman scattering line in graphene can be used to deduce local strain using TERS. Figure f (right) shows a TERS image of a deformed CNT which has been constructed from the peak positions of the G⁺-mode. The image has a spatial resolution better than 20 nm which is about 25 times finer than the diffraction limit of the excitation wavelength of light (488 nm). The colour variation (as indicated by the scale bar) corresponds to the local peak position of the G⁺-mode and represents the variation of strain along the CNT.

Stimulated Raman Scattering Microscopy

Unlike CARS, SRS microscopy does not contain a nonresonant background signal that degrades image contrast. However, SRS can be affected by cross-phase modulation (where light at one wavelength modulates the refractive index in the medium affecting another wavelength of light), transient-absorption (which is characteristic of femtosecond light pulses) and photo-thermal effects which can modify the vibrational energy levels and reduce hyperspectral image contrast. SRS is quantified by the amount of energy transfer from the pump light to the Stokes light when the difference frequency between the pump and Stokes light matches a specific vibrational frequency; ω_{osc}. In addition, the resulting signal from SRS is strongly sensitive to the incident polarisations when the orientation of the probed vibrating species is ordered. This polarisation dependence can be exploited to probe the orientational order of vibrational modes in samples. However, currently developed techniques are not able to perform large-field fast time scale dynamics instantaneously due to the requirement of point-wise scanning over the sample space. Conventional polarisation-resolved techniques take minutes because each point of the scanning area must be polarisation tuned sequentially.

Multi-lamellar myelin plays a crucial role for efficient transmission of nerve impulses as an electrical insulator. The lipids and proteins in myelin self-assemble into a highly ordered and stable structure to form a tightly packed membrane. In neurological disorders, this compact structure is highly perturbed leading to dysfunctions of the central nervous system. As these biological processes are highly dynamic, researchers seek to observe the dynamics of molecular order with sufficient resolution and frame rate. Hofer et al. have recently demonstrated fast-polar-SRS by exploiting high-speed amplitude- and polarisation modulation with an acousto-optic modulator (AOM) and electro-optical polarisation modulation, respectively, to read out the molecular order and orientation at a fast rate. They therefore obtain both amplitude and phase information. The linear polarisation direction of the pump beam is rapidly rotated while the Stokes polarisation is circularly polarised to avoid polarisation dependence from the Stokes beam. The polarisation is further modified by a quarter-wave plate. The polarisation modulation leads to an α-dependant response of the signal intensity given by the following:

$$I(\alpha) \propto a_0 + S_2 \cos 2(\alpha - \varphi_2)$$

Where, α is the rotating pump polarisation direction in the sample plane, a_0 is the total measured intensity, and S_2 and φ_2 are the amplitude and phase of the second-order induced modulation.

Figure below a) i) shows a comparison of conventional polarisation SRS with that from Hofer's fast-polarisation SRS on a multi-lamellar lipid vesicle (MLV). The fast-polarisation SRS image in the bottom of Figure a) i) was obtained in 1 s which is two orders of magnitude faster than the conventional-SRS image (top) using the same incident powers, number of pixels and dwell time per pixel. Figure a ii shows sub-second frame-rate imaging of a MLV using double EOM-AOM modulation SRS at two instances in time. The measurement technique was remarked to have little effect on the lipid order properties during the measurement. Hofer et al. were able to observe second-timescale dynamics in thin lipid membranes down to the cell plasma membrane using fast-polarisation-resolved SRS as shown in figure a) iii).

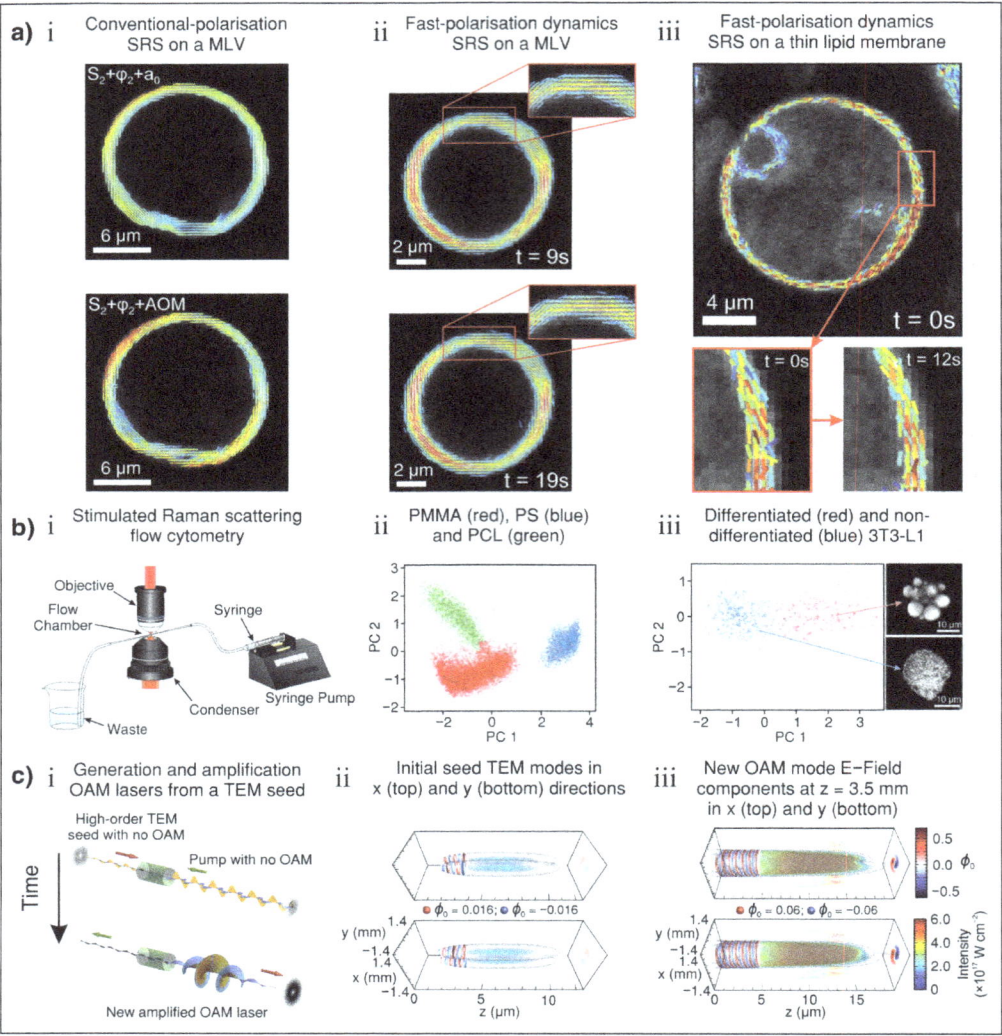

Figure above shows a) Comparison of conventional and fast-polarisation SRS on multi-lamellar lipid vesicles (MLVs). a) i–iii) the composite images show absolute local molecular order (S_2) and mean molecular orientation (φ_2) values represented as coloured sticks. a) i) (top) Conventional polarisation SRS on a MLV using step angles of 5° (acquisition time 112 s); the total measured intensity (a_0) is represented as a grey-scaled background. a) i) (bottom), fast-polarisation SRS on the same MLV (acquisition time 1 s); the acousto-optic modulation (AOM) is shown as a grey-scaled background. a) ii) Fast-polarisation dynamics SRS images of lipid order in a MLV taken at different times of the observation sequence shown; the AOM is shown as a grey-scaled background. Zoomed regions at the upper part of the MLV contour show no change in lipid order during the measurement over tens of seconds. a) iii) Fast-polarisation dynamics of lipid order in a thin lipid membrane. Coloured sticks show S_2 and φ_2 and the grey-scale background shows the AOM amplitude. b) i) SRS flow cytometry (SRS-FC) setup. b) ii) Colour-coded constrained principle component analysis (CPCA) scatter plot of SRS-FC spectra from mixed PMMA (red), PS (blue) and PCL (green) beads. The principle components (PC 1 and PC 2) are distinguished subpopulations of mixed polymer beads according to the distinct Raman spectra. Data were acquired in 6 s using a bead mixture with a concentration of 2% solids. The beads were 10 μm in diameter. b) iii) (left)

Colour-coded CPCA scatter plot from SRS-FC analysis of lipid amount in 3T3-L1 cells with principle components differentiated through quantification of distinct chemical compositions inside single cells. Data were acquired in 3 s. b) iii) (right) SRS images of the two 3T3-L1 cell types. c) i) Illustration of the generation and amplification of a new orbital angular momentum (OAM) laser in a configuration with no initial OAM using a Hermite-Gaussian (TEM) laser. c) ii), iii) Simulation of the generation and amplification of a new OAM mode from initial configurations with no net OAM. c) ii) The initial seed TEM modes in x- and y-directions (top and bottom, respectively). c) iii) The new OAM mode electric field components at z = 3.5 mm in x- and y-directions (top and bottom, respectively). The new mode is linearly polarised in the x- and y-directions with $l_{lx} = l_{ly} = -1$ from an initial seed polarised in the x-direction with a TEM01 mode and in the y-direction with a TEM10 mode that is $\pi/2$ out of phase with respect to the TEM01 mode polarised in x. The pump is a Gaussian laser polarised at 45°. Projections in the (x, y) plane (blue-white-red) show the normalised vector potential ($\varphi 0$) field envelope of the new OAM mode at the longitudinal slice where the laser intensity is maximum. The envelope of the 3D laser intensity is also shown in blue-green-red colours and normalised vector potential isosurfaces in blue and red. The values of the laser vector potential illustrated by the isosurfaces are shown in c (ii) and c (iii).

Flow cytometry (FC) is one of the most important technologies for high-throughput single-cell analysis. FC is a technique used to measure physical/chemical characteristics of a population of cells or particles suspended in a fluid. The fluid suspension flows through the instrument detectors for fluorescent labelling which is the primary approach for cellular analysis in FC. Figure b) i) shows an optical FC setup. However, for small molecules, the fluorescent tags can perturb the biological function of the species. In addition, non-specific binding of fluorescent labels as well as cellular autofluorescence can also reduce the clarity of the result. SRS flow cytometry (SRS-FC) non-invasively detects chemical cell content but conventional techniques suffer slow acquisition rates.

Zhang et al. have recently demonstrated label-free high-throughput single-particle SRS-FC with a 32-channel multiplexing technique. Their technique measured single-particle chemistry at a rate of 5 μs per SRS spectrum, approaching that of standard fluorescence-based FC. The SRS-FC technique was based on broadband laser excitation and a multiplex spectral detection system. The systems allowed the acquisition of 200,000 spectra per second, more than 11,000 particles per second. The subpopulations of species, such as mixed polymer beads and 3T3-L1 cells, could be separated and distinguished through compositional principle component analysis (CPCA) of the SRS signals. The principle components were designated according to their Raman spectra. An agglomerative clustering procedure was performed on the resulting CPCA spectral matrix. This procedure assumed the number of cluster groups (κ) was known to separate the clusters of principle components in the CPCA analysis. Figure b) ii) shows the CPCA of the SRS spectra for a mixture of three types of beads: poly-methyl-methacrylate (PMMA), polystyrene (PS) and polycaprolactone (PCL), all with a 10-μm mean diameter, mixed at a 2:1:1 ratio of PMMA:PS:PCL and a final concentration of 2% solids in the fluid. The flow speed was $\approx 0.16 \, ms^{-1}$, the SRS-FC data was acquired in 6 s. The CPCA plot (Figure b ii) shows three distinct clusters of principle components. The agglomerative clustering procedure ($\kappa = 3$) allowed the quantification of ~ 7100 PMMA bead (red), ~ 3400 PS beads (blue) and ~ 3600 PCL beads (green) as shown in Figure b ii. Their measurement demonstrated the ratio of $\approx$ 2:1:1 (PMMA:PS:PCL) at a throughput rate of ~ 2350 particles per second and that their multiplex SRS-FC system, paired with the CPCA analysis, could distinguish different chemical components with small spectral differences. Zhang et al. were able to detect beads as small as 1 μm and were even able to detect single Staphylococcus aureus bacteria flowing through the laser focus highlighting the potential to characterise subcellular organelles with SRS-FC.

Zhang et al. also demonstrated the discrimination of 3T3-L1 cells at different stages of cell differentiation according to their difference in lipid amount using SRS-FC. After insulin-induced differentiation, 3T3-L1 cells acquire an adipocyte-like phenotype with a significantly increased amount of triglycerides which aggregate to form large lipid droplets. This aggregation of triglycerides causes the intensity of the methylene symmetric vibration at $\Delta\tilde{v}$ = 2850 cm^{-1} from fatty-acid acyl chains to become stronger compared to that of non-differentiated cells which provides the means for CPCA analysis. Figure b) iii) shows the CPCA scatter plot of the cell mixture measured by Zhang et al. which were separated using the agglomerative clustering approach. The insert SRS images (Figure b) iii), right) show a non-differentiated 3T3-L1 cell and a differentiated cell with the formation of large lipid droplets.

Twisted Laguerre-Gaussian lasers, with orbital angular momentum (OAM) and characterised by doughnut-shaped intensity profiles, are of great interest to a number of growing research fields such as ultra-cold atoms, microscopy and imaging, atomic and nanoparticle manipulation, ultra-fast optical communication, quantum computing, astrophysics and plasma accelerators. Spiral phase plates or computer-generated holograms are usually used to generate visible light with OAM. Spiral phase plates can produce light with predefined OAM modes. By using plasma as an optical medium to generate and amplify laser pulses with OAM and relativistic intensities, well above the damage threshold of optical devices, could provide for high-energy-density science and applications. Plasmas also allow for greater flexibility in the level of OAM in the output laser beam than conventional optics. Vieira et al. have shown that SRS in nonlinear optical media with Kerr nonlinearity (e.g. plasmas, optical fibres and nonlinear optical crystals) can be used to generate and amplify OAM light. Figure a) i) illustrates the process in which the pump EM fields can have different OAM components in both transverse directions x and y (blue and orange in Figure a) i) l_{ox} is the pump electric field component of OAM in the x direction. Likewise, l_{oy} is the pump electric field component of OAM in the y direction. The initial seed electric field component has an OAM component l_{1x}. After interacting with the plasma, the pump is depleted, and a new electric field component appears in the seed with OAM $l_{1y} = l_{1x} + l_{ox} - l_{oy}$.

Stimulated Raman backscattering in plasma is a three-wave mode coupling process in which a pump pulse decays into an electrostatic plasma wave as well as a counter-propagating seed laser. The plasma can be viewed as a high-intensity mode converter. The presence of OAM in the pump and/or seed results in additional matching conditions that ensure the conservation of angular momentum of the pump when the pump decays into a scattered electro-magnetic wave and a Langmuir wave.

Particular superpositions of Hermite-Gaussian modes TEM modes are mathematically equivalent to Laguerre-Gaussian modes. Vieira et al. therefore explore the use of Stimulated Raman backscattering to generate and amplify light with OAM using TEM laser beams with no initial net OAM. Each Hermite-Gaussian beam in the simulation is described by TEM$_{m,n}$, where m and n represent the Hermite-Gaussian mode index. Figure c ii and iii show the 3D simulation results from the setup shown in Figure c i. The simulations show that SRS results in a new OAM mode with $l_1 = 1$ linearly polarised at 45°. The field topology of the seed normalised vector potential changes from plane isosurfaces in Figure c ii, to helical isosurfaces in. Hence, light with OAM has been generated from light with no net OAM. In the case of plasma, the interaction between the seed light and the pump light occurs via an electron Langmuir wave. This interaction ensures that the frequency, wavenumber and OAM matching conditions are conserved.

Coherent Anti-stokes Raman Scattering Microscopy

CARS results from an induced anti-Stokes scattering of radiation, ω_{AS} which is enhanced when $\omega_p - \omega_S = \omega_{OSC}$. One of the main challenges with CARS microscopy is the nonresonant background. The existence of the nonresonant background can either distort or even saturate the resonant signal of Raman peaks, which reduces the image contrast. Qin et al. have recently demonstrated multi-colour background-free coherent anti-Stokes Raman scattering microscopy using an all-fibre, low-cost, multi-wavelength time lens source. A time lens, in analogy to a spatial lens, is simply a quadratic optical phase modulator in time, which can be approximated by a portion of a sinusoidal phase modulator. Three different wavelength picosecond pulse trains were provided by the time lens source, at 1064.3 nm (stable), 1052–1055 nm (tuneable) and 1040–1050 nm (tuneable). The time lens was used to apply temporal quadratic phase modulation to a continuous-wave laser to broaden its spectral bandwidth. In this instance, the time lens was applied with fibre-integrated electro-optic radio-frequency phase modulators. The phase modulation and pulse synchronisation were derived from a mode-locked Ti:Sapphire laser that provided synchronised multi-colour picosecond pulses with dispersion compensation. Electronic tuning of the pulse delay was used to achieve temporal overlap between the pump and Stokes laser pulse trains, which is a convenient substitution for mechanical optical delay paths. Two of the three wavelengths of light from the time lens source were used for two-colour on-resonance imaging and the third wavelength for off-resonance (nonresonant background subtraction) imaging. Pixel-to-pixel wavelength switching was achieved, which provided simultaneous two-colour CARS imaging with real-time non-resonant background subtraction. Qin et al. demonstrated the technique with an excised fresh tissue sample from a mouse ear and imaged molecular stretching vibrations at 2845 cm^{-1} (CH_2) and 2940 cm^{-1} (CH_3) and non-resonance background at $\Delta\tilde{v}$ = 2770 cm^{-1}. Figure a) i–iii) shows the process applied to the Raman peak of CH_3 stretching vibration from the mouse ear tissue sample.

The nonresonant background signal in CARS can also be suppressed by applying an external static electric field to the sample known as electro-CARS. Capitaine et al. demonstrate this electro-optical technique on n-alkanes in solution with broadband multiplex coherent anti-Stokes Raman scattering spectroscopy. The nonresonant background is suppressed due to the orientation response of the molecules to the electric field. The molecular orientation is related to the induced electric dipole moment. The enhancement of the CARS signal-to-noise ratio was achieved in the case of the CH_2 and CH_3 symmetric/asymmetric stretching vibrational modes.

Conventional CARS provides information about the chemical nature but not about the molecular organisation or symmetry in the system. The Cartesian components of the nonlinear susceptibility tensor $\chi^{(3)}$ represent the vibrational symmetry properties of the material. These tensor elements can be extracted with polarisation-resolved coherent Raman scattering schemes. However, these schemes often involve the acquisition of multiple images from different polarisation angles requiring long acquisition times due to limits imposed by polarisation tuning and time-consuming post-processing.

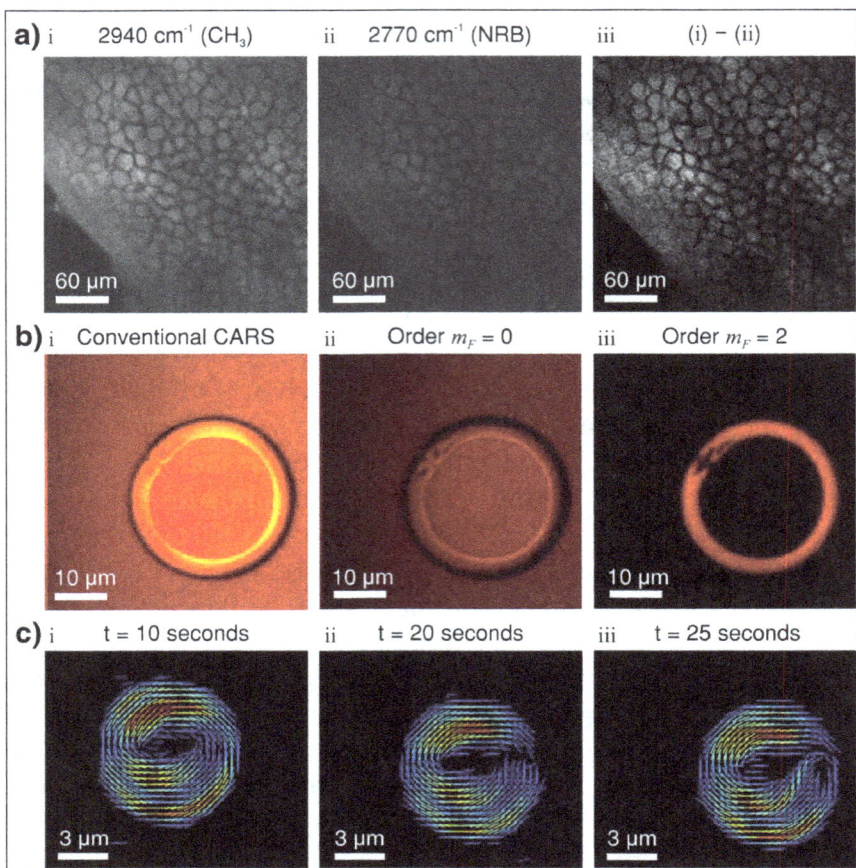

Figure a) Simultaneous two-colour CARS imaging with real-time nonresonant background subtraction from a mouse tissue sample at a surface depth of 45 μm. a) i) CARS image acquired at 2940 cm⁻¹ (CH3 stretching vibration). a) ii Off-resonance background CARS image at 2770 cm⁻¹. a iii Background-free image of (i) at 2940 cm-1 after subtraction of the nonresonant background (ii). b Multi-lamellar lipid vesicle (MLV) imaged with conventional (i) and symmetry-resolved CARS (SR-CARS) at 1133 cm-1 (ii) and (iii). ii The incident circularly polarised pump, Stokes, probe and anti-Stokes light have co-rotating handedness ($m_F = 0$). iii the incident circularly polarised pump, probe and anti-Stokes light have co-rotating handedness and the Stokes light has counter rotating handedness ($m_F = 2$). c High-speed polarisation-resolved CARS image sequence on a MLV moving over the sample surface taken at different times of the observation sequence shown as a composite image of S_2 and φ_2 as coloured sticks and with the acousto-optic modulation (AOM) as a grey background.

Cleff et al. have recently demonstrated a label-free microscopy technique that uses circularly polarised light to probe the symmetry as well as the chemical fingerprint of the probed sample in a single acquisition. This symmetry-resolved CARS (SR-CARS) depends on both the presence of (ro-)vibrational modes as well as their local organisation. By switching between combinations of left- and right-handed circular polarisation states for the involved fields, the individual symmetry contributions of the sample can be imaged. This technique offers a straightforward means to access the local organisation of (ro-)vibrational bonds with improved image contrasts (with 1 to 2 orders of magnitude) for anisotropic samples, as well as improved chemical selectivity without post-processing and independently of sample orientation in the transverse plane. In addition, SR-CARS provide higher chemical selectivity with the contrast in symmetry characteristics, which are not accessible with conventional spontaneous Raman or SRS microscopy.

Multi-lamellar lipid vesicles (MLVs) are made of a tight packing of lipid layers forming a ring of highly ordered matter with twofold symmetry and a lipid orientation distribution close to a Gaussian angular shape.

Figure b) i) shows a conventional CARS image of an aqueous MLV at $\Delta \tilde{v} = 1133\,cm^{-1}$ (C-C stretching vibration) which illustrates the expected poor contrast due to the nonresonant background. Figure b) ii) and iii) show the zeroth and second-order $m_{\bar{F}}$ -value image of the same MLV as in Figure b) ii). The $m_{\bar{F}}$ -value is the summation of the light circular polarisation handedness quantum numbers of the incident light beams:

$$m_{\bar{F}} = m_p - m_s + m_{pr} - m_{as}$$

When light with field tensor $\bar{F}$ probes matter with nonlinear susceptibility tensor $\chi^{(3)}$, in a CARS process, the light probes only the parts of the matter with identical rotational invariant symmetries (i.e. identical $m_{\bar{F}}$). Hence, by engineering the field tensor of the light, specific sample symmetries can be directly read out, creating a symmetry-based image contrast mechanism. The aqueous solution surrounding the MLV is only visible in the $m_{\bar{F}} = 0$ image due to its purely isotropic nature. Background-free imaging of the MLV with superior contrast with respect to conventional CARS is shown in Figure b) iii) at $m_{\bar{F}} = 2$, which results from the symmetric microscopic organisation of the lipids in the MLV. Imaging at $m_{\bar{F}} = 4$ (not shown) lacked sufficient signal strength to provide an image of the MLV due to the lack of anti-symmetry in the lipid organisation.

As with SRS, the CARS signal is sensitive to the polarisation of the incident light when the orientation of the scattering species is ordered. Polarisation-resolved CARS (PR-CARS) requires monitoring of the CARS signal response depending on the relative rotation of the incident light polarisations (pump and Stokes) to the sample, in species with ordered orientations. Provided that the molecular bonds are oriented, the detected intensity of the anti-Stokes signal is maximised when the incident polarisations lie along the averaged direction of the bonds. The ability to monitor lipid order without the need for fluorescent labels can provide information on lipid packing properties. PR-CARS schemes often involve long acquisition times due to limits imposed by polarisation tuning and time-consuming post-processing.

In addition to fast-polar-SRS, Hofer et al. have recently demonstrated fast-polar-CARS imaging with combined electro-optic polarisation and acousto-optic amplitude modulations. Figure c shows fast-polarisation CARS with similar sensitivity to that of SRS. Despite the requirement of lock-in amplification for the detection of low modulation over a large nonresonant background, the fast-polarisation technique demonstrated by Hofer et al. can considerably improve the signal-to-noise ratio in CARS imaging. Despite the robustness of MLVs, occasional alteration of molecular order in MLVs could be observed at the time scale accessible in Hofer's experiment (0.25–1 s per image). MLVs could detach from the sample surface, inducing motion or shape change. The modifications observed in Figure c) were attributed to a local membrane disruption, followed by its spontaneous reformation. Hofer et al. demonstrated the possibility of visualising local modification during MLV displacement that was not accessible using the minute-time-scale conventional polarisation Raman experiments.

There have also been a number of developments in CARS flow cytometry (CARS-FC). However, these techniques were shown to be much slower than fluorescence-based FC. Out-of-focus microparticles can randomly impede CARS-FC and the fluid often generates a strong nonresonant background limiting CARS-FC from achieving high-throughput single-cell analysis. Recently, however, O'Dwyer et al. have demonstrated that it is possible to significantly enhance the fraction of unambiguously and instantly recognised in-focus microparticles, in unconstrained flows by co-monitoring CARS-FC with linear scattering of light.

CARS is invariably performed with two synchronised picosecond laser sources owing to the coherence life time of Raman resonance. Ti:Sapphire oscillators or optical parametric oscillators pumped by a picosecond frequency-doubled Nd:Vanadate laser are the instruments of choice, which are generally very expensive and the synchronisation mechanisms can be challenging. In addition, the spectral drift in the pump wavelength can introduce errors in the calculation of ω_{osc}. Langbein et al. have demonstrated CARS micro-spectroscopy using a single Ti:Sapphire laser oscillator and simple passive optical elements. Vibrational excitation, tuneable over a large spectral range with adjustable spectral resolution, was achieved by spectral selection with dichroic mirrors and linear chirping by glass elements.

Tip-enhanced Dual Wavelength Coherent Anti-stokes Raman Scattering Microscopy

TERS offers spatial resolution far beyond the diffraction limit of the probing light. The more conventional technique is to directly illuminate the tip-sample cavity. This technique achieves the desired resolution (beyond the diffraction limit) by forcing the evanescent light into the far field image formation. However, the far field light presence in the tip-sample cavity generates an unfavourable background light source. It is possible to perform TERS by coupling the far-field excitation light to the tip a few tens of micrometres from the tip apex. Femtosecond laser pulses can be coupled to the tip surface by shining the light on a grating fabricated on the tip surface. The SPPs then propagate to the tip apex and generate background-free localised optical excitation.

Toma et al. previously demonstrated selective excitation of a single Raman mode and its CARS imaging of CNT using ultra-fast SPP pulse nanofocusing using an Au tapered tip. In a more recent publication, seminal work by Tomita et al. demonstrated simultaneous nanofocusing of ultra-fast SPP pulses at 440 and 800 nm, which were coupled with a common diffraction grating structure. Figure a i, illustrates the scheme. The Al-tapered tip had an apex radius of ≈ 35 nm. Selective CARS microscopy that combined an 800 nm (ω) SPP pump pulse and a 440-nm (2ω) SPP probe pulse was achieved. Figure a ii illustrates the energy level process of ω - and 2ω -CARS. The pump pulse achieves selective vibrational excitation by spectral focusing. Raman shift intensities with this 2ω-CARS scheme were reported to increase by as much as 4 compared with that of ω -CARS for monolayer graphene. The selectivity of vibration band excitation and background noise suppression were confirmed on the CARS intensity probed by a 2ω-SPP plasmon pulse for a monolayer graphene sample. Venezuela et al. reported the Raman lines in graphene associated with both phonon-defect processes (such as the D line at $\Delta\tilde{\nu} \sim 1350\,cm^{-1}$) and two-photon processes (such as the 2D line). The 2D-band intensity in graphene was reduced monotonously when the defect concentration was increased, contrary to the D-band. Tomita et al. applied their multi-vibrational-mode 2ω-CARS imaging method to a multi-walled CNT (MWCNT) at the D, G and 2D bands. This dual-wavelength nanofocusing technique could open new nanoscale micro-spectroscopy and optical excitation schemes in SPM, such as sum frequency mixing, two-photon excitation $(\omega+2\omega)$ and pump-probe schemes.

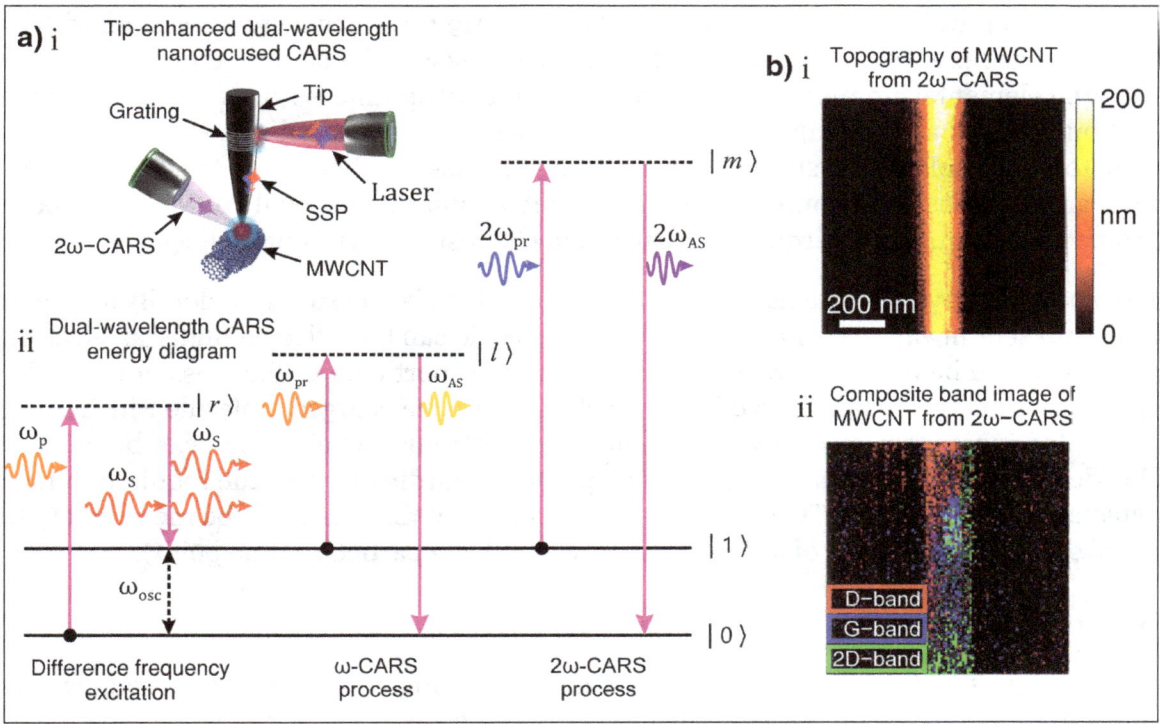

Figure shows a) i) Illustration of tip-enhanced dual-wavelength nanofocused CARS on a multi-walled carbon nanotube (MWCNT). a) ii) Energy diagram for ω-CARS and 2ω-CARS. When the difference frequency between pump and Stokes light matches the vibrational mode of the molecule, it is resonantly excited. When ω and 2ω-probe light is simultaneously incident within the dephasing time, ω- and 2ω-CARS photons are respectively generated. b) i) Simultaneous topographical 2ω-CARS imaging of a MWCNT. b) ii) Composite image of three 2ω-CARS images of the MWCNT using the 2ω-CARS spectrum from D- (red), G- (blue) and 2D- (green) bands.

Figure b) i) shows a topography image of a MWCNT with a diameter of $\sim 175\,nm$ measured by the Al-tapered tip. Figure b) ii) shows a composite image of three 2ω-CARS images of the MWCNT using the 2ω-CARS spectrum from D- (red), G- (blue), and 2D- (green) bands. In ref., the D- and 2D-band showed a negative correlation (in agreement with ref.) except for the central part of the MWCNT. The 2D- and G-bands were intense near the central part of the MWCNT. Tomita et al. indirectly estimated the spatial resolution of their technique to be less than 90 nm by taking the profile of the 2D-band signal across the axis of the MWCNT.

Mass Spectrometry

Study of various molecules based on its mass is called mass spectrometry (MS). A spectra of different masses can be obtained which can be matched with known masses or pattern of masses obtained from the theoretical breakdown of the whole compound. The concept is so universal that a wide variety of compounds can be identified and characterized. The first research on mass analysis goes back in the early decades of 20th century. J. J. Thomson was awarded Nobel Prize in 1906 for his creation of prototype to measure m/z (ratio of mass to charge). The concept that m/z can be studied was first time applied as tool in scientific experiments to measure the mass of elements by Francis Aston, a physicist from Cambridge in the year 1919 for which he was awarded with Nobel Prize in the year 1922. Initially attempts were made to measure the natural abundance of specific isotopes and elemental atomic weight and the abundance of specific isotopes. Later on the field of application widened and first application in biological science appeared as measurement of trace heavy isotopes in biological systems. More recently, this has also been used to sequence oligonucleotides and peptides. With continuous development in multidimensional knowledge, today mass spectrometry is the most powerful analytical tool in chemistry, physics or biology.

In Mass spectrometry (MS) the mass-to-charge ratio of ions is measured to identify and quantify molecules present in various samples. A charged molecule can be deflected under an electromagnetic field and can be forced to travel or spin which is proportional to the mass of the molecule. Thus, a heavier mass which is charged will travel slowly than a charged molecule which is lighter. The separated ionic mass can convert into multiple electronic signals which can be detected by suitable detector. Different masses thus can be plotted according to the time required to reach to the detector (time of flight or TOF) and relative frequency of that mass can be measured in terms of the height of peak (number of unique signals received at a particular time point).

Methodology

Doing MS requires that mass should be in the appropriate range. The Molecular weight can be determined up to 5 ppm level or 0.0005% which is extremely accurate for confirming the molecular formula from mass alone. The MW for large molecules can be determined with an accuracy of 0.01% (i.e. within 5 Da for a 50 kD protein). The molecule must be ionized so that it can travel under electric field. The mass analyzer separates ions according to differences in m/z. A detector detects the ionic mass and the software interprets the received signals in terms of mass to charge ratio.

Ionization of a Molecule

A molecule to be identified by MS must possess charge on it. The charged molecule can be accelerated in electromagnetic field and hence can be separated based upon different mass to charge ratio. The process of transferring charge on molecules is called ionization. It can be achieved either by transferring free electrons to the molecule or knocking out electrons from the molecule. Some common methods for ionization are:

1. Electron Impact (EI - Hard method): It is suitable for small molecules i.e. 1-1000 Da. The sample enters into the system in the form of spray after its evaporation on hating at high voltage. The sample in the gas phase is bombarded with electrons generated from tungsten or rhenium filament

(energy = 70 eV). Under the impact of such high energy molecule is "shattered" into fragments (70eV ionization energy is too high in comparison to 5 eV of bond energy).

We can understand the process of ionization with an example of methanol by electron impact ionization.

$$CCH_3OH \rightarrow CH_3OH^+$$

$$CH_3OH \rightarrow CH_2O = H^+ + H$$

$$CH_3OH \rightarrow CH_3 + OH$$

$$CH_2O = H+ \rightarrow CHO = H^+ + H$$

EI is not suitable for proteomics because such high energy breaks all the bonds of protein leading to very short peptides (2, 3, 4 amino acids long). Even amino acids are broken down. In such case the signals obtained are too simple to differentiate between proteins. It results in large number of small ions from a single protein which is too complex to analyze.

2. Fast Atom Bombardment (FAB – Semi-hard): Beam of high energy atoms are used to strike the surface a molecule to produce ions. The atoms are typically from an inert gas such as argon or xenon. It is a kind of soft ionization which results in MH+ or MH- ions similar to that of ESI and MLADI. FAB can be used to study peptides and sugars up to 6000 Daltons.

3. Electrospray Ionization (ESI - Soft): This method is suitable for the ionization of peptides and proteins up to 200,000 Daltons. The concept of Electro-spray ionization was first conceived by Malcolm Dole in 1960's and was practiced by John Fenn in 1980's. It is a soft ionization technique which keeps the molecule of interest fully intact.

Solvent containing analytes is forced to pass through a glass/steel capillary needle at high voltage to disperse them electrostatically. The proton donors like formic acid or Trifluoroacetic acid(TFA) present in the solvent imparts charge to the rapidly evaporating liquid. Samples are usually dissolved in polar, volatile buffer (no salts) and pumped through a stainless steel capillary (70 - 150 mm) at very slow flow rate (Nano liter (nl) to micro (ul) per minute). The solvent containing analytes are sprayed in the form small mist-like droplets by using nebulizing gas usually inert Nitrogen. This process is therefore called nebulization. Strong voltage (3-4 kV) at tip aerosolizes the solvent and guides the flow of ions ahead. The aerosols move towards the regions of higher vacuum and evaporate to near atomic size which still carry the charges. Off late nano spray has been designed which allows a flow rate of in nl amount per minute. The ESI technique is highly sensitive which hardly needs the target molecules in picomoles. Salt and detergents present in the sample can badly affect the ionization of target molecules. Ionization can be of two types. The positive ion mode operates in the presence of 0.1% formic acid in the solvent which yields $(M+H)^+$ while negative ion mode measures $(M-H)^-$ if ammonia is in the solvent. Positive ion mode detection method is used if the experimental samples contain functional group which readily accepts H^+ ions (amide and amino groups present in peptide and proteins). If sample has functional groups that readily lose a proton (such as carboxylic acids and hydroxyls as found in nucleic acids and sugars), then negative ion detection is used.

4. Matrix Assisted Laser Desorption (MALDI-Soft): Franz Hillen kamp and Michael Karas introduced MADI in 1985. Large molecules such as proteins (up to 500 kDa), peptides, and DNA molecules can

easily be analyzed by MADI using quadrupole, ion trap and TOF. Unlike ESI, MALDI generates spectra that have a singly charged ion. Sample preparation for MALDI is done by mixing a UV absorbant matrix (4-hydroxycinnaminic acid for peptides and sinapinic acid for proteins). Sample is then ionized by bombarding the sample with multiple lasers light. Laser wavelength supplies energy equivalent to of the absorbance maxima of matrix. The matrix transfers some of its energy to the analyte leading to ion sputtering. Positive ion mode generates ions of M+H ions while negative ion mode generates ions of M−H ions. MALDI is generally more robust than ESI because former is not affected by salts and nonvolatile components in the sample. It requires approximately 10 uL of 1 pmol/uL sample.

Mass Analysis with Mass Analyzer

Once the molecules under study have been ionized they need to be separated based on the difference in their m/z ratio. All instruments are designed to carry at least two functions; one for selecting the m/z range of interest and one for measuring the m/z values and intensities of the fragments ions. Different instruments have been invented which works on different principle taking into consideration the time of flight, their electromagnetic properties, orbital frequency and the path stability. Some of the instruments are mentioned below with their principle.

Matrix Assisted Laser Desorption Ionization - Time of Flight Instrument (MALDI-TOF)

It is capable of differentiating molecules based on its molecular weight. The ions liberated from ionizer and accelerated through vacuum in mass analyzer are separated based on m/z ratio. It is because different size molecules take different time to reach to the detector. The whole system operates under intense vacuum during the entire procedure. The signals which hit the detector are amplified, the ions are subsequently analyzed. The relative abundance of the ions is interpreted on the basis of their mass by charge ratio (m/z) and respective intensity. It is a high throughput technology and there is no upper m/z limit of detection. TOF has been modified according to the type of required application e. g. TOF in triple quadrupole is good for amino acid sequencing, MALDI QqTOF is good for amino acid sequencing as well as molecular weight determination. QqTOF is good for amino acid sequencing and any protein modification.

How TOF information is converted into m/z ratio?

Any charge particle (ions) in a region of potential difference (voltage) possesses potential energy which can be represented as:

Potential Energy (PE) = charge (q) * potential difference (V)

The accelerated ions travel in the field of potential difference and gain kinetic energy which can be equated as:

$$qV = \frac{1}{2} mv^2$$

Assuming the velocity remaining constant on the path of TOF, velocity can be expressed as a function of distance and time. The distance in TOF instrument is specific and known. Therefore, kinetic energy can be written as:

$$qV = \frac{1}{2} mv^2 = \frac{1}{2} m \left(d / t \right)^2$$

$$t = \frac{d}{\sqrt{2V}} \sqrt{\frac{m}{q}}$$

$\frac{d}{\sqrt{2V}}$ being constant can be written as K. therefore,

$$t = K \sqrt{\frac{m}{q}}$$

$$m/q = (t/K)^2$$

Other variants of instruments are:

- Magnetic Sector Analyzer (MSA): High resolution, exact mass, original MA.

- Quadrupole Analyzer (Q): It has low (1 amu) resolution but fast and cheap.

- Ion Trap Mass Analyzer (QSTAR): Good resolution, all-in-one mass analyzer.

- Ion Cyclotron Resonance (FT-ICR): Highest resolution, exact mass, costly.

Other types of Mass analyzers are:

- Triple Quadrupole: It is good for amino acid sequencing.

- MALDI-QqTOF: It is suitable for amino acid sequencing and Molecular weight determination.

- QqTOF: It is suitable for amino acid sequencing and identification of protein modification.

- ESI-QTOF: Electrospray ionization source + quadrupole mass filter + time-of-flight mass analyser.

- MALDI-QTOF: Matrix-assisted laser desorption ionization + quadrupole + time-of-flight mass analyzer.

- LC-MS - Liquid Chromatography MS: separates delicate compounds in HPLC column and ientifications by mass.

- LC/LC-MS/MS-Tandem LC and Tandem MS: Separates by HPLC, ID's by mass and AA sequence.

Tandem MS-MS/MS

It is a combination of more than one mass analyzer in sequence so that the resolution could be improved or targeted ions could be selected and studied in detail. It is very useful approach in the study of complex mixtures. It induces fragmentation of selected ions coming from first MS and then mass analyzes the fragment ions. This is done by placing a collision cell filled with Argon or Xenon in between the two mass analyzers. Depending on the intended goal tandem mass analyzers can be configured as:

- Quadrupole-quadrupole (low energy),

- Magnetic sector-quadrupole (high),

- Quadrupole-time-of-flight (low energy),

- Time-of-flight-time-of-flight (low energy).

Data Analysis

The final part of whole experiment is data analysis and interpretation. Data analysis mainly is of two types a) identification of the compound and b) differential quantitation of the compounds in two or more samples. The detected masses in the form of actual m/z ratio are plotted as mass spectrum. A typical mass spectrum looks like as given in the figure below. The list of all M/z obtained in experiment is finally compared with a theoretical list of putative mass of compounds using comprehensive algorithm. Data recorded in first mass analyzer is called as MS1 data. These are also called as precursor masses. In MS2 mode the precursor masses are further fragmented by high collision energy dissociation which yields typical fragments specific to the precursor ions. Thus MS2 reveals structural information of the precursor ions making the identification more confident. Inferring meaning from the generated compounds m/z is done on sophisticated software platform which enables accurate and specific search on the known database. For example, MASCOT, SEQUEST, X!Tandem, Phenyx, Protein Prospector etc. performs statistical matching between observed and known peptides m/z.

Intensity of the ions recorded on y axis can be used to compare the relative amount of the analyte in two samples. This is what we call differential analysis of biological samples. Again many free and paid dedicated software programs are available to do the comparison between samples. For example, MaxQuant is a freeware for quantitative proteomics developed by Jirgen Cox at Max Planck Institute of Biochemistry in Germany. It is important to know some of the definitions and terminologies used in reading the mass spectrum.

Resolution and Resolving Power

Resolution is the characteristics exhibited by the instrument by virtue of which it is able to plot different m/z as distinct and separate entity. If various m/z are very close to each other, probably the instrument will show spectra of overlapping peaks. In such case the peak would not specify to actual mass identified. Or some masses may not be identified at all. Therefore, the resolving power of instrument should be as high as possible. One peak should address only one specific m/z at its best accuracy.

In proteomics, resolution is defined for a single peak. It is calculated by dividing the particular m with delta "m". Delta m is the width of the peak at a height which is a specified fraction of the maximum peak height. The recommended heights are, 50%, 5% or 0.5%. In general resolution can be defined as the ratio of mass upon full width at half its maximum height (FWHM 50%).

Resolving power = m/delta m

Mass spectrometer instruments can be categorized on the basis of their resolution capacity. g. low, middle and high. Commonly a resolution up to 2000 is denoted as low and those over 20,000 as high.

Quadrupole Mass Analyzer

A quadrupole mass filter consists of four parallel metal rods with different charges. Two opposite rods have an applied positive potential and the other two rods have a negative potential. The applied voltages affect the trajectory of ions traveling down the flight path. At specific voltage specific entities reach to detector while others collide to the rods and fail to reach. Therefore, it is possible to select different m/z entities by scanning the sample over a wide voltage range. It is different from TOF in the sense that here the ions are separated in mass analyzer by voltage while in TOF the ions are separated by difference in the time of flight.

In triple quadrupole instrument, three quadrupoles are added in tandem. The middle one is for fragmentation (usually based on CID). The first and the second act as mass analyzers. With triple Quad, ESI is usually the ion source although atmospheric pressure chemical ionization (APCI) can be used.

Q-TOF Mass Analyzer

It is combination of two mass analyzers, one being Quadrupole and the other being TOF. Functionally it is identical to a triple quad except that Q3 is replaced by a TOF analyzer. It is faster and resolution is better than triple quad. Both MALDI and ESI can be used with it.

Ion Trap Mass Analyzer

Wolfgang Paul invented ion trap mass analyzer which won him Nobel Prize in 1989. In an ion trap mass analyzer, ions are trapped by a three-dimensional quadrupole field. It has good mass resolving power. It is used to trap ions in a confined space so that MS and MS-MS analysis both can be used in same space. Different ion trap devices have been developed like 3D ion trap, linear ion trap (LIT), Orbit rap etc. 3D ion trap is a modification of triple quad in that two quadrupoles serve as end cap and the other two rods of triple quad has been modified to serve as ring electrode at the center. The centre hole serves as ion trap. The ions coming from first end cap is trapped in the center where collision gas is introduced to fragment the precursor ions.

Different analyzers can be combined to create hybrid type analyzers. For example, ESI-QQQ, MALDI QQ TOF, Q TRAP, etc.

References

- Spectroscopic-techniques-i-spectrophotometric-techniques, principles-and-techniques-of-biochemistry-and-molecular-biology: cambridge.org, Retrieved 08, August 2020

- Analytical-aspects-of-uv-visible-spectroscopy, pharma-analysis: pharmatutor.org, Retrieved 08, July 2020

- Circular-dichroism, biomolecular-Interactions: sbs.ntu.edu.sg, Retrieved 03, Feb 2020

- Modern-spectroscopic-techniques-and-applications, atomic-spectroscopy: intechopen.com, Retrieved 13, March 2020

- Spectroscopy-elemental-isotope-analysis, industrial: thermofisher.com, Retrieved 28, March 2020

Electrophoretic Techniques

Electrophoretic techniques are based on the motion of charged particles under the influence of electric field. There are four types of electrophoretic techniques: polyacrylamide gel electrophoresis, agarose gel electrophoresis, capillary electrophoresis and microchip electrophoresis. The chapter closely examines the key concepts of these electrophoretic techniques to provide an extensive understanding of the subject.

Electrophoresis is defined as movement of charged particles in an electric field leading to their separation. In other words, it is separation technique which is based on differences in mobility of charged particles in an electric field. It is essential to understand factors which determine mobility of particles and these are being explained with the help of following statements.

When voltage (potential difference) is applied across the electrodes, it results in generation of potential gradient (E) which is equal to applied voltage (V) divided by distance between the electrode (d). Thus,

$$E = V/d.$$

It is required to be noted that 'd' will vary from equipment to equipment and thus for the same applied voltage, potential gradient will vary.

If charge on molecule is 'q' coulombs, then force exerted on particle is equal to 'Eq' newtons. This force causes the charged particles to move towards electrode. Particles carrying positive charge are referred as cations and will migrate towards cathode whereas particles carrying negative charge are called as anions and these will move towards anode. Thus, cathode is negatively charged and anode is positively charged.

It is also essential to be noted that particles under movement experiences frictional resistance which retards movement of charged particles. The frictional force is dependent on size, shape, and hydrodynamic volume of particle as well as on pore size of support matrix used for separation and viscosity of medium (buffer) used in electrophoresis. Thus, velocity 'v' is function of 'E' and frictional coefficient 'f' and is defined by following equation:

$$v = E / f = Eq / f$$

Electrophoretic mobility 'μ' of an ion is defined as ratio of velocity to potential gradient and is expressed as:

$$\mu = v / E = q / f$$

Molecules with identical or similar charges can be also separated if these molecules experiences different frictional force. Thus, separation of particles differing in charge as well as of similar charge is feasible in electrophoresis. In fact, separation of charged particles based on differences in frictional forces is routinely used in laboratories.

In electrophoresis, electric field must be switched off before ions reach respective electrode to allow their separation. Failure to do so will result in their neutralization at respective electrodes, a process called 'electrolysis'.

During electrophoresis, heat is generated in supporting medium and is given by following equation:

$$W = I^2 R$$

Where, W is power generated in Watts, I is current which increases with increase in voltage (V / I = R) and R is resistance. Power generated is dissipated as heat.

Heating of medium (support) will cause following problems:

- Sample and buffer ions will diffuse resulting in broadening of bands.

- Viscosity of buffer will decrease and thus, resistance of medium will also decrease. Thus, as per Ohm's law, it will result in increase in current at a given voltage. Heat can denature proteins and enzymes which may lose activity.

- If during electrophoresis, constant voltage is used, current will continue to slowly increase with time because of heat generated. More is the value of current, more is heat generated. Thus, this increased amount of heat will further increase current. As such amount of heat will be continuously increasing with time. Some workers prefer to run electrophoresis at constant current. Thus it should be clear that higher is the current, higher is heat generated and higher is velocity. In order to minimize heat generation, low voltage could be an option. However, this will end in long separation time which ultimately will result in diffusion of separated bands. Effects of heat can be minimized if support is maintained at low temperature. This is achievable if electrophoresis is performed at low temperature or cold water is circulated through buffers in electrode chambers or support material is layered over surface with good heat conducting properties or through Peltier cooling. It is also repeated here that force exerted on charged particle is dependent on voltage, not on current. It is therefore advised to perform electrophoresis on constant voltage which when applied should not result in generation of excess heat and also should not lead to diffusion of separated bands.

Charge or Net Charge on Macromolecule

For the students of biochemistry or life-science subject, electrophoretic separation of protein is more frequently encountered and therefore this macromolecule is discussed here. Now we will explain how charge on protein is contributed. To make this clear, it is essential to understand concept of pH, buffer, dissociation constant, pKa and different amino acids which acts as monomeric unit in protein. pH is minus logarithmic of hydrogen ion activity (concentration), i.e. pH = -log [H^+].

Water is weak electrolyte. It ionizes to small extent to give [H$^+$] and [OH$^-$]. Its dissociation and ionic product are summarized in following equations:

$$H_2O \rightleftharpoons \left[H^+\right] + \left[OH^-\right]$$

$$K = Equilibrium\,Constant = \frac{\left[H^+\right]\left[OH^-\right]}{\left[H_2O\right]} = 1.8 \times 10^{-16}$$

$$Kw = Ionic\,product\,of\,water = K.\left[H_2O\right] = \left[H^+\right]\left[OH^-\right]$$

$$= 1.8 \times 10^{-16} \times \left[H_2O\right]$$

$$= 1.8 \times 10^{-16} \times 55.6 = 1 \times 10^{-14}\,mol^2\,dm^{-6}$$

Concentration of H$^+$ and OH$^-$ ions from water is equal and is equal to 10^{-7} mol^2 dm^{-3}. Thus, pH of water is 7. For alkaline solution, pH is greater than 7 while for acidic solution, pH lower than 7. Also, it may be noted that pH values for strong acids will be less than 0 and that of strong bases greater than 14. For weak acids such as acetic acid, extent of dissociation and concentration of acetic acid will determine pH of solution.

$$HA \rightleftharpoons \left[H^+\right] + \left[A^-\right]$$

$$Ka = Dissociation\,Constant = \frac{\left[H^+\right]\left[A^-\right]}{\left[HA\right]}$$

$$\left[H^+\right] = \frac{Ka.\left[HA\right]}{\left[A^-\right]}$$

$$-Log\left[H^+\right] = Log\,Ka - Log\frac{\left[HA\right]}{\left[A^-\right]}$$

$$pH = pKa + Log\frac{\left[A^-\right]}{\left[HA\right]}$$

Previous equation is referred as Henderson–Hasselbalch equation. This equation can be used for calculating extent of dissociation of group at a given pH. pKa which is defined as negative log-arithmic of dissociation constant is constant for a given group. Thus by changing pH, extent of dissociation can be changed and consequently the charge contribution. Carboxylic group on dissociation will give one negative charge whereas un-dissociated will be charge-less. Contrary to it, amino group in undissociated state will carry one positive charge while dissociated state will be charge-less.

Proteins are polymers of amino acids which are linked together by peptide bond. Amino acids contain amino group and carboxylic groups. Carboxylic group of one amino acid is linked with amino group of other amino acid and thus the groups involved in peptide bond formation will not contribute to charge. However, α-amino group of first amino acid and α- carboxylic group of last amino acid in protein or peptide will contribute to charge on molecule. Different amino acids contain different side chains and some of them do contain groups which can contribute to charge on protein. For example side chains of aspartic acids and glutamic acid have carboxylic group while that of lysine has amino group. Similarly imidazole of histidine can contribute to charge. Thus (i) pKa of α-amino, ε-amino, α-carboxylic, ßcarboxylic, γ- carboxylic and imidazole, (ii) number of these groups in protein and (iii) pH will determine net charge on protein. It is again emphasized that dissociation of carboxylic group will lead to one negative charge while dissociation of amino group will abolish one positive charge. pKa values of groups in protein is given table.

Group	pKa	Group	pKa
α-amino (all amino acids)	8.80 to 10.6 (Different values for different amino acids)	Guanidino (Arg)	12.48
ε-amino (Lys)	10.53	Imidazole (His)	6.00
α-carboxylic (all amino acids)	1.83 to 2.83 (Different values for different amino acids)	Sufhydryl (Cys)	8.18
ß- carboxylic (Asp)	3.65	Phenolic (Tyr)	10.0
γ- carboxylic (Glu)	4.25		

pH of solution and number of dissociable groups in protein and their pKa values will decide the charge on protein molecule. Extent of dissociation can be determined by use of Henderson–Hasselbalch equation. When pH is equal to pKa, there will be 50% dissociation. When pH is one unit higher than pKa, there will be about 90% dissociation. When pH is two units higher than pKa, there will be about 99%.When pH is one unit lower than pKa, there will be about 10% dissociation. When pH is two units higher than pKa, there will be about 1% dissociation.

Isoelectric point (pI) of protein is pH at which net charge on protein is zero and at this pH, protein in electric field will not migrate. At pH value higher than pI, protein will have net negative charge and move towards anode. However, at pH value lower than pI, protein will have net positive charge and move towards cathode. Most of proteins have PI in the range of 6.0 to 6.8 and therefore at alkaline pH, such proteins will have negative charge while at pH lower than 6.0, the net charge will be positive. As the pH is increased from say from 2.0 to 10.0, net positive charge will gradually decrease and will reach zero and then acquire negative charge. Also, it is made clear that net charge on protein will continue to increase as one move away from pI. From these discussions, it is also clear that pH is required to maintained constant during the electrophoresis so that electrophoretic mobility remains dependent only on composition of protein. pH can be maintained by using buffer during electrophoresis. Let us know more about buffers.

Buffers

Buffers are solutions which resist change in pH on addition of acid or alkali. Contrary to it, if acid or alkali is added to water, it will bring large changes in pH. Buffers are mixtures of weak acids and salts or weak base and its salts. For example, mixture of acetic acid and sodium acetate will constitute buffer. Molar ratio of sodium acetate and acetic acid will determine pH of buffer. Addition of acid or alkali to buffer brings the changes in ratio of sodium acetate and acetic acid. If you see Henderson– Hasselbalch equation, log of ratio of dissociated state (eg. CH_3COO^-) to undissociated state (eg. CH_3COOH) will determine pH of buffer. Because of log of ratio, buffer resists change in pH. Alternate way of explaining how buffer functions requires consideration of dissociation of acetic acid as given below:

$$CH3COOH \rightleftharpoons CH3COO^- + H^+$$

When acid is added, the equilibrium is shifted to the left, in accordance with Le Chatelier's principle. Thus, hydrogen ion concentration increases by less than the amount expected for the quantity of strong acid added. Similarly, when alkali is added to the mixture, the hydrogen ion concentration decreases by less than the amount expected for the quantity of alkali added. Buffers exhibits better buffering in pH range of one unit either side of pKa values of weak acid. Buffering is very poor or negligible if pH of buffer is more than one unit away from pKa value. The pKa value of most common buffer is given in table.

Group	pKa	Group	pKa
Phosphate	1.97, 6.82, 12.5	Acetate	4.76
Citrate	3.09, 4.75, 5.41	Tricine	8.05
Boric acid	9.23, 12.74, 13.80 T	Tris (hydroxymethyl) amino methane	8.06
Carbonic acid	6.35, 10.33		

Preparation of Buffer

Preparation of Buffer is illustrated with the help of preparation of 0.1 M acetate buffer of pH value 5.76. Use Henderson–Hasselbalch equation in determining ratio of sodium acetate and acetic acid. pKa of acetic acid is 4.76.

pH-pKa= 5.76-4.76 = 1.0 = logarithmic of ratio of sodium acetate and acetic acid molar concentrations. Antilog of 1.0 = 10 = ratio of sodium acetate and acetic acid molar concentrations.

Mixing of 100 ml of 0.1 M sodium acetate (8.203g CH_3COONa is dissolved in water in total volume of 1000 ml) with 10 ml of 0.1 M acetic acid (6.005 g CH_3COO H diluted with water in total volume of 1000 ml) will make acetate buffer of pH 5.76 and of 0.1 M molarity. Check pH of buffer in pH meter calibrated with standard buffer. Adjust the pH with 0.1 M CH3COOH or 0.1 M CH3COONa if required.

Alternately, 90.90 mMoles of sodium acetate (7.456g) and 9.09 mMoles of acetic acid (0.5458g) can be dissolved in 900 ml water. pH is checked and adjusted with HCl or NaOH if required. Total volume is then made to 1000 ml with water.

Equipment Required for Electrophoresis

Performance of electrophoresis requires two basic equipments; power supply and an electrophoresis assembly. Power supply can be used in 'constant voltage' as well as 'constant current' mode. It has meter to display voltage and current and two slots for connecting lids. Power supply to electrophoresis assembly was through lids. The colour of lid is usually colour-coded. Red lid and red marked slot for lid is positively charged and is required to be connected to anode. Similarly black colourlid is connected to cathode.

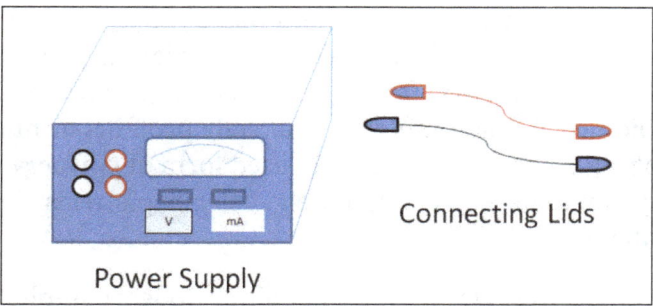

Assembly is different for Horizontal Gel Electrophoresis and Vertical Gel Electrophoresis. In Vertical Gel Electrophoresis, buffer chamber containing cathode is connected to buffer chamber containing anode through support gel. The assembly as well as support gel is kept in vertical position.

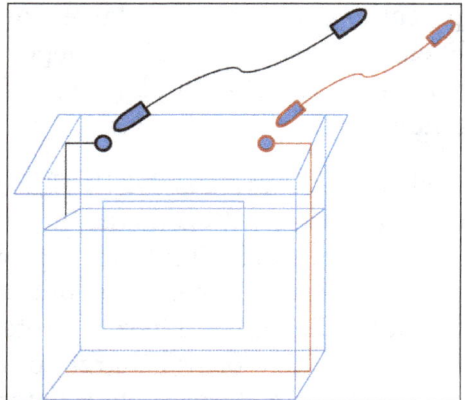

Figure: Vertical Gel Electrophoresis Unit.

On the other hand, horizontal electrophoresis is being usually done in one buffer system and support gel is submerged in buffer during electrophoresis. Agarose gel electrophoresis is done in horizontal electrophoresis.

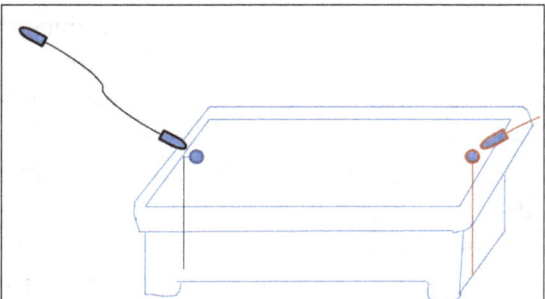

Figure: Horizontal Electrophoresis Unit.

Polyacrylamide Gel Electrophoresis

Polyacrylamide gels are chemically cross-linked gels formed by the polymerization of acrylamide with a cross-linking agent, usually N,N'-methylenebisacrylamide. The reaction is a free radical polymerization, usually carried out with ammonium persulfate as the initiator and N,N,N',N'-tetramethylethylendiamine (TEMED) as the catalyst. Polyacrylamide gel electrophoresis (PAGE) is a technique widely used in biochemistry, forensic chemistry, genetics, molecular biology and biotechnology to separate biological macromolecules, usually proteins or nucleic acids, according to their electrophoretic mobility. The most commonly used form of polyacrylamide gel electrophoresis is the Sodium dodecyl suplhate Polyacrylamide gel electrophoresis (SDS- PAGE) used mostly for the separation of proteins.

Polyacrylamide Gel Electrophoresis (PAGE) and sodium dodecyl sulphate – polyacrylamide gel electrophoresis (SDS-PAGE): In both the techniques, gel is prepared by polymerization of acrylamide. Mixture of proteins can be separated in both PAGE and SDS-PAGE. In PAGE, electrophoretic mobility is dependent on net charge on each of protein which of course is dependent on composition of protein and pH of buffer used in electrophoresis. Since all proteins in mixture are separated simultaneously, potential gradient and to large extent frictional coefficient will be constant. However in SDSPAGE, electrophoretic mobility is dependent on molecular size or mass of protein. In this technique, charge to mass ratio for all proteins is made constant and frictional forces are different for different molecules.

Sodium Dodecyl Sulphate (SDS) chemical formula is CH_3-$(CH_2)_{10}$-$CH_2OSO_3^-Na^+$) which is an anionic detergent. When SDS comes in contact with proteins in presence of a reducing agent like β-mercaptoethanol or dithiothreitol, it binds to hydrophobic regions of protein molecule and provides net negative charge on protein molecule. The binding of SDS to per-unit-length of protein molecules is almost constant for large number of different proteins. For every two residues of amino acids, one molecule of SDS binds. This brings charge-to–mass ratio almost constant for most proteins. Mobility of charged particles in electric field is determined by ratio of force exerted on charged particle to frictional coefficient. The net charge on protein after binding of SDS is negative and therefore, protein-SDS complex will migrate towards anode in an electric field. Further, charge to mass ratio is constant and thus electrophoretic force exerted on all protein molecules in mixture will be same. The difference in electrophoretic mobility will be as a result of difference in frictional force. More is frictional force, lower will be mobility and vice versa. Electric field provides movement of Protein- SDS complex. The complex moves through pores of acrylamide gel and friction is experienced by complex while moving. The quantum of friction is different for different molecules. Relatively large size molecules will experience high friction and thus move slowly. On the other hand, small size molecules will experience low friction and will therefore move speedily. Depending on the size of molecule, the value of friction will vary and ultimately it will be reflected in velocity. Also, it is to be noted that if size of protein-SDS complex is larger than pore of gel, the particle will not move and particles will remain stagnated at entry point and these will not be separated. Similarly if size of all particles is much smaller than size of gel-pore, all particles will move speedily and close to dye front. These particles will also be not separated. The ideal situation demands that protein molecules in mixture must experience frictional force different in magnitude to allow their separation. In case either desired protein is not separated from rest of protein or all proteins are not resolved in gel, gel of altered pore size is required to be used. Protein molecules after binding to SDS unfolds and SDS-PAGE is carried out in unfolded (denatured) state to have all molecules in unfolded state. β-mercaptoethanol or dithiothreitol reduces disulphide bonds and therefore assists in complete unfolding of proteins. Oligomeric proteins after reduction with β-mercaptoethanol or dithiothreitol in presence of SDS will exist in monomeric form.

SDS-PAGE is carried out in discontinuous buffer system. In this system, buffers used in tank and for preparation of gels are different. Further, separation requires migration of particles first through large pore gel (stacking gel) and then in small pore gel (separating or running gel). Stacking gel is layered over separating gel. The upper end of stacking gel and lower end of separating gel remain in contact with buffer during electrophoresis. Lower molecular weight proteins move faster than high molecular weight proteins. The method described by Laemmli is widely used. In this method, discontinuous buffer system is employed. For preparation of stacking gel and separating gel, different buffers are used and also tank buffers are different from gel buffers. When electrophoresis is started, ions from stacking gel buffer (Tris-HCl), ions from tank buffer (Tris-glycine) and protein-SDS complex start moving into stacking gel. Cl from Tris-HCl leads while glycinate ion trails and protein-SDS complexe moves between thes two ions. This leads to concentrate protein in a thin zone referred as stack. The pH of stacking gel is 6.8 and at this pH, net negative charge on glycine is low. Therefore, glycinate ion trails in stacking gel. The protein molecules continue to move in the stack until they reach the separating gel. When glycinate ion enters separating gel of which pH is 8.8, it acquires more negative charge resulting in enhanced mobility. In separating gel, glycinate ion will cross Protein-SDS complex. The purpose of stacking gel is to concentrate proteins before their entry into separating gel. Acrylamide concentration in stacking gel is kept low around 4 % while higher concentration of acrylamide from 7.0 to 15% is used separating gel.

Chemistry of Polyacrylamide Gel

Acrylamide gel is prepared by polymerization of mixture of acrylamide and methylene bis acrylamide using free radical reaction. Free radicals are generated by decomposition of ammonium persulphate (APS) by tetramethyl ethylenediamine (TEMED). TEMED exists in free radical form and it catalysis conversion of persulphate to sulphate-free radical. Aqueous solution of APS can also generate sulphate- free radical but this process is fastened by TEMED.

$$S_2O_8^{2-} + e^- \rightarrow S_2O_8^{2-} + SO_4^-$$

Or

$$S_2O_8^{2-} \rightarrow 2SO_4^-$$

Sulphate-free radical snatches one electron from carbon-carbon double bond of acrylamide ($CH_2=CH-CONH_2$) or bisacrylamide ($CH_2=CH-CONH-CH_2-NHCO-CH=CH_2$) and latter becomes free radical. Acrylamide or bis-acylamide-free radicals will act on new molecule of acrylamide or bisacrylamide and in this way long chains of polymers incorporating many molecules of acrylamide and some molecules of bis-acrylamide are synthesized. Bisacrylamide forms cross links between linear chains of acrylamide and therefore bisacrylamide is used in low amount in reaction mixture.

1. $CH_2 = CH - CONH_2 + 2SO_4^- \bullet \rightarrow CH_2 - C \bullet H - CONH_2$

2. $-CH_2 - C \bullet H - CONH_2 + CH_2 = CH - CONH_2 \rightarrow -CH_2 - CH - CONH_2$
$$| $$
$$CH_2 - C \bullet H - CONH_2$$

3. $-CH_2-CH-CONH_2$
   |
   $CH_2-C\cdot H-CONH_2 + CH_2 = CH-CONH_2 \rightarrow -CH_2-CH-CONH_2$
   |
   $CH_2-CH-CONH_2$
   |
   $CH_2-C\cdot HCONH_2$

Bisacrylamide (CH_2=CH-CONH-CH_2-NHCO-CH=CH_2) can similarly linked to two linear chains of growing acrylamide polymer. Bisacrylamide provides rigidity and is added in low concentration in comparison to acrylamide so that gel remains elastic and does not form cracks in handling.

Experimental Procedure

Equipments and Chemicals

Mini-vertical gel electrophoresis dual model with glass plates, spacer, comb and powersupply; orbital shaker; acrylamide, N,N' methylene bisacrylamide; ammonium persulfate; β- mercaptoethanol; sodium dodecyl sulfate; molecular weight markers; coomassie brilliant blue R-250; TEMED; tris; glycine; dithiothreitol.

Stock Solutions

- Acrylamide/Bisacrylamide (30%): 29.2 g acrylamide and 0.8 g bisacrylamide are dissolved in distilled water and total volume was made to 100 ml. The solution is filtered and filtered solution can be stored at 40C in dark bottle up to 3 months.

- 4 x Running Gel Buffer (1.5 M Tris-HCl, pH 8.8): 18.15 g Tris is dissolved in about 80 ml distilled water. pH is adjusted to 8.8 with 1 N HCl and total volume is made to 100 ml with distilled water. Prepared buffer can be stored up to 3 months at 40C in dark bottle.

- 4 x Stacking Gel Buffer (0.5M Tris-HCl, pH 6.8): 3.0g Tris is dissolved in about 40 ml distilled water. pH is adjusted to 6.8 with 1 N HCl and total volume is made to 50 ml with distilled water. Prepared buffer can be stored up to 3 months at 40C in dark bottle.

- 10% SDS: 10 g sodium dodecyl (lauryl) sulfate is dissolved in distilled water and total volume is made to 100 ml with distilled water. Prepared solution can be stored at room temperature.

- 5 x Electrode Buffer (125 mM Tris, 960 mM Glycine, 0.5 % SDS, pH 8.3: 15 g Tris, 72 g glycine and 5 g SDS are dissolved in distilled water and total volume is made to 1 litre with distilled water. The pH of buffer should be 8.3 ± 0.2. Stock electrode buffer is diluted five times with distilled water before use. The stock buffer can be stored at room temperature up to 1 month. The diluted stock buffer is 25 mM tris, 192 mM glycine, and 0.1% SDS.

- 10% Ammonium Persulfate: 100 mg ammonium persulfate is dissolved in 1.0 ml distilled water. The solution is always prepared fresh.

- 2 x Sample Buffer (0.125 M Tris, 4% SDS, 20% glycenol 0.2 M DTT, .02% bromophenol blue, pH 6.8: 2 x sample buffer is prepared by mixing following solutions/chemical.

- 4 x stacking gel buffer-2.5ml.

- Glycerol-2.0ml.

- 10% SDS-4.0ml.

- Bromophenol blue-2.0mg.

- Dithiothreitol (DTT)-0.31g.

- Distilled water-1.5ml.

2 x sample buffer can be stored in small aliquots at - 20 °C up to 6 months. Instead of DTT, 1.0 ml of β-mercaptoethanol can be used but the volume of water is reduced to 0.5 ml.

- Overlay Buffer (0.375 M Tris, 0.1%, SDS, pH 8.8): Overlay buffer is prepared by mixing 25 ml running gel buffer, 1ml 10% SDS and 74 ml distilled water. This buffer can be stored up to 3 months at 4 °C in the dark bottle.

Preparation of Glass Sandwich

One notched glass plate is placed on a flat surface. One spacer (1.0 mm) each is then placed along the each of two edges so that spacer aligns with the notch. Subsequently, rectangular glass-plate is placed over it. The sandwich is held firmly between thumb and fingers. Side-ways of both spacers were sealed with appropriate tape to overcome any possible gel-leak during gel plate preparation. There is always the possibility of leakage at the bottom of the plate. This is taken care by placing molted agar (1% in water) up to 5 mm height in trough of gel-casting unit. The plate in standing position is then quickly placed in casting unit and screws are finger tightened.

Preparation of Running Gel

Running gel of desired concentration is prepared by mixing appropriate volumes of solutions as shown below in table.

Final Gel Concentrations				
Solutions	7.5%	10%	12.5%	15%
Acrylamide / bisacrylamide (30%)	5.0ml	6.7ml	8.3ml	10ml
4 x Running gel buffer	5.0ml	5.0ml	5.0ml	5.0ml
10% SDS	0.2ml	0.2ml	0.2ml	0.2ml
Distilled water	9.7ml	8.0ml	6.4ml	4.7ml
10% Ammonium persulfate	0.1ml	0.1ml	0.1ml	0.1ml
TEMED	6.7 µl	6.7 µl	6.7 µl	6.7 µl

Acrylamide/bisacrylamide, running gel buffer, SDS and distilled water are added to conical flask and degassed. Then ammonium persulfate and TEMED are added and contents mixed gently. With the help of dispensor, the running gel solution is delivered to sandwich to a level about 3 cm below the top of rectangular plate. Air should not be trapped while filling sandwich with running gel solution. A small volume of water or overlay-buffer (~200 µl) is layered over gel solution. This prevents exposure to oxygen and also keeps gel surface flat.

Preparation of Stacking Gel

Stacking gel of 4% concentration is prepared by mixing appropriate volumes of solutions as shown in table.

Solutions	Volume
Acrylamide / bisacrylamide (30%)	1.3 ml
4 x Stacking gel buffer	2.5 ml
10% SDS	0.1 ml
Distilled water	6.1 ml
10% Ammonium persulfate	50 µl
TEMED	10 µl

The preparation of stacking-gel solution is similar to preparation of running-gel solution. After removal of water or overlay buffer, stacking-gel solution is layered over running gel. Appropriate comb is inserted into the stacking gel to make wells for sample application. Comb is removed after polymerization of gel.

Sample Preparation

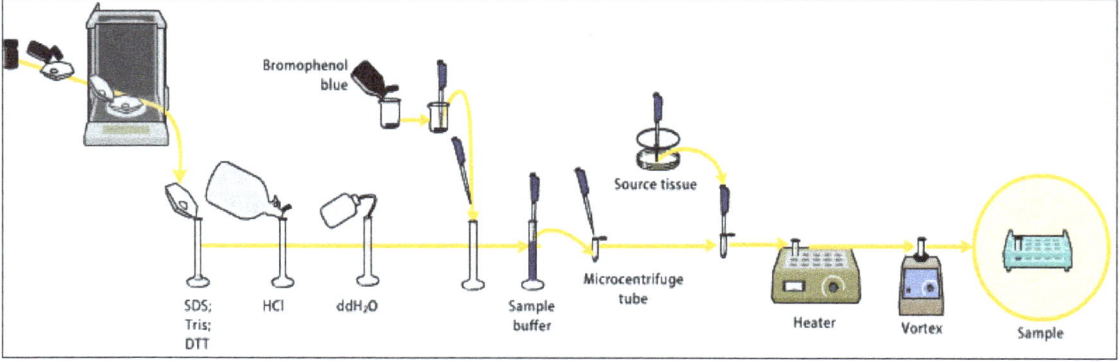

Samples may be any material containing proteins or nucleic acids. The sample to analyze is optionally mixed with a chemical denaturant if so desired, usually SDS for proteins or urea for nucleic acids. SDS is an anionic detergent that denatures secondary and non–disulfide–linked tertiary structures, and additionally applies a negative charge to each protein in proportion to its mass. Urea breaks the hydrogen bonds between the base pairs of the nucleic acid, causing the constituent strands to anneal. Heating the samples to at least 60 °C further promotes denaturation. A tracking dye may be added to the solution. This typically has a higher electrophoretic mobility than the analytes to allow the experimenter to track the progress of the solution through the gel during the electrophoretic run.

Preparation of Polyacrylamide Gel

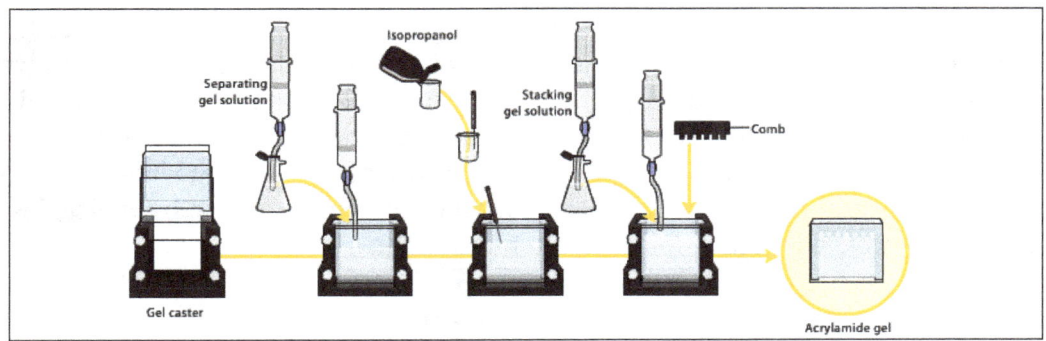

The gels typically consist of acrylamide, bisacrylamide, the optional denaturant (SDS or urea), and a buffer with an adjusted pH. The ratio of bisacrylamide to acrylamide can be varied for special purposes, but is generally about 1 part in 35. The acrylamide concentration of the gel can also be varied, generally in the range from 5% to 25%. Lower percentage gels are better for resolving very high molecular weight molecules, while much higher percentages of acrylamide are needed to resolve smaller proteins. Gels are usually polymerized between two glass plates in a gel caster, with a comb inserted at the top to create the sample wells. After the gel is polymerized the comb can be removed and the gel is ready for electrophoresis.

Electrophoresis

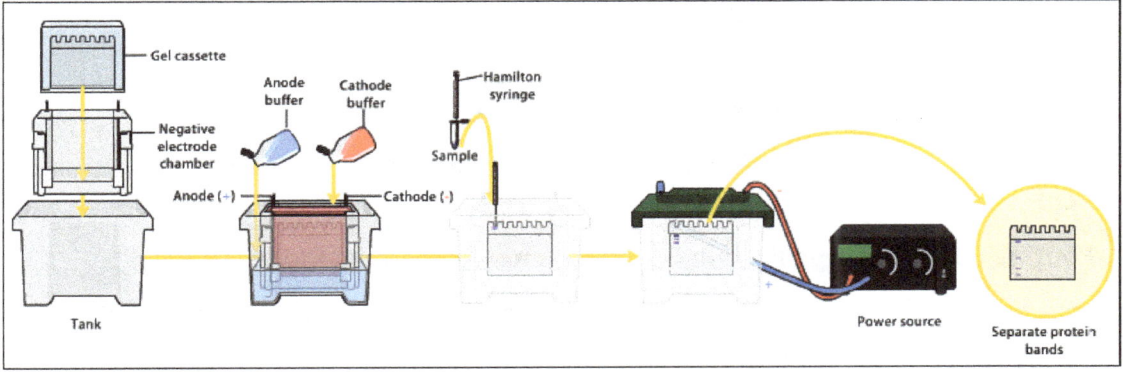

Various buffer systems are used in PAGE depending on the nature of the sample and the experimental objective. The buffers used at the anode and cathode may be the same or different. An electric field is applied across the gel, causing the negatively charged proteins or nucleic acids to migrate across the gel away from the negative and towards the positive electrode (the anode). Depending on their size, each biomolecule moves differently through the gel matrix: small molecules more easily fit through the pores in the gel, while larger ones have more difficulty. The gel is run usually for a few hours, though this depends on the voltage applied across the gel. After the set amount of time, the biomolecules will have migrated different distances based on their size. Smaller biomolecules travel farther down the gel, while larger ones remain closer to the point of origin. Biomolecules may therefore be separated roughly according to size, which depends mainly on molecular weight under denaturing conditions, but also depends on higher-order conformation under native conditions.

Detection

Following electrophoresis, the gel may be stained (for proteins, most commonly with Coomassie Brilliant Blue or autoradiography; for nucleic acids, ethidium bromide; or for either, silver stain), allowing visualization of the separated proteins, or processed further (e.g. Western blot). After staining, different species biomolecules appear as distinct bands within the gel. It is common to run molecular weight size marker sof known molecular weight in a separate lane in the gel to calibrate the gel and determine the approximate molecular mass of unknown biomolecules by comparing the distance traveled relative to the marker.

Applications of Polyacrylamide Gel Electrophoresis (PAGE)

- Measuring molecular weight.

- Peptide mapping.

- Estimation of protein size.

- Determination of protein subunits or aggregation structures.

- Estimation of protein purity.

- Protein quantitation.

- Monitoring protein integrity.

- Comparison of the polypeptide composition of different samples.

- Analysis of the number and size of polypeptide subunits.

- Post-electrophoresis applications, such as Western blotting.

- Staining of Proteins in Gels with Coomassie G-250 without Organic Solvent and Acetic Acid.

- Pouring and Running a Protein Gel by reusing Commercial Cassettes.

- Selective Labelling of Cell-surface Proteins using CyDye DIGE Fluor Minimal Dyes.

- Detection of Protein Ubiquitination.

Advantages of Polyacrylamide Gel Electrophoresis (PAGE)

- Stable chemically cross-linked gel.

- Greater resolving power (Sharp bands).

- Can accommodate larger quantities of DNA without significant loss in resolution.

- The DNA recovered from polyacrylamide gels is extremely pure.

- The pore size of the polyacrylamide gels can be altered in an easy and controllable fashion by changing the concentrations of the two monomers.

- Good for separation of low molecular weight fragments.

Disadvantages of Polyacrylamide Gel Electrophoresis (PAGE)

- Generally more difficult to prepare and handle, involving a longer time for preparation than agarose gels.

- Toxic monomers.

- Gels are tedious to prepare and often leak.

- Need new gel for each experiment Stable chemically cross-linked gel.

Agarose Gel Electrophoresis

Agarose is one of the two principal components of agar which was discovered in 1658 and coined as a Kanten by Mino Tarōzaemon in Japan. Agar is mixture of polysaccharides isolated from sea weeds such as Gelidium and Gacilaria. Agarose is a linear polysaccharide which contains agarobiose as a repeating unit. Agarobiose is a disaccharide and consists of D-galactose and 3,6-anhydro-L-galactopyranose. Its molecular weight is about 12,000 Daltons.

Figure: Structure of agarose.

Agarose solubility in water is low. However, when aqueous suspension of agarose is boiled, it results in clear solution. When boiled and clear solution of agarose is slowly allowed to cool, it forms gel. Gel formation is ascribed to formation of intra and inter-molecular hydrogen bonding. Intra hydrogen bonds are within long agarose chains while intra hydrogen bonds are formed between chains. Pore size of gel is dependent on agarose concentration. Gel formed from higher agarose concentration will be of low pore size and vice-versa. Although agarose is supposed to be free from charge, sulphate groups are present at alternate sugars to a variable degree. Support matrix for electrophoresis should be free from charge. Sulphate group in agarose interferes in electrophoresis. This is explained in terms of electroendosmosis phenomenon. Cations from buffer ions or salts can bind to sulphate and make electric double layer. During electrophoresis, when voltage is applied, cations from electrolyte layer moves towards cathode and in this process carry electrolyte solution towards cathode. This is referred as electroosmotic flow in the direction of cathode. The quality of agarose is determined by level of sulphate and agarose with minimum level of sulphate is used in agarose gel electrophoresis for better separation.

Agarose gel is used for separation of nucleic acids and proteins in electrophoresis. Only large size proteins can be separated in agarose gel electrophoresis. Immnoelectrophoresis is good example for separation of antibodies on agarose gel prior to diffusion of antigen and antibody through agarose gel. In isoelectric focusing also, agarose gel is used. It is true that electrophoretic separation of nucleic acid is mostly achieved on agarose gel. It is pertinent to discuss the pore size of acrylamide gel and agarose gel. Although pore size of acrylamide gel and agarose gel is dependent on their concentration, in general pore size of agarose gel is larger than acrylamide gel. Also, in general size of nucleic acid is much larger than size of protein. Both in acrylamide gel electrophoresis and agarose gel electrophoresis, protein or nucleic acid is required to move through pores of support and pores must create differential frictional force in movement of charged particles. This is the reason why proteins are mostly separated on acrylamide while nucleic acids on agarose gel. Of course, small size nucleotides can be separated in acrylamide gel electrophoresis. Comparatively, elasticity of agarose gel is poor in comparison to acrylamide. It is difficult to remove agarose gel from small tube or glass plates after vertical gel electrophoresis. Thus, agarose gel electrophoresis is not performed on vertical electrophoresis unit and is carried out in horizontal electrophoresis unit.

In both agarose gel electrophoresis and SDS-PAGE, separation is based on size of size of nucleic acids and protein respectively. In both the case, charge to mass ratio of respective macromolecules is constant. In SDS-PAGE, this is achieved by binding of SDS to protein which acquires net negative charge. The binding of SDS to protein is almost constant for wide range of proteins and for every two residues of amino acids in protein, one molecule of SDS binds. SDS binds to hydrophobic pockets in protein and in doing so, sulphate groups are exposed out. Nucleic acids do not require any treatment for making their charge to mass ratio constant. Nucleic acids have negative charge and amount of net negative charge is proportional to size of nucleic acids. So, charge to mass ratio of nucleic acids is constant. Large size nucleic acids move slowly whereas small size molecules move faster in agarose gel electrophoresis. Size of nucleic acids and size of pore in agarose gel determine magnitude of frictional force exerted on nucleic acids in their movement. This situation is similar to SDS-PAGE. Pore size of agarose gel is dependent on concentration of agarose. Size of nucleic acids should be such that these are able to enter pores of gel and all components in mixture should experience variable frictional force to allow their separation.

Different types of agarose are available in market. These include agarose with characteristics of low electroendosmosis (EEO), medium EEO, and high EEO, high gelling temperature, low gelling temperature and high gel strength. High gelling temperature gives higher thermal stability with elasticity and flexibility to the gels. Low gelling temperature agarose is preferred for cloning experiments for extracting small size nucleotides.

Factors affecting Separation on Agarose Electrophoresis

The separation of targeted DNA fragments in agarose electrophoresis depends: (i) Size and shape of the nucleic acids, (ii) concentration of the agarose (iii) voltage and electric field, (iii) concentration of the agarose and (iv) buffer.

Size and Shape of the Nucleic Acids

Size of the molecule is most important factor in resolution of the sample. Larger size molecules or fragments migrate slower because of higher frictional force generated by the gel. Smaller molecules or fragments move faster through the pores. Since, nucleic acids are negatively charged at pH of buffer, these moves towards anode.

Shape of the molecule also effects the migration of the molecules in the electric field. Molecules having globular shape will move faster than linear or fibrous shape. When plasmid subjected to resolution in electrophoresis, super coiled plasmid will move faster followed by coiled plasmid and then by linear plasmid.

Concentration of the Agarose

The concentration of agarose depends upon the size of the DNA fragments which are to be separated. Agarose concentration vis a vis range of size of DNA fragments is given below in table.

Agarose (%)	Fragment Size (base pairs)
0.5	10,000-15,000
1.0	500-10,000
1.5	200- 5000
2.0	100-2000

Molecular weight of nucleic acids is in general much larger than proteins. Size of protein is represented in terms of its molecular weight and denoted by Dalton or kilo Dalton (kD). However for nucleic acid, size is represented in terms of base pairs (bp) or nucleotides (nt). Size of double stranded nucleic acid is denoted by base pairs 'bp' whereas size of single stranded nucleic acid is expressed in terms of nucleotide 'nt'. One kilo base pair or '1kb' is equivalent to 6,20,000 Daltons. Commonly used plasmid PBR322 is of 24,00,000 Daltons and size of it is referred as 4.32 kb.

Pore size or the sieving effect of the agarose depends upon the concentration of the agarose. Generally molecular size of above 50 bp can be separated easily. Mobility of the nucleic acids slows down in gels prepared with higher concentration of agarose. Sample resolution time is inversely promotional to concentration of the agarose.

Voltage and Electric Field

Migration of ion in an electric field depends on potential gradient which is equal to applied voltage divided by distance between two electrode, charge on particle and frictional force exerted by medium on particle. Higher voltage will increase current which will results in more heat generation. Generated heat is to be efficiently dissipated otherwise it will lead to poor resolution and broadening of bands.

Buffer

The buffer affects the mobility of the resolving molecules by decreasing its migration in the electric field. Most commonly Acetate or citrate buffers are used in DNA separation. If the ionic strength of the buffer is more, migration of the molecule will be slow. In low ionic strength of the buffer DNA fragments carry the current to migrate faster than the buffer ions.

Methodology for Performing Agarose Gel Electrophoresis

Materials Required

Power Supply, horizontal electrophoretic unit, casting unit, comb, agarose, DNA sample, DNA ladder, tris base, glacial acetic acid, boric acid, ethylenediaminetetraacetic acid (EDTA), disodium EDTA, ethidium bromide, xylene cyanol dye, bromophenol blue dye, sodium dodecyl sulfate, glycerol, transilluminator.

Electrode Buffer

Low ionic strength as well as relatively high ionic strength buffers is used in agarose gel electrophoresis. Low ionic strength buffer is popularly referred as Tris-Acetic Acid-EDTA of TAE which is comprised of 4 mM Tris and 1mM EDTA. pH of buffer is adjusted with acetic acid to 8.0. This buffer is used for high molecular weight double stranded DNA. EDTA sequesters divalent ions which are essential for nucleases action. This buffer is also preferred for DNA recovery. Other buffer which is common is referred Tris-borate-EDTA (pH-8.3) and popularly called as TBE. It is comprised of 89 mM Tris, 89 mM Boric Acid and 2 mM EDTA Na2. Ionic strength of TBE buffer is higher than TAE buffer. Borate is inhibitor for many enzymatic reactions and thus TBE buffer is to be avoided for such application. Both buffers inhibit the protonation of DNA and protect DNA molecule from the enzymatic degradation.

Sample Buffer and Sample Preparation

Stock solution of 6 X sample buffer (loading buffer) contains 30% glycerol in distilled water, 0.25% bromophenol blue and 0.25% xylene cyanol and is prepared in sample buffer which should be of lower ionic strength than the electrode buffer. High ionic strength of loading buffer causes bands to be fuzzy and results in migration of DNA at unpredictable rates. Alkaline loading buffer is used in alkaline gel electrophoresis. Amplified or DNA fragment is dissolved in loading buffer. The purpose of loading buffer is to increase density of the sample and its visibility on loading. Movement of dyes present in loading buffer provides status of electrophoretic run. Loading dyes are negatively charged in buffer and hence these will migrate in the direction of DNA migration.

Agarose Gel Preparation and its Casting in Gel Casting Unit

Agarose is dispersed in 1X TAE/TBE buffer without clumps in clean glass conical flask to obtain required concentration of gel. The contents are heated in microwave oven or on hot plate. When solution becomes clear and transparent, it is allowed to cool to approximately 550C. At this stage, ethidium bromide (EtBr) is added to obtain 0.5µg/ml concentration in gel and contents are swirled without forming any air bobbles in melted agarose. Immediately, transparent agarose gel (without further cooling) is gently poured in gel casting unit avoiding trapping of air bubbles. Depth of gel can be kept to 3 mm. Ethidium bromide is a carcinogenic compound and thus it requires careful handling which includes use of lab coat and gloves. Comb is then placed gently in gel at appropriate location. When gel is solidified, gel is ready for electrophoresis.

Molecular Marker or DNA Ladder

Molecular weight markers or referred as DNA ladder is also dissolved in loading buffer. These essentially contain mixture of DNA molecules whose molecular size is known. Ladders are available in size difference of 100 bp or smaller. Size of bacteriophage λDNA is 49kb which on action of restriction enzyme generates fragments of 21.80 kb, 7.52 kb, 5.93 kb, 5.54 kb, 4.80 kb and 3.41 kb. These fragments can also use as molecular markers. These are loaded to one of the well in agarose to allow correct determination of size of DNA in sample.

Method

Electrophoretic Run

When gel and samples are ready to use, electrophoresis unit tank is filled with 1X TAE or TBE buffer to a height to allow merging of gel in buffer. Then, gel is placed in electrophoretic unit. Comb is removed without any damage to the well. Sample is loaded in wells created by the comb in gel by using dispenser. Molecular marker or the DNA ladder are usually loaded in first and last well. Electrophoretic chamber is then covered with lid. Electrophoresis is done at voltage of approximately 5 Volts per centimeter of distance between electrodes. It is essential to connect lids to power supply correctly. The movement of DNA and markers will be towards anode. The edge effect will be observed when the high voltage is applied. Run is continued till loading dye migrates to appropriate distance.

Visualization

The nucleic acids are visualized by using ethidium bromide. Ethidium Bromide intercalates the major grooves present in the DNA. Its density enhances wherever DNA in gel is there. When gel is exposed to UV light, ethidium bromide emits florescence and DNA appears as orange red.

DNA can also be stained with ethidium bromide after electrophoretic run and this requires placing of gel in ethidium bromide solution for 5-10 minutes. 10 ng of DNA can be detected. Other alternative dyes to visualize the DNA are also available. This includes SYBR Green which is used for quantification of nucleic acid. This dye can also label DNA within the cells. Other dyes 'GelRed' is closely related to the EtBr in its structure and binding but it is less toxic and more sensitive. Once the gel is stained with ethidium bromide, DNA fragment can be visualized under ultraviolet transilluminator. Under ultraviolet light aromatic ring of ethidium molecules activate and release energy in the form of light. Generally ethidium bromide absorbs a range of 302/312 nm known as UV-B rays. Longer time exposure of the DNA to a UV light can damage DNA and thiamine dimmers can form. If DNA is to be recovered, it should be minimally exposed to UV light. Visualization at 365 nm (called UV-A) causes less damage to the DNA but visualization is poor. It is essential to protect one's eye by wearing goggles when UV light is used. While viewing gel exposed to UV light, UV shields must be placed over gel for protecting eyes.

Applications of Agarose Gel Electrophoresis

The agarose gel electrophoresis is widely employed to estimate the size of DNA fragments after digesting with restriction enzymes, e.g. in restriction mapping of cloned DNA. It has also been a routine tool in molecular genetics diagnosis or genetic fingerprinting via analyses of PCR products. Separation of restricted genomic DNA prior to Southern blot and separation of RNA prior to Northern blot are also dependent on agarose gel electrophoresis.

Agarose gel electrophoresis is commonly used to resolve circular DNA with different supercoiling topology, and to resolve fragments that differ due to DNA synthesis. DNA damage due to increased cross-linking proportionally reduces electrophoretic DNA migration.

In addition to providing an excellent medium for fragment size analyses, agarose gels allow purification of DNA fragments. Since purification of DNA fragments size separated in an agarose gel is necessary for a number molecular techniques such as cloning, it is vital to be able to purify fragments of interest from the gel.

Increasing the agarose concentration of a gel decreases the migration speed and thus separates the smaller DNA molecules makes more easily. Increasing the voltage, however, accelerates the movement of DNA molecules. Nonetheless, elevating the currency voltage is associated with the lower resolution of the bands and the elevated possibility of melting the gel (above about 5 to 8 V/cm).

Capillary Electrophoresis

Capillary Electrophoresis technique is known for its high performance and also has been mentioned as high performance capillary electrophoresis or free solution capillary electrophoresis, capillary zone electrophoresis or capillary electrophoresis. In this technique, the reagents used are in microliter scale, sample detected are in nanoscale and the time taken is in minutes. Techniques such as PAGE, SDS-PAGE, isoelectric focusing and pulsed field gel electrophoresis can be carried out in capillary systems. This technique is very useful for the separation and analysis of wide variety of compounds such as amino acids, peptides, proteins, oligonucleotides, DNA fragments, nucleic acids, metal ions, drugs.

During the performance of different methods of capillary electrophoresis, the analyte is separated on the basis of its migration in the applied electric field. The main advantage of using CE over the conventional methods using the slab gels is to overcome the heating effects generated during the electrophoresis. In electrophoresis, high voltage generates heat which has deleterious impact on the resolution as well as on the structural integrity of biopolymer being separated. This problem can be resolved by performing the electrophoresis in thin capillaries having small internal volumes and large surface to volume ratio. Therefore, heat generated in such capillaries during electrophoresis can be dissipated efficiently.

Theoretical Plates: The numbers of theoretical plates are calculated by following equation:

$$N = \mu V/2D$$

Where,

- μ = electrophoretic mobility.
- V = voltage applied.
- D = Diffusion coefficient of the sample in the electrophoresis medium.

Number of theoretical plates is dependent on electrophoretic mobility, applied voltage and diffusion coefficient. With increase in electrophoretic mobility or voltage,, number of theoretical plates will increase while with increase in diffusion coefficient, number of theoretical plates will decrease.

Retention Time

Time (t) taken for sample to pass through capillary is presented by following equation.

$$t = L^2 / \mu V$$

Where, L is length of tube, μ is electrophoretic mobility and V is applied voltage.

From these above two equations, it is clear that column length does not influence theoretical plates and hence resolution. However, it does influence time required to pass the sample through capillary tube. High voltage will result in increase of number of theoretical plates and hence resolution. Essentially it is to be also understood that higher voltage also results in more generation of heat and ultimately it results in increase in diffusion coefficient. In electrophoresis, increase in voltage will result in increase in current and later will result in more generation of heat. Generated heat must be dissipated fast for minimizing its effect on diffusion. In capillary electrophoresis, internal volume is very low i.e. in microlitres while surface area is very large. Thus, generated heat is very quickly dissipated. This is great advantage seen in capillary electrophoresis. Because of quick dissipation of heat in capillary electrophoresis, no support such as agarose or acrylamide is required to minimize diffusion. Thus in capillary electrophoresis, molecules move freely in solution.

Length of capillary tube does not influence number of theoretical plates and thus longer capillary tubes are not advised. However, it is to be noted that shorter capillary tubes will result in decreased resistance and thus increase in heat. In shorter capillary, surface area available for heat dissipation will also decrease and can result in increase in diffusion coefficient. Also, there will be physical limit to reduce the size of capillary tube. Capillary length is kept usually in the range of 10 to 100 cm. internal diameter of tube is between 10 and 100 µm while external diameter is usually 300 µm.

Instrumentation for Capillary Electrophoresis

The basic instrumentation for capillary electrophoresis is shown here and includes a power supply for applying the electric field, anode and cathode compartments containing reservoirs of the buffer solution, a sample vial containing the sample, the capillary tube, and a detector. The sample and the source reservoir are switched when making injections.

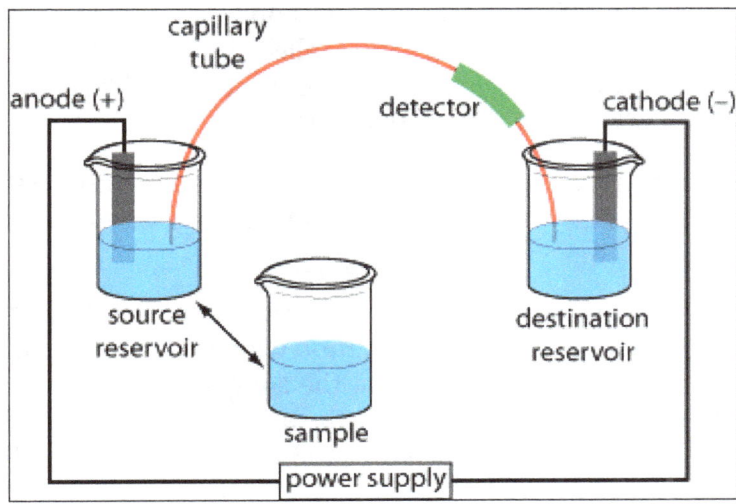

The illustration below shows a cross-section through a typical capillary tube. Most capillary tubes are made from fused silica coated with a 15–35 μm layer of polyimide to give it mechanical strength. The inner diameter is typically 25–75 μm—smaller than the internal diameter of a capillary GC column—with an outer diameter of 200–375 μm.

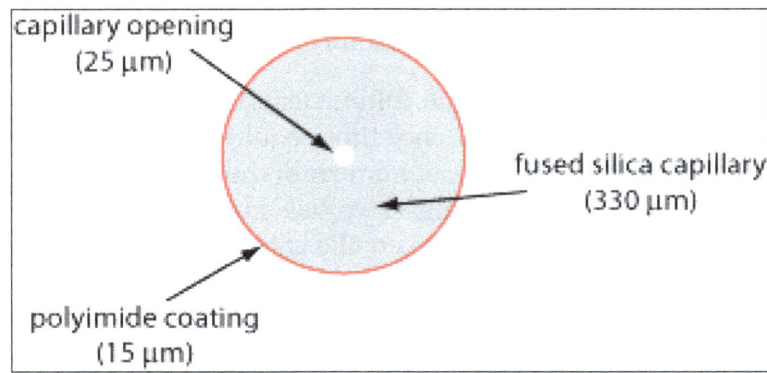

The capillary column's narrow opening and the thickness of its walls are important. When an electric field is applied to the buffer solution within the capillary, current flows through the capillary. This current leads to the release of heat—what we call Joule heating. The amount of heat released is proportional to the capillary's radius and the magnitude of the electrical field. Joule heating is a problem because it changes the buffer solution's viscosity, with the solution at the center of the capillary being less viscous than that near the capillary walls. Because a solute's electrophoretic mobility depends on viscosity, solute species in the center of the capillary migrate at a faster rate than those near the capillary walls. The result is an additional source of band broadening that degrades the separation. Capillaries with smaller inner diameters generate less Joule heating, and capillaries with larger outer diameters are more effective at dissipating the heat.

Instrumentation of capillary electrophoresis is essentially consists of two reservoirs, capillary tube, detector, integrator and power supply. Therefore, instrumentation is very simple in design. These two reservoirs are called anode reservoir or cathode reservoirs. In anode reservoir, anode electrode is placed whereas in cathode reservoir, cathode electrode is placed. Capillary ends are placed in these reservoirs. End of capillary placed in anode reservoir is called anodic end while other end of capillary placed in cathode reservoir is called cathodic end. Electrodes are connected to power supply. Detector is connected towards cathodic end of capillary. Power supply should be able to provide voltage in the range of 10 to 100 kV.

The sample injection is very critical in CE where 5-30 μm^3 sample solution (1-10 nL) is introduced from the anode end of the capillary by one of the two methods: pressure injection or by high voltage injection.

- Pressure injection: Capillary is removed from the buffer reservoir present at the anodic end and carefully placed in the sample solution which is present in air tight or sealed container. Another tube provides pressure to the sample solution which forces the sample to enter the capillary from one end. Once the sample is introduced into the capillary, it is returned back into the buffer reservoir. And then, voltage is applied to start the electrophoresis.

- High Voltage injection: Firstly, when the voltage is not applied, the anodic buffer reservoir is replaced by the sample solution reservoir. Once the sample reservoir is present in the position of anodic buffer reservoir, then high voltage is applied which introduces the sample into the capillary tube. Again by turning off the voltage, the sample solution reservoir is removed, anodic buffer reservoir is placed at its original position and the voltage is switched on to start the electrophoresis.

In CE, the high voltage power supply which is commonly used should be capable of supplying up to 50 kV direct current. The high voltage applied across anodic and cathodic end of the capillary tube moves the sample components at the different rates along the length of the capillary tube.

Electroendosmosis or electroosmotic flow is of much significance in capillary electrophoresis. Capillaries made up of glass contain negatively charged silica on the surface. These negative charges attract the positive charges present in the buffer and form a layer on the inner surface of the capillary. This positively charged layer further attracts the hydration layer of water and forms a double layer. When electric field is applied the positive charges along with layer of water moves towards the cathode. The volume of the capillary is very small and the major fraction of volume is occupied by the hydration layer of the water, therefore, it effects the separation. Electroendosmotic flow is affected by the pH of the buffer used. Electroendosmotic flow is approximately ten times faster at low pH buffers than at high pH buffers.

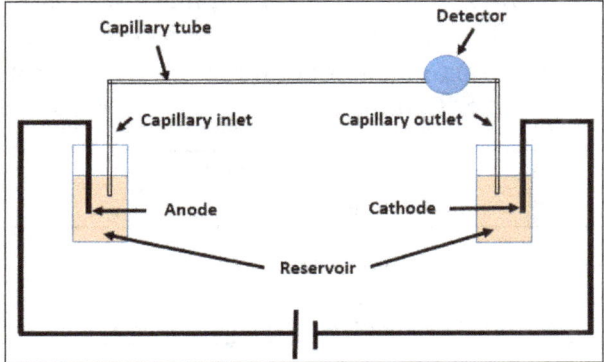

In addition to electroendosmotic flow, the samples also move according to the electrophoretic flow. Electrophoretic flow is the movement of positive ions towards cathode and movement of negative ions towards the anode. Combining the effects of the electroendosmotic flow and electrophoretic flow, the positively charged ion will move towards the cathode, where the effect of flow is additive. Uncharged sample will move towards the cathode because of the electroendosmotic flow. Negatively charged ions will move towards the anode because of the electrophoretic flow but also experience the flow towards the cathode as consequence of electroendosmotic flow and therefore the net movement depends upon the difference between the electrophoretic and electroendosmotic flow.

In another version where electroendosmotic flow is undesirable and not required, the inner surface of the capillaries is coated a polymer (polyimide or teflon) which blocks the interaction of positively charged components with the negatively charged silanol groups. In such capillaries, when electrophoresis is conducted, the positively charged ions will move towards cathode, negatively charged ions will move towards the anode and the neutral ions will remain immobile. Since, detection is possible at only one end at a time therefore, only few components are detected and these components are selected depending upon the practical utility. CE is modified depending upon the version.

A typical run time in CE is 10 to 30 minutes. As the separated components reach the cathodic end of the capillary tube, detection system at this end is installed for the detection and analysis of the sample components. Transmitted signals are then recorded by the recorder. Detection is possible via electrochemical detection, mass spectroscopy which allows two dimensional detection of sample components, laser induced fluorescence detection in UV-visible range in case of peptides, proteins and nucleic acids. Many commercial CE systems use UV or UV-visible absorption detection system. The fluorescence detection system can also be applied when the separated analyte is a fluor and if not then extrinsic fluor is attached to the analyte. This method of detection increases the sensitivity and selectivity of the separated components. Laser induced fluorescence detection in CE detects the analytes at the level of as low as 10^{-18} to 10^{-21} M. The detection system which detects the analyte while moving towards the cathodic end gives results with good resolution.

Thermostatting or heat dissipation mechanism is very critical for the CE. There should be minimum heat gradient between the interior and exterior capillary wall. This depends on the heat transfer rate between the outermost surface of the exterior wall and its surrounding medium. The temperature gradient between the interior and exterior capillary wall is 125 °C/mm when air is the surrounding medium and this temperature is decreased to 51 °C/mm with the forced air cooling setup is installed, Further, with the arrangement of temperature controlled aluminium plates , the temperature gradient decreased to 7 °C/mm.

Sample	CE mode	Detection
Catecholeamines	Free solution	Electrochemical
Metals	Free solution	UV
Propanolol	Free solution	UV
Amino acids	Free solution	UV
Peptides	Free solution	UV
Insulin	SDS-PAGE	UV
Serum proteins	Discontinuous	free solution UV
DNA	Free solution	Fluorescence
DNA (mutation detection)	Pulsed field in ultradilute sieving solutions	UV
DNA (restriction fragments)	Agarose	Fluorescence
Oligonucleotides	Free solution	UV
Virus particles	Free solution	UV
Bacteria	Free solution	UV

Advantage of Capillary Electrophoresis over the Slab Gels

- The main advantage of using capillary systems is that high voltage (5-50kV) can be applied without affecting the resolution and structural integrity of the samples as the set up for these systems is very effective in heat dissipation.

- Requires less samples volumes and therefore increases the sensitivity (can detect as little as 10^{-15} M).

- Exhibits high resolution. It can also separate the chiral mixtures. The separation efficiency of components in CE is greater than the HPLC.

- It separates the complex mixtures within minutes. This technique is known for its rapidity.

Limitations of Capillary Electrophoresis

- In free solution, where electroendosmotic flow is advantageous, it can also be a shortcoming as the positively charged proteins or peptides or any other sample component can be adsorbed on the inner surface of the capillary wall. This can lead to the tailing or the complete loss of component during detection.

- Small volume of the sample used in the capillary electrophoresis. If the sample is diluted, even the highly sensitive detectors cannot detect the sample.

- High voltages used in CE may cause lethal damage if necessary heat dissipation methods are not applied.

Applications of Capillary Electrophoresis

- Point mutations which are responsible for the disease can be identified using CE.

- CE is used for the quantitative analysis of DNA and RNA.

- It can be applied for the purity detection of the synthetic oligonucleotides.

- Wide range of metals, drugs, small molecules and other compounds are being detected in the biological samples.

- It helps in the separation of chiral mixtures. The separation is normally carried out using cyclodextrins as chiral selectors.

- Isotachophoresis has wide application in CE. Similarly, as in SDS-PAGE, the sample ion is stacked between the leading and the terminating ion. Under the constant current leading ion has the high electrophoretic mobility than the sample ion, and sample ion has the high electrophoretic mobility than the terminating ion. (Electrophoretic mobility: Leading ion > sample ion > terminating ion). Therefore, when the electric field is applied, the ion with low electrophoretic mobility experiences high field strength and the ion with high electrophoretic mobility experiences low field strength (Field Strength: Leading ion < sample ion < terminating ion) and all the ions move with the same velocity causing the sample ion to stack between leading ion and the terminating ion, also known as zone sharpening effect. This effect relies on the principle that the ions having the higher electrophoretic mobility will diffuse in the zone where these ions experiences low field strength and vice versa. Isotachophoresis is conducted in quartz and teflon cuvettes at voltages up to 30 kV.

- Capillary Isoelectric Focusing: The buffer used in this type of capillary electrophoresis have pH gradient is which provided by commercially available carrier ampholytes. The pH gradient of different range (narrow or broad) can be setup with these ampholytes having their isoelectric points in this range. These ampholytes are very small, therefore in an applied electric field they move very quickly and pH gradient is formed. The capillary in which electrophoresis will be conducted is filled with ampholytes and analytes which are to be separated. When the voltage is applied all the ampholytes starts to migrate on the basis of their isoelectric point and establishes the pH gradient. As already discussed that being very small, these ampholyte molecules move faster than the analyte molecule. Each analyte molecule will move according to their isoelectric point. For the identification of analytes chemical makers are available with known isoelectric points and can be identified when placed in the electropherogram. These markers can be can be added along with the markers.

- Ligand affinity of protein or DNA can be applied with electrophoresis. Depending upon the electrophoretic mobility, the uncomplexed ligand and uncomplexed protein can be separated from ligand-protein complex as all the three species have different electrophoretic mobility.

- Separation by capillary electrophoresis is in line with HPLC system, which facilitates 2-D analysis of sample. In can be performed by the directing the eluted samples from HPLC to the CE system where it can be analysed rapidly.

- Micellar electrokinetic capillary electrophoresis: In this type of capillary electrophoresis, separation can be performed by including SDS in the system. Here, SDS when suspended in an appropriate buffer and above certain concentration it form micelles and act as a stationary phase, therefore, the separation occurs on the basis of the interaction of the sample with the detergent as well as on their electrophoretic mobility in electric field.

- In capillary electro-chromatography, C-18 stationary phase used in reverse-phase chromatography is packed into capillaries; therefore electrophoretic mobility is based upon the affinity of the sample towards C-18 stationary phase and their own electrophoretic mobility.

Microchip Electrophoresis

Microchip electrophoresis is a separation method performed by applying a high voltage along a channel inside a small microfluidic chip or lab-on-a-chip. All ions (positive and negative) are pulled through the channel in the same direction by electro-osmotic flow (EOF). The analytes separate as they travel along the channel because of differences in their ionic mobility and they are detected by the C4D electrodes integrated on the chip near the outlet.

Contactless conductivity detection can be used for virtually all charged species: inorganic anions and cations, as well as organic ions, such as carboxylic acids, amines, amino acids, peptides, proteins, DNA fragments, antibiotics and many other pharmaceutical compounds. Tagging or other modification of the analytes is usually NOT required, while limits of detection are often comparable to, or sometimes even better, than UV-visible absorption techniques.

Microchip electrophoresis with C4D has been used for the analysis of cation, anions and for the study of electro-osmotic flow. The ER455 Microchip Electrophoresis Kit is a set of equipment for microchip electrophoresis experiments. It includes:

- C4D Data System

- High Voltage Sequencer

- Chip Platform

- Standard Test Solutions

- PowerChrom Software

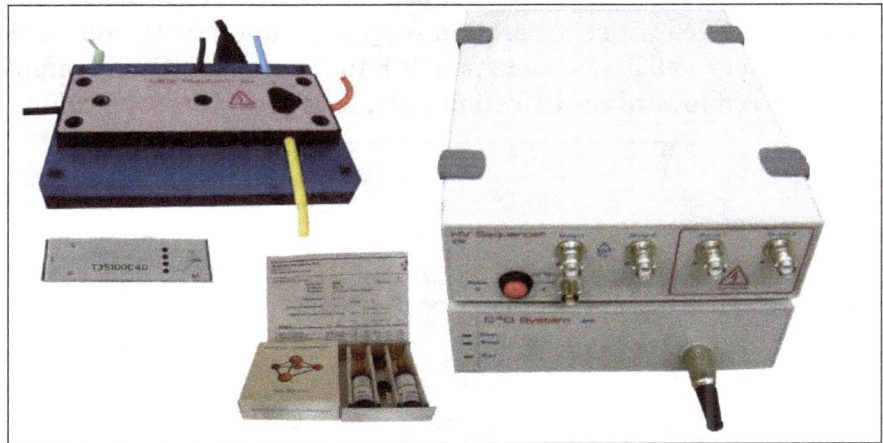

For microchip electrophoresis researchers with existing systems, the ER430 High Voltage Sequencer can be integrated into your setup. The sequence of high voltages is fully controlled using either the eDAQ QuadSequencer software or third-party software such as LabVIEW, WinWedge, or HyperTerminal. The analog signal from UV and fluorescence detectors can be recorded using the C4D Data System, PowerChrom or e-corder systems, and the data analysed using PowerChrom or Chart software.

In modern disease diagnosis (health examination), objective and accurate judgments are made on the basis of quantified body conditions obtained by the measurement of biomarkers in blood, urine, etc. These measurements are referred to as clinical tests. Determination of clinical test items requires expensive equipment and highly skilled personnel. Recent fundamental and clinical studies have revealed new indicators of biological changes (biomarkers) related to diseases, which adds to the number of test items used for disease diagnosis. Since clinical tests take considerable time, detailed results are seldom available on the same day. Both medical workers and patients desire inexpensive and quick (simple) measurement of various biomarkers included in clinical test items.

In this context, microchip electrophoresis (ME) has been increasingly used for DNA analysis. This method is based on microchannels of micrometer size formed on a glass or plastic substrate by nanoprocessing using semiconductor technologies. The microchip capillary electrophoresis has the advantages of low sample consumption, shorter analysis time, and high sensitivity. At the moment, several manufacturers offer commercial ME systems with disposable microchips. In recent years, the scope of analysis has gone beyond DNA to include RNA, proteins, sugar chains, cellular functions, and other items. Implementation of reactions and detection on microchips several centimeters in size results in simple operation and low dependence on operator skills, which assures accurate and consistent data acquisition. In combination with the compact size of the equipment, this promises wide use for on-site analysis in medical treatment and clinical tests.

Features of Microchip Electrophoresis

Structure of Microchip and Principle of Detection

Microchip electrophoresis is carried out using microchannels 10 to 100 μm wide and 5 to 50 μm deep fabricated by semiconductor nanoprocessing on a chip made of quartz or poly(methyl methacrylate) (PMMA), polydimethylsiloxane (PDMS), polycarbonate, or other plastics. The conceptual configuration of ME for DNA analysis used in our experiments on sugar testing is shown in figure. The crossing channels are filled with a separation support solution (gel), and an electric charge is applied successively to every well as shown in panels B to D. As a result, only samples in the crossing portions are transferred toward the detection zone, thus providing separation and detection.

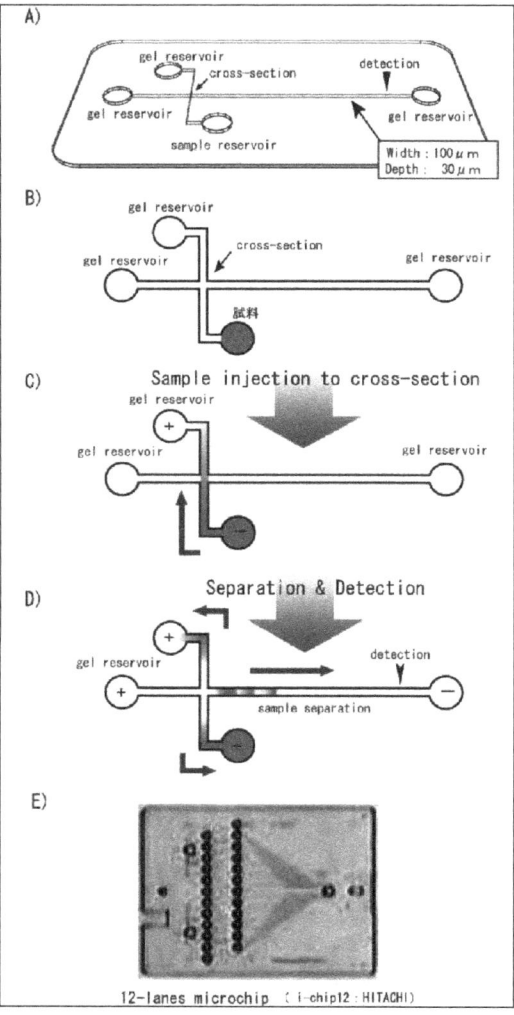

Figure: DNA sample application, separation, and analysis on microchip.

Since the channels are very small, an operation cycle from sample injection to detection takes about 5 minutes. In addition, high-speed high-throughput systems can be implemented by designing microchips for multisample analysis (panel E).

In DNA analysis, presently the most popular ME application, polymer solutions (non-cross-linked polyacrylamide, cellulose derivatives, polyethylene oxide, etc.) are employed for separation support. When the concentration of polymeric solutions such as methyl cellulose reaches a certain level, the polymer molecules entangle with each other, and "dynamic pores" are generated (unlike the conventional gels used in electrophoresis, such as agarose or acrylamide, the pores formed by the polymer molecules have no fixed size; however, the average mesh size depends on the polymer concentration). Thus, DNA undergoes molecular sieving, as when conventional gels are used, which results in separation by molecule size. There are reports of optimal polymers for different DNA sizes; the polymers and electrophoresis conditions are selected as necessary. Ethidium bromide and other fluorescent indicators are added to the support solutions to detect the separated DNA samples by fluorescence using semiconductor lasers or light-emitting diodes. Conventional agarose electrophoresis with ethidium bromide stain allows a DNA detection sensitivity of about 1.0 ng/μl; the detection sensitivity of commercial ME systems such as the Hitachi SV-1100 is 0.1 ng/μl, an order of magnitude higher, even by conservative estimate. In addition, the sample volume required for analysis is of picoliter order, and high-sensitivity analysis can be implemented with very small samples. Packaged software provides display of migration patterns, analysis of DNA chain length, presentation of pseudo-gel images and other data in the course of electrophoresis. As confirmed by actual results of DNA analysis using conventional gel electrophoresis and ME (SV-1100), microchip electrophoresis outperforms gel electrophoresis in detection sensitivity. A difference of just a few base pairs can be detected reliably. These results indicate that microchip electrophoresis is a very efficient analysis tool with the advantages of small sample volume and short analysis time.

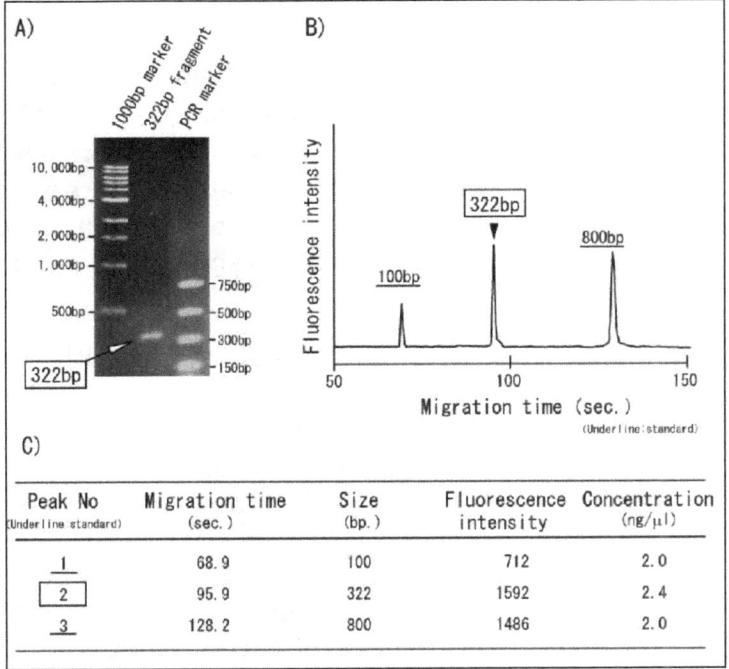

Figure: Data output of assay of 322-bp DNA fragment using agarose gel and microchip electrophoresis.

Application of Microchip Electrophoresis to Biomarker Analysis

Commercial ME systems can rapidly measure the molecular weight of DNA, RNA, and proteins using small samples, while providing a resolution of a few base pairs. However, from another standpoint, the use of this advanced equipment is limited to measurement of the molecular weights of such substances as DNA and RNA. Considering the equipment cost, conventional electrophoresis using agarose and acrylamide might be preferable in cost–performance terms.

Thus, we considered the possibilities of completely different applications using commercial ME systems (including analysis chips) in their present form, while modifying the reaction and detection methods. In particular, we focused on the glycans, which have been attracting attention recently as a "third chain of life," alongside proteins and nucleic acids, and attempted to use microchip electrophoresis for biomarker analysis in medical treatment and clinical tests. The method proposed in this study has the advantages of rapid measurement with small samples of the blood biomarkers, namely, sugar level and sugar-related enzyme (amylase), at the same accuracy as conventional clinical tests.

Blood Sugar Measurement by Microchip Electrophoresis

In current clinical practice, blood sugar levels are determined mainly by absorbance measurement of sugar modification enzymes such as glucose oxidase and glucose dehydrogenase. However, techniques based on enzymes require many samples and reagents as well as calibration controls. In some cases, accurate measurement is impossible without deproteination and other complicated procedures. In addition, in the case of patients undergoing intravenous infusion, overestimated results may be obtained because of reducing sugars such as lactose and galactose that are included in the IV solution. Consequently, direct measurement of blood sugar using high-speed separation of glucose and other monosaccharides by ME has been investigated.

The main principle is that glucose (Glc) in human blood is labeled directly with fluorochrome (2-aminoacridone, AMAC), and then subjected to electrophoresis. Direct labeling of blood plasma involves reductive amination so that only reducing substances are labeled. In addition, 0.2 M boric acid buffer is used for electrophoresis: that is, electrophoresis is driven only by the negative electric charge of sugar–borate ion complexes, and almost all biological substances labeled by this method are carbohydrates. As a result, one can expect specific detection of blood sugar.

We labeled unprocessed human plasma by reductive amination with AMAC, and performed electrophoresis using a commercial microchip (Hitachi i-chip). As a result, we obtained a single peak at a migration time of about 140 s. The peak was also obtained at nearly the same migration time when a standard Glc preparation (10 μM) was used. Thus, the Glc-specific reaction was confirmed. When glucokinase (GK) was added to the reaction solution so as to promote phosphorylation, the electrophoresis time changed from 140 s to 70 s. Thus, we may conclude that Glc in blood is phosphorylated by GK and converted to glucose-6-phosphate (Glc-6-P). Hence, the negative charge increases and electrophoresis accelerates. That is, the proposed method makes possible the direct measurement of Glc in blood, and produces Glc-specific peaks.

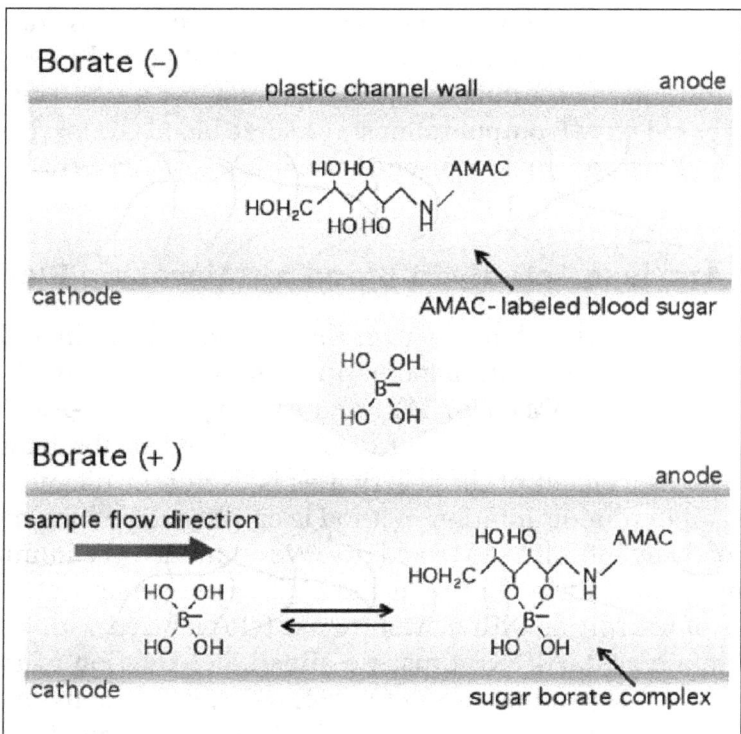

Figure: Reducing sugar electrophoresis in presence and absence of boric acid.

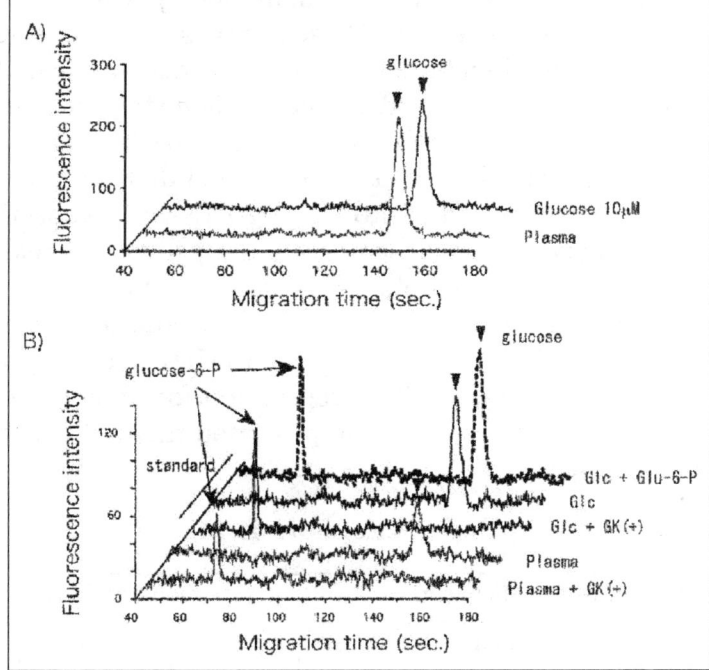

Figure: Electropherogram of reducing glucose in human plasma.

We also showed that the peak intensity varies almost linearly with the peak concentration in the range of 1.0 to 300 mM (the data are omitted here because of space limitations). Such direct measurement of concentration is very likely to be useful in clinical tests.

We compared the proposed ME method with the conventional enzymatic method used in clinical tests. Simple regression analysis showed a strong correlation between the two measurement methods with a regression formula $y = 0.9138x + 8.584$ ($r = 0.999$, $p < 0.05$). This result indicates that the proposed method based on ME outputs almost the same blood Glc level as the current clinical tests; thus, the proposed method appears useful for clinical tests in terms of simplicity and low sample consumption.

Measurement of Amylase Activity in Blood by Microchip Electrophoresis

α-Amylase (AMY) is an enzyme in blood serum that has long been included in clinical tests. Hyperamylasemia is observed in acute pancreatitis. There are two amylase isozymes in human blood, namely, pancreatic AMY (P-AMY) and salivary AMY (S-AMY). The activity fluctuations of both isozymes reflect the state of diseases related to the pancreatic and salivary glands, and accurate measurement of these activities is useful for diagnosis and screening. In current clinical tests, an immunoinhibition method is employed for such measurement: monoclonal antibodies selectively inhibit S-AMY activity. We combined the immunoinhibition method using S-AMY monoclonal antibodies with the high-speed separation of glucose by ME to determine the degree of hydrolysis with maltohexaose (G6) labeled with fluorochrome (8-aminopyrene-1,3,6-trisulfonic acid, APTS); thus, we aimed at reduction of the time and sample consumption required for measurement of total AMY (T-AMY) activity and P-AMY activity in human blood. First, samples were prepared with a final concentration of 50 mM HEPES (pH 70 mM NaCl, 1.0 mM $CaCl_2$, and 60 μM APTS-labeled maltohexaose (APTS-G6). In addition, 10 μl of standard human amylase (about 300 U/L) or human blood plasma was added. The resulting 100 μl of mixture was kept at 30 °C for 10 min. These samples were subjected to electrophoresis with a commercial microchip (Hitachi ichip). A single peak was obtained at a migration time of 83 s for the control samples without standard amylase; when standard amylase (0.003 U) was added, the peak at 83 s decreased and shifted to 64 s. Comparison with electrophoresis data for polysaccharides (G2 to G7) labeled with APTS confirmed that the peaks at 83 and 64 s correspond to maltohexaose (APTS-G6) and maltotriose (APTS-G3), respectively. Thus, the peak shifted because G6 was hydrolyzed by amylase to G3. Nearly identical peaks were also detected when human plasma was added to the samples. We examined the correlation between amylase activity (U/L) and the fluorescence intensity of APTS-G3 produced by hydrolysis. In the amylase activity range of 5 to 5000 U/L (detection limit: 4.38 U/L), a strong correlation was found between the two measurement methods, with a regression formula $y = 3.5512x + 26.485$ ($r = 0.999$, $p < 0.05$). These results indicate that the proposed method can accurately determine amylase activity in blood samples.

Table: Comparative accuracy of microchip electrophoresis.

Sample No	Microchip method		Conventional method	
	(mM)	(mg/dl)	(mM)	(mg/dl)
1	5.40	97	5.44	98
2	6.18	111	5.94	107
3	9.96	179	9.71	175

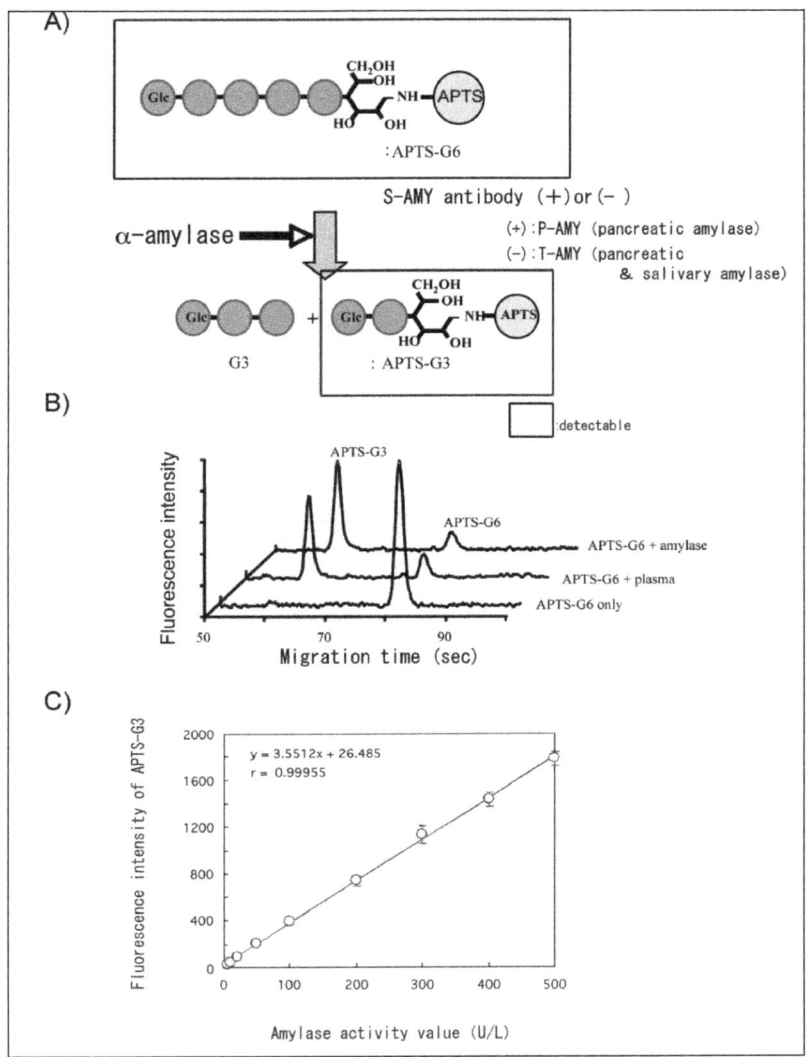

Figure: Determination of AMY activity in human plasma. (A) Reaction scheme;
(B) electropherogram; (C) standard curve.

We next confirmed that P-AMY (pancreatic amylase) activity can be measured by the proposed method when S-AMY (salivary amylase) activity is inhibited by the immunoinhibition method using S-AMY monoclonal antibodies. Actually, the S-AMY activity is about 98% inhibited. The amylase activities related to pancreatic and salivary glands can be calculated accurately from the S-AMY inhibition rate and the measured P-AMY and T-AMY activities. Exact quantification of the activities of the two isozymes may be very useful in medical treatment. In order to examine this possibility, we compared the proposed method with blood amylase activities determined by the existing clinical test method. Specifically, we drew a blood sample and divided it into two halves; one half was used for microchip electrophoresis, and the other half was sent to a clinical laboratory to determine the T-AMY and P-AMY activities. In Figure, the AMY activities obtained by the proposed method and by conventional clinical testing are represented by the horizontal and vertical axes, respectively. As is evident from the graphs, a strong correlation with conventional clinical tests occurs for both TAMY and P-AMY; the regression formulas are $y = 1.000x + 4.7174$ ($r = 0.995$) and $y = 1.032x + 4.3305$ ($r = 0.989$), respectively. Simple regression analysis showed no statistically significant difference in the T-AMY and P-AMY activity values between the proposed method and the conventional method (p < 0.01). Considering all the results presented above, the proposed microchip measurement of amylase activity in human blood is quite comparable to conventional clinical tests. Furthermore, the proposed method would be very useful in clinical practice due to its rapidity and low sample consumption.

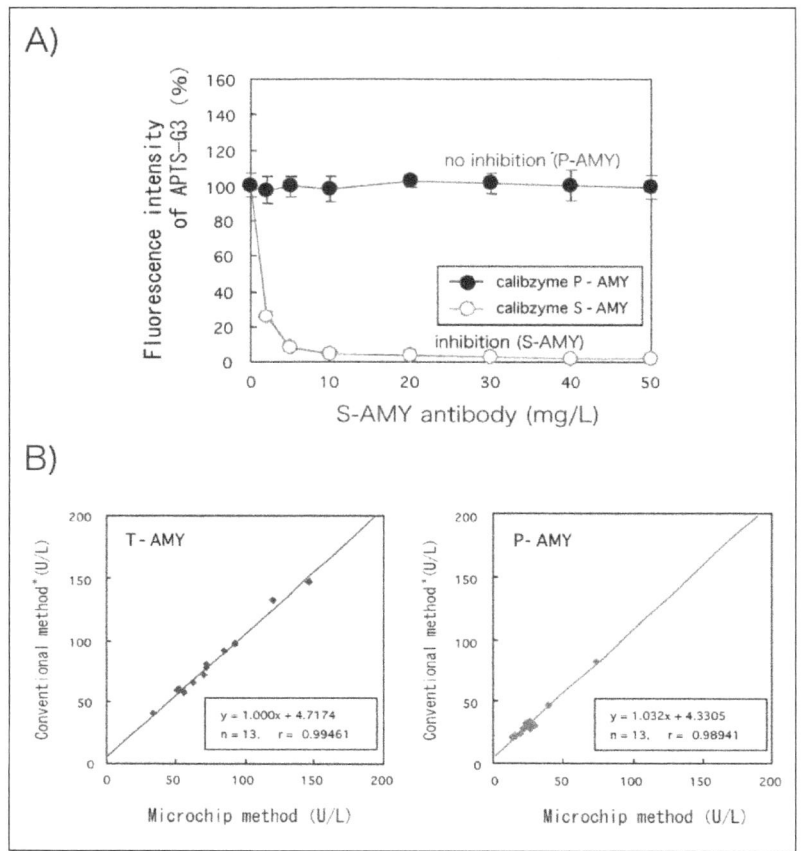

Figure: Determination of two types of amylase activity. (A) Examination of S-AMY antibody dose effects; (B) accuracy of values obtained by microchip electrophoresis. * clinical assay.

Future of Microchip Electrophoresis

In this study, we showed that blood biomarkers can be measured by microchip electrophoresis instead of conventional clinical tests. We used a commercial microchip system for DNA analysis, modifying only the reaction and detection methods. The use of microchip electrophoresis is advantageous because the analysis can be completed rapidly with small samples. We have also studied other applications of microchip electrophoresis in molecular biological analysis, such as simultaneous DNA analysis and restriction enzyme digestion, or gene expression analysis and protein–protein interaction analysis by labeling. Such molecular biological analysis is a basic part of clinical tests related to pathogen detection and SNPs analysis. We reported measurement of the mitochondrial membrane potential with an Agilent Bioanalyzer 2100 provided with flow cytometry functions. Mitochondria are not only "cellular power plants": they are also known to be closely related to apoptosis. In clinical practice, measurement of the mitochondrial membrane potential is required for definite diagnosis of mitochondrial diseases and other diseases with mitochondrial dysfunctions. FACS and other techniques have been used conventionally for membrane potential measurement. However, microchip analysis allows an inexpensive simple operation, and the required cell mass is about one-tenth as great as in FACS, which opens the prospect of its increasing adoption in clinical diagnosis. Hence, we may expect expanding applications of microchip electrophoresis, in addition to nucleic acid analysis, such as the measurement of blood sugar levels and other biomarkers.

Permissions

Index

A

Acrylamide, 33, 77, 87, 205-211, 214 219, 227-228
Affinity Chromatography, 2, 50, 79-83 102-105
Agarose Concentration, 213, 215, 218
Albumin, 3, 11, 83, 91-92, 105
Ampholytes, 223-224
Amylase Activity, 230, 232
Analytical Technique, 5, 14, 92
Annealing, 17-18, 26-29, 33-35
Annealing Temperature, 18, 27-29 33-35
Anode, 2, 10, 200, 203-206, 211, 215 217, 219-222
Atomic Spectroscopy, 5, 144-145 149-151

B

Beam Splitter, 153-155, 176
Biochemical Analysis, 1, 3
Bioinformatics, 26
Biomarkers, 6, 10, 225, 228, 232-233
Biosensor, 11
Boric Acid, 204, 216, 228-229
Buffer System, 205, 207

C

Capillary Electrophoresis, 6, 12-14, 38 218-225
Capillary Tube, 219-222
Cathode, 2, 10, 146, 200, 203-205, 211 213, 219-221
Chromatographic Column, 7, 53, 66, 79 100
Circular Dichroism, 132-133, 135-143

D

Denaturation, 17, 19-20, 26, 28, 31-34 36, 46, 76, 90, 130, 210
Diffusion Coefficient, 61, 218-219

E

Electro-osmotic Flow, 224-225
Electrode Buffer, 208, 216
Electroendosmotic Flow, 221, 223
Electrolyte, 10, 71, 202, 213
Electropherogram, 224, 229, 231
Electrophoresis, 2-3, 6, 12-14, 27, 31 36-38, 47, 100, 200-201, 203-208 211-233
Electrophoretic Mobility, 200, 203, 206 210, 218-220, 223-224

E

Elongation, 17-18, 20, 27
Enhancing Agent, 85
Ethidium Bromide, 30, 33, 36-37, 41-42 127, 212, 216-217, 227

F

Fluorescence Detectors, 59, 225
Fourier Transform, 151-152, 160
Frictional Force, 200, 206, 214-215

G

Gel Electrophoresis, 2-3, 12, 27, 36-37 47, 205-206, 208, 212-214, 216-218, 227

H

Heat Dissipation, 219, 222-223
Human Genome, 2-3
Hydrolysis, 23-25, 33, 230

I

Immune Complex, 14, 97
Immunoassay, 9, 14, 45, 93-99, 131
Infrared Spectroscopy, 9, 151-152 157-158, 160
Instrumental Techniques, 5
Isoelectric Point, 70, 203, 224
Isopsoralen, 32

L

Lymphocytes, 4, 7, 86

M

Maltohexaose, 230
Mass Spectrometry, 3, 6, 9-10, 15, 58 104, 194
Microchip Electrophoresis, 224-228 230-233
Mitochondrial Membrane, 232
Molecular Biology, 2-3, 36, 206
Monoclonal Antibodies, 84, 86, 230-231
Monomers, 1, 212-213

N

Nonlinear Susceptibility, 167, 189, 191
Nucleic Acids, 1, 3, 11, 40, 61, 66 71-72, 77, 195, 206, 210-212, 214-215 217-218, 222, 228
Nucleoside Triphosphate, 20-22
Nucleotides, 18-20, 23-24, 66, 71, 77 214-215

P

Peptide, 11, 130, 137, 195, 202, 212

Plasmonic Nanoparticles, 170-171

Polyacrylamide Gel Electrophoresis 205-206, 212-213

Polymerase Chain Reaction, 10, 16 18-19, 27

Polymerization, 19, 23, 26, 205-207, 210

Polysaccharides, 68, 213, 230

Purification Procedure, 31, 104

R

Regression Analysis, 230-231

Rule Of Thumb, 29, 113

S

Sodium Dodecyl Sulfate, 103, 208, 216

Spectrometer, 59, 133, 135, 138-140 145, 151, 153, 173-178, 198

Spectrophotometry, 5, 8

Surface Plasmons, 163, 170-171, 178 180

T

Taq Polymerase, 18-20, 28-29, 31-35

Theoretical Plates, 50, 52-56, 218-219

Therapeutic Proteins, 76, 105

Thermal Cycler, 18, 26, 29, 32-34, 45

Transferrin, 4, 91, 94, 96

Transilluminator, 32, 216-217

U

Ultracentrifugation, 2, 92

Ultrafiltration, 8, 32-33

V

Viscosity, 75, 200-201, 220

X

Xylene Cyanol, 33, 216

CPSIA information can be obtained
at www.ICGtesting.com
Printed in the USA
BVHW062121290822
645775BV00003B/122